Pvt. Ltd.

Industrial Engineering & Operation Management

Application of Decision Making Techniques

by

Arvind Jayant

Mechanical Engineering Department

Sant Longowal Institute of Engineering and Technology

(University under MHRD Govt. of India)

Longowal, Dist. Sangrur, Punjab – 148 106

2019

Studium Press (India) Pvt. Ltd.

Industrial Engineering
&
Operation Management

Application of Decision Making Techniques

©2019

This book contains information obtained from authentic and highly regarded sources. Reprinted material is quoted with permission, and sources are indicated. A wide variety of references are listed. Reasonable efforts have been made to publish reliable data and information, but the author and the publisher cannot assume responsibility for the validity of all materials or for the consequences of their use.

All rights are reserved under International and Pan-American Copyright Conventions. Apart from any fair dealing for the purpose of private study, research, criticism or review, as permitted under the Copyright Act, 1956, no part of this publication may be reproduced, stored in a retrieval system or transmitted, in any form or by any means–electronic, electrical, chemical, mechanical, optical, photocopying, recording or otherwise–without the prior permission of the copyright owner.

ISBN: 978-93-85046-45-2

Published by:

Studium Press (India) Pvt. Ltd.
4735/22, 2nd Floor, Prakash Deep Building
(Near Delhi Medical Association)
Ansari Road, Darya Ganj, New Delhi-110 002
Tel.: + 91-11-43240200-15 (15 lines); Fax: 91-11-43240215
E-mail: pubdir@studiumpress.in

Printed at India

About the Author

Dr. Arvind Jayant is presently working as an Associate Professor in the Department of Mechanical Engineering and formerly he was Head of the Department of Disabilities Studies at Sant Longowal Institute of Engineering & Technology, Deemed University (CFTI under Ministry of HRD, Govt. of India), Longowal, Punjab for a period of three years (2012-2015). He received B.E. (Production & Industrial Engineering) degree from Delhi College of Engineering, Delhi and and M.Tech. (Industrial Engineering) from IIT, Delhi. He obtained his Ph.D. degree in the field of Reverse supply chain management under joint research collaboration of SLIET Deemed University & DTU, Delhi. Dr. Jayant having more than 22 years teaching and research experience with 02 years industry experience. His research area includes Industrial Engineering & Operation Management, Supply Chain Management, Reverse Logistics, CIM, ERP, TQM, and Modelling & Simulation of Manufacturing System. He has completed 03 sponsored research projects of Rs. 90 Lacs funded by MHRD, CSIR and MSJE, New Delhi. He has been published more than 130 research papers in International/ national journals and proceedings of International and national conferences. He is listed in the panel of reviewers of 50 International high impact Journals including, IEEE Transaction on Engineering Management, *OMEGA: International Journal of Management Science*, Production, Planning and Control, International Journal of Resource, Conservation and Recycling, Pervasive and Mobile Computing, Sustainable Production and Consumption, Journal of Small Business Management and many other reputed high impact International journals of Elsevier, Inderscience, John Wiley, Taylor & Francis and Springer publications. He serves on the editorial boards of several reputed international journals. He has attended 47 International/ National conferences and STTP in India & Abroad. He has been organized 11 Conferences and AICTE/ISTE Sponsored STTP. He has been delivered 15 Keynote/Invited Lectures in India and abroad. Dr. Arvind Jayant biography had been published in MARQUIS Who's Who in Science and Engineering (USA), 31th Edition, 2014. He is recipient of prestigious Dewang Mehta National Education Leadership Award (22nd Dewang Mehta & BSA Awards), Nov. 15, 2014 in the capacity of HOD (DDS). He visited USA, Japan, England, Spain, Germany, Thailand, Nepal and Singapore to disseminate his research among academic fraternity and interact with faculty. He has guided more than 75 students for their final year PG and UG Industry oriented projects. At present he is supervising 05 Ph.D. students. He is the members of different professional bodies like ISTE, SMIIE, ISME, IWS, IIE (USA), IACSIT (Singapore), IAEngg. (Hongkong) etc.

Acknowledgements

During preparation of this reference book, literature from various journals, magazine, books and online articles have been referred. I express my sincere gratitude to all worthy authors, specially my research and post graduate students, publishers, institutions and small medium enterprise (SMEs) of northern part of India for providing data and required information's for the case studies development. Many of them are listed in the references. If some are left out unfortunately, I seek their pardon.

I express my sincere gratitude to Professor PremVrat (Ex-Director, IIT Roorkee), His excellence in Industrial Engineering and Operation management. He is a role-model and inspiration for many like me. I wish to acknowledge my sincere thanks to Prof. S.K. Garg (Pro-VC, DTU-Delhi) and Prof. R.P. Mohanty (Ex-Vice Chancellor, Siksha 'O' Anusandhan University, Bhubaneswar), another role- models for excellence in teaching and research of Industrial Engineering. I have learned a lot from my professors from IIT Delhi and Delhi Technological University, Delhi, I thank them all.

A conducive ambiance was needed to pen a book. SLIET Deemed to be University, Longowal with its calm beautiful scenic campus and infrastructure provided the right elements for that. I express my sincere thanks to our institute Director Prof. Shailendra Jain, Prof. P.K. Singh, Dean (R&C), Prof. P. Gupta, former head, Mechanical Engineering Department and Prof. Rajesh Kumar, current head of the department for their all kind of support and for providing a terrific environment in which to teach and conduct research successfully.

Other acknowledgement goes to the editorial and production staff at Studium press, they have had much patience with me over the years and contributed greatly towards the success of this book. Especially, I express my deep sense of gratitude to Dr. J.N. Govil, Publication Director of Studium Press for making many valuable contributions by his careful checking of the manuscript and proof materials. Special Thanks goes to my sincere and diligent students who are always motivation for me. Support of wife Neelam Jayant and inquisitiveness of my genius daughters Amisha and Navya encouraged me to complete the book. I remember and adhore all the teachers in my life including my great parents' blessings and good wishes are counted.

Arvind Jayant
Ph.D. (Industrial Engineering & Operation Management)
Associate Professor
SLIET Deemed to be University, Longowal, Sangrur, Punjab
(CFTI under Ministry of HRD, Govt. of India)

Preface

Foreword: *This reference book is a useful reminder of what good business practice is and how it should be applied within domain of industrial engineering and operation management practices in Indian business environment.*

This reference book is written as a self-teaching tool to assist every technical teacher, engineering student, MBA students, mathematician, or non-mathematician for educating themselves or others, and to support their understanding of the elementary concepts of assessing the performance of a set of homogenous firms. The book can be used as a reference for undergraduate or graduate level courses in industrial engineering, operation management, economics, management, business, statistics, or There is no pre-course required to start reading the book, except a basic and elementary knowledge of mathematics and computer.

The content is relevant; concepts are clearly explained and supported by case studies that bring the concepts to life. The language used is clear and contemporary; visualisations re-enforce the concepts well. Organisations operating on a global stage must get this stuff right, in both process and physical terms: it is an essential element to delivering profitable growth. This book offers an invaluable guide to understanding the specific dynamics of your supply chain and the fundamentals underpinning it. It provides the framework for delivering a supply chain strategy based upon recognised best practice.

Research works often require one or more analytical tools for problem solving, optimisation, prediction, system modelling, data analysis or interpretation etc. In a typical academic research work or any professional research assignment, at some point the researcher looks for a suitable analytical strategic decision modelling tool to progress the work before arriving at some inferences. However, a wrong selection of the tool for a specific problem can be counter–productive and mislead the hard research efforts. The fact is that for each research problem or situation, one must be judiciously select an analytical tool that is best suitable for it.

The rationale of this book is to guide the researcher in the field of Industrial Engineering and management for selecting the best tool for the research problem by briefly introducing to various tools along with their typical features and applications in Indian business environment.

Some illustrative examples using popular MCDM techniques (AHP, ANP, DEA, VIKOR, Fuzzy-AHP, GRA, TOPSIS and PROMETHEE) and discrete event simulation applications in manufacturing environment are included to help the readers in tool selection by bridging the gap between theory and application. The book attempts to contain most of the contemporary analytical multi criteria decision making tools but giving exhaustive description for each tool is neither feasible nor the objective. The unique book can serve as a reference source for various academic courses, dissertations, research projects and assignments. The hypothetical illustrative examples cited in the book are solved either through software or otherwise. The book is penned over a period of two years while keeping the needs and circumstances of the researchers in mind.

This is the first edition of this book and it is possible there might be some typos. Author will be grateful to hear from the reader about the usefulness, queries or factual errors and suggestions for improvements at the following email address, if there are any points, correction, or questions, arvindjayant@gmail.com. SLIET Deemed to be university, Longowal, Sangrur, Punjab, India, Arvind Jayant Wishing the reader, a successful research endeavour.

Assoc. Prof. (Dr.) Arvind Jayant
Author

Table of Contents

List of Abbreviations

ANP	:	Analytic Network Process
BSC	:	Balanced Score Card
DEA	:	Data Envelopment Analysis
ECM	:	Environmentally Conscious Manufacturing
GRE	:	Grey Relational Analysis
ISM	:	Interpretive Structural Modeling
EOL	:	End of Life
VIKOR	:	VIšekriterijumsko Kompromisno Rangiranje
AGV	:	Automated Guided Vehicle
FAHP	:	Fuzzy Analytical Hierarchy Process
SC	:	Supply Chain
SCM	:	Supply Chain Management
GSC	:	Green Supply Chain
EA	:	Environmental Aspects
GSCM	:	Green Supply Chain Management
GSS	:	Green Supplier Selection
DEMATEL	:	Decision Making Trial & Evaluation Technique
AHP	:	Analytical Hierarchy Process
MCDM	:	Multiple Criteria Decision Making
TOPSIS	:	Technique for Order of Preference by Similarity to Ideal Solution
PROMETHEE	:	Preference Ranking Organisation Method for Enrichment Evaluations
ELECTRE	:	ELimination Et Choix Traduisant la REalite
FMS	:	Flexible Manufacturing System

List of Figures

List of Tables

1

Vendor Evaluation in Supply Chain Management (SCM) Through AHP and VIKOR Technique

1.1. INTRODUCTION

Supply chain management is basically flowing of product and services from the origin of product to end of the product. Supply chain management main objective is to monitor production, distribution, shipment, and service. A supply chain consist all parties involved directly or indirectly in fulfilling customer requirement. In supply chain Vendor, manufacturer, distributor, retailer, and shopper are involved. The vendor, manufacturer, distributor, retailer, and shopper play vital role in efficient supply chain management.

In the Fig. 1.1, we see that product, information, finance flow Vendor to the manufacturer, manufacturer to distributor, distributor to retailer, retailer to the shopper and reverse logistics also occur.

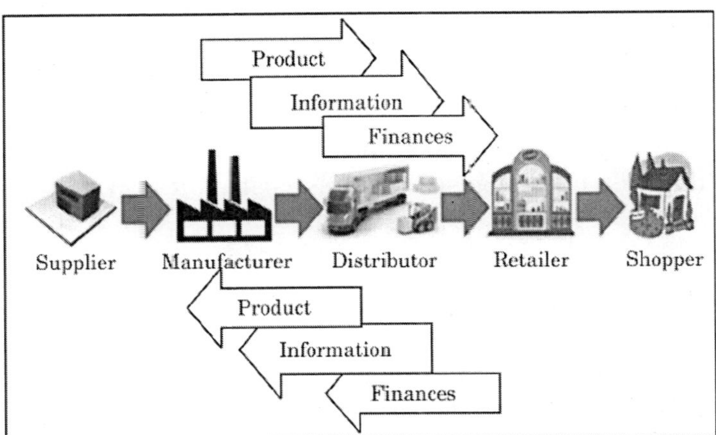

Fig. 1.1: Supply chain process flows.

1.2. SUPPLY CHAIN MANAGEMENT PROCESS FLOW

In Supply chain management process flow is used to make sure to efficient and cost-effective. In supply chain retailer used a collection of steps from Vendor to consumers. The basic five components of the supply chain are discussed below.

1.2.1. Plan

The first stage of the supply chain is planning stage. In the planning stage, we must make a strategy to fulfill demand and necessities of the consumer. In plan the main objective to design strategy to get maximum profit. In planning retailer also focus on cost, quality, resources utilization and risk factor.

1.2.2. Develop (Source)

After planning next step is sourcing. In this stage the main objective is a too strong relationship between vendors. In this stage we analysis stock, make to order, scheduling deliveries, receiving, authorizing Vendor payment, identify and selecting Vendor, assessing Vendor performance, managing incoming inventory and Vendor agreement

1.2.3. Make

Production units are identified mostly with their decision to make or buy. In other words, do we wish to produce the desired product on their own or do we want to purchase it from the foreign market? This decision is critical because of the third-party Vendors especially in countries like Eastern Europe, China.

1.2.4. Deliver

On time, delivery is very important aspects for customer satisfaction. Delivery depends upon transportation, geographical location, product capacity etc. Delivery depends upon distribution center and retailer inventory.

1.2.5. Return

The return also referred as reverse logistics. When products do not reliable in its warranty, period then product returns from customer to the retailer, retailer to distributor and distributor to manufacturer. Fast return makes a good market reputation and customer satisfaction.

1.3. SUPPLY CHAIN FLOW

There are three flow of supply chain: materials, information and financial flow

- ***Material flow:*** These are all physical product flow along the chain from manufacturer to distributor, distributor to retailer, retailer to consumer.
- ***Information flow:*** All the data related to demand, shipment, orders, returns, schedule are information flow.
- ***Financial flow:*** Financial flow includes all transfer of money, payments, credit card information etc.

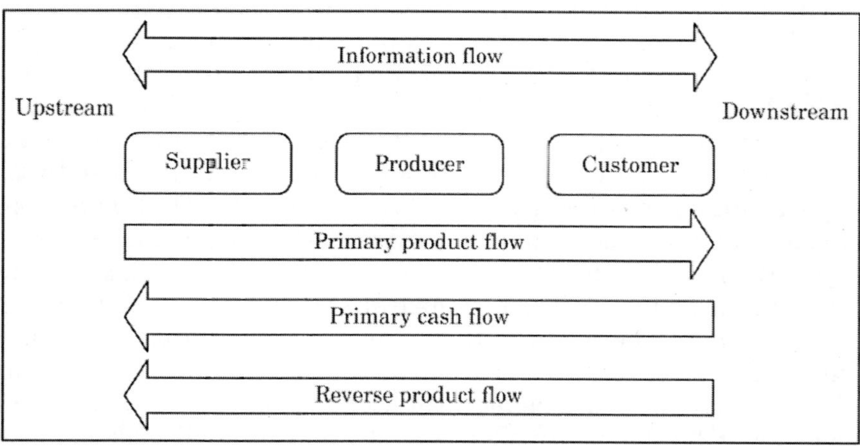

Fig. 1.2: Supply chain flows.

In the Fig. 1.2 shows that there are three stage Vendor, producer, customer play important role in supply chain management. Information flow is bi-directional. In SCM information flow from Vendor to customer and customer to Vendor. Primary product flow from Vendor to the customer. Primary cash flow from customer to Vendor. Reverse product flow also from customer to Vendor.

1.4. VENDOR SELECTION IN SUPPLY CHAIN MANAGEMENT

Vendor of all organizations realizes that obtaining products at the right cost, in the right quantity, with the right quality at the right time from the right source is crucial for their growth and survival today. If the buyer chooses the right vendor, the buyer can promote its competitive advantage in the market. Therefore, an efficient vendor selection process needs to be in place and of paramount importance for successful supply chain management. However, this decision becomes complicated in case of multiple Vendors, multiple conflicting criteria, and indefinite parameters. In addition to the uncertainty and vagueness of the experts' opinion is the prominent characteristic of the problem. The solution to this problem begins with the realization of the need for a vendor with specific requirements and therefore the need of a robust vendor selection process. Then determination and formulation of decision criteria are done with the help of a survey of purchase experts. This helps in reducing prejudices and biases. These criteria depend on the product and environment of purchase. Pre-qualification (initial screening and drawing up a shortlist of potential vendors from a large list) is also done with the help of Purchase experts. Finally, the vendor selection is done by evaluating the shortlisted vendors on the listed criteria. It is a multi-criteria decision-making problem which involves multiple criteria which can be tangible as well as intangible. In this study a very comprehensive application of Analytic Hierarchy process. The AHP is designed to solve complex multi-criteria decision problems. It is based on the human ability to make sound judgments about small problems and comparisons. The judgment made by an experienced individual between two vendors evaluated on one criterion is comparatively more accurate than the judgments made between multiple vendors evaluated on multiple criteria. AHP also facilitates decision making by organizing perceptions, feelings, judgments, and memories into a framework. In the end, it is suggested to develop a formal, systematic and rational selection model for Vendor Selection.

XYZ enterprises are the manufacture of various types of automobile product like a piston, gear, helical spring etc. As a part of its strategy, division outsources the noncore mechanical jobs to Private vendors due to capacity constraints and facility constraints. The outsourcing department of XYZ Enterprises is responsible for the techno-commercial function of registering new vendors, selecting the appropriate vendors for the job from the vendor list, finalization, sending enquiries, tendering, making comparative statement, negotiations, purchase proposal, placing purchase order, quality inspections, receipt of goods and final payment to vendors.

Few of the problems described below

a) Default by vendors in supplying quality goods and meeting delivery schedules is a major concern for management.

b) Lots of cases of Re tendering are reported where the vendors are sent the inquiries but they don't respond. Increasing Administrative cost per order is a concern for management.

c) Quality of products is being compromised by many vendors and there is nothing much that can be done after placement of purchase order.

d) Cases are observed where inquiries were also sent to vendors who don't manufacture the parts mentioned in the inquiry.

e) Increasing dependency on very few vendors is obstructing in-house production schedules. Delays are causing cost escalations in many products.

f) Some vendors are taking the order just to get associated with XYZ Enterprises and with no intention to fulfill its terms.

g) Some vendors are taking excess orders with respect to their capacities whereas some are having lots of spare capacities unutilized.

h) The lack of classification of vendors according to a class of components is resulting in an unfair competition between high-quality vendor and low-quality vendor.

i) Presently Vendors are not given reasons for not qualifying in the preliminary shortlisting.

All these above points indicate that there is no consistent approach to select the vendors. The evaluators are performing the vendor selection procedures without full consideration of all the factors. There are general restrictions imposed on the selection procedure based on the lowest price, which may not completely fulfill what is required in terms of quality, delivery, service etc. Therefore, this research takes into consideration the whole criteria and sub-criteria that control the vendor selection No method for selecting vendors by bringing consensus within the department process. This research will try to develop a model for selecting the best vendors who full fill all satisfying criteria. Present research has been conducted for highly critical mechanical jobs outsourced in XYZ Enterprises. It is an attempt to improve the selection process in general and develop a model that can be used regularly to restrict these issues and problems at the early stage of vendor selection and achieve fair competition among vendors. Vendor selection of any manufacturing firm is done on selected criteria and its sub-criteria. Criteria are selected for vendor selection: quality, price, service, business overall performance, technical ability and on time delivery. Each Criterion has various sub-criteria *e.g.,* Reliability and durability are sub-criteria of quality.

1.4.1. Quality

The quality parameter of Vendor has been established as a primary concern in the vendor selection process for decades. Quality is no longer simply applies to the product itself but also applies to the service and other aspects of the vendor-manufacturer relationship. The quality factor was measured in terms of vendor ability to provide inputs that are reliable (consistency over a range of past job works) and durable (measure of useful life of the product/no deviation in design parameters with time), possessing the required quality systems, adherence to quality tools, percent rejection and vendor reputation and position in the market.

1.4.2. Price

Best pricing is not just the lowest one but a Competitive Pricing which is a win-win price for both and doesn't poses any hurdle in supplies after placement of Order, understanding and willingness to follow financial security clauses of XYZ company without any defaults, and compliance with payment terms (suitability of terms and conditions regarding payment of invoices, open accounts, credit letter, bank guarantee, bond, insurance and payment schedule), and payment procedures understanding.

1.4.3. Service

The service factor includes the vendor's attitude to handling complaints, the ability to maintain product/service, and the ability and willingness to provide technical support and training, and the vendor flexibility (the ability and willingness of the vendor to change order volumes and a mix of products and delivery schedules).

1.4.4. Business Overall Performance

The business overall performance factor has been measured on the basis of the importance of the following dimensions: the vendor's financial stability, Quality Certifications, knowledge of the market and if the vendor is using information systems, management capability (includes management's commitment, and willingness to develop a closer working relationship with the buyer) and performance history of the vendor which relates to the vendor reputation for performance.

1.4.5. Technical Ability

The ability of the vendor to provide technical support, the use of current technology, technical know-how, understanding of technology and experience in the related class of jobs, the responsiveness of the vendor to changes in purchase quantities and due dates and personnel capabilities (the overall skills and abilities of the workforce).

1.4.6. On Time Delivery

If a vendor submits the lowest price, it doesn't mean much to the XYZ Company. The vendor is late in deliveries as it delays XYZ Company's production schedules. The delivery factor has been measured based on the ability and willingness to expedite an order, how quickly a vendor can deliver (lead time), spare capacity available to meet XYZ Company's requirement, upcoming delivery commitments, safety and security of components during the transportation and geographical location of the vendor which can affect the deliveries.

1.5. ANALYTICAL HIERARCHICAL PROCESS

In AHP, we structure the problem in the form of a hierarchy and then capture the criteria, sub-criteria, and alternatives. All the criteria are compared fairly to determine their relative weights. Then, the alternatives are compared fairly with respect to each criterion. The final out-come of the procedure is a score for each alternative AHP avoids the main drawback of the traditional linear scoring model, which assigns weights and scores arbitrarily. At the same time, it can make a trade-off between the quantitative and qualitative criteria. The important advantages of AHP are its simplicity, robustness, and the ability to incorporate intangibles into the decision-making process.

For a selection of a vendor for critical mechanical jobs at XYZ Company, the data of previous purchases are not recorded in the form required for the neural network. Also, it is a technique which needs investment and it will not be appropriate to the extent of Outsourcing activities done at present. Mathematical programming will be quite complex as quantification of some factors is difficult and we are considering many criteria. The scoring model seems quite simple and the replica of what is being practiced at present. Statistical analysis will also require an extensive survey and the number of expert purchasers dealing with these problems was less. Hence it was judged that AHP will provide the required ability to deal with intangible criteria, the simplicity of formulation, the better condition for the purchasers to make choices and it will be best suited for this case. AHP is a multiple criteria decision-making approach based on the reasoning, knowledge, experience, and perceptions of experts in the field. It is a robust technique that allows managers to determine preferences of criteria for selection purposes, quantify those preferences, and then aggregate them across diverse criteria. AHP is particularly used in this study to overcome the inconsistency associated with the vendor selection problem and to deal with subjective decision criteria. The AHP provides an effective structure for group decision-making by imposing a discipline on the groups thought processes and that is what we have done by asking 5 purchase experts to fill the pair-wise questionnaire. The consensual nature of AHP influences group decision-making and improves the consistency of the judgments.

1.5.1. AHP Steps

All the steps for AHP is shown in Figs. 1.3 & 1.5.

a) Specify the set of criteria for evaluating the vendors' proposals, then find the sub-criteria within each criterion and construct a decision hierarchy as shown in Fig. 1.4.

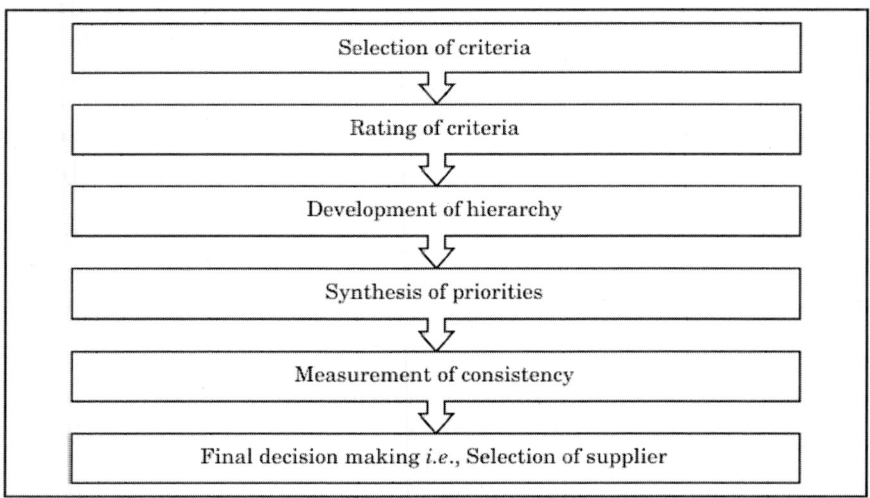

Fig. 1.3: Vendor selection using AHP.

b) Obtain the pair-wise comparisons of the relative importance of the criteria in achieving the goal, determine whether the input data for the importance of criteria and sub-criteria satisfy a consistency test and compute the priorities or local weights of the criteria and sub-criteria based on this information.

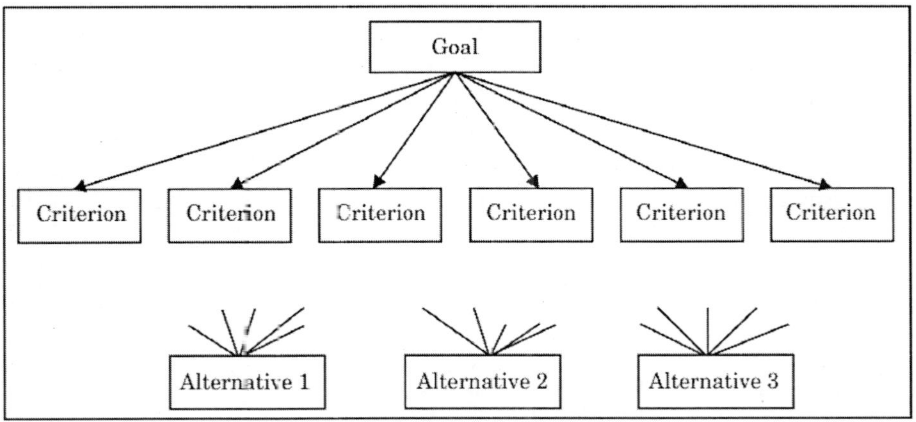

Fig. 1.4: Simple AHP hierarchy.

c) Obtain the pair-wise comparisons to obtain the priorities of the alternatives with respect to each criterion as given in Table 1.1. Local priorities of the alternatives are then multiplied by the weight of the respective criteria and the results are summed up to produce the overall priority of each alternative (vendor). Then determine whether the input data for vendors satisfy a consistency test; if not, re-do the pair-wise comparisons and obtain corresponding priorities of vendors in the form of local weights.

Table 1.1: Level of importance of vendor.

Preference weights/ level of importance	*Definition*	*Explanation*
1	Equally preferred	Two vendors are equal in this criterion
3	Moderately	Experience and judgment slightly favors' vendor a over vendor b
5	Strongly	Experience and judgement strongly and essentially favors vendor a over vendor b
7	Very strongly	Vendor a is strongly favored over b and its dominance is well demonstrated
2, 4, 6, 8	Intermediate values	Used to represent compromise between the preferences listed above

d) A final priority vector of each vendor is obtained by synthesizing all the priority vectors *i.e.,* the local weights of criteria, sub-criteria and that of vendors for each sub criteria. The preferences are quantified in Pair-wise comparisons by using a nine-point scale.

The hierarchy can be visualized through the Fig. 1.2, with the goal at the top, the alternative at the bottom, and the filling criteria up to the middle. In such diagrams, each box is called node. The boxes descending from any node are called its children. The node from which a child node descends is called the parent. Applying these definitions to the diagram below, six criteria are children of the goal, the goal is the parent of each six criteria. Each alternative is the child of each criterion. In practice, many criteria have one or more layer of sub-criteria. These

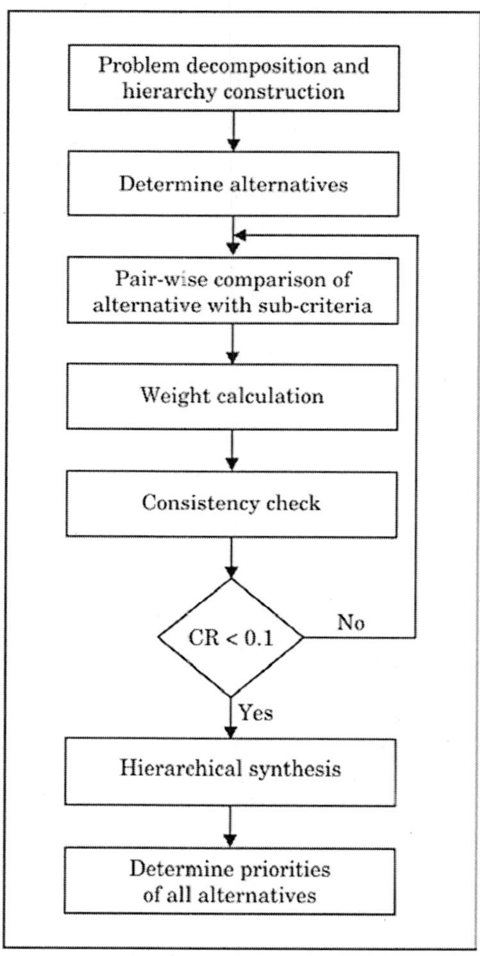

Fig. 1.5: Flow chart of AHP (Analytic Hierarchy Process)

are not shown in the simplified diagram; the lines between the alternatives and criteria are often omitted or reduced the number. Regardless of any such simplifications in the diagram, in the actual hierarchy, each alternative is connected every one of its covering criteria has the lowest level of criteria, sub-criteria, etc.

More and more researchers realizing that AHP is an important generic method and are applying it to various manufacturing areas. In addition to the wide application of AHP in manufacturing areas, recent research and industrial activities of applying AHP on other selection problems are also quite active. AHP can thus apply any selection problem in manufacturing firms.

1.6. STRUCTURING VENDOR SELECTION PROBLEM: DEVELOPING A HIERARCHICAL MODEL

This phase involves formulating an appropriate hierarchy of the AHP model consisting of the goal, strategic factors, criteria and sub-criteria and the alternatives. The goal of our problem is to prioritize vendors for critical mechanical jobs of XYZ Company which can meet the quality and delivery requirements for Production and at the same time follows the strategic factors of Purchase policy in terms of fairness in selection and getting the product at the best competitive price. Here we want to rationalize the process of vendor selection and make it analytical and consensual so that there is no scope for personal bias. This goal along with the strategy is placed on the first level of the hierarchy. The second level of the hierarchy occupies the criteria defining the 6 criteria quality, price, service, business overall performance, technical capability and on time delivery. Criteria and sub-criteria used in level one and two of the AHP hierarchy can be assessed using the basic AHP approach of pair-wise comparisons of elements in each level. The pair-wise comparison judgments for criteria and sub-criteria are obtained from the group of 5 purchase experts who were given the questionnaire in the form of matrices and asked to find the extent of importance of factors at both the levels of criteria and sub-criteria and rate them on the scale. This approach is found to be very useful in collecting data. This evaluation team of five members is those who are frequently involved in vendor selection of critical jobs within the organization. Of these five evaluators, one is a senior manager from the outsourcing department. Two evaluators are deputy managers with good experience in dealing with wide range of Vendors. Other Two evaluators are senior technicians with good experience in critical mechanical jobs of XYZ company. Each one of them has more than five years of experience in vendor selection projects for XYZ. Thus, the evaluators have sufficient experience in vendor selection for the critical mechanical job and, hence, are qualified to assign pair-wise comparison judgments for the criteria and sub-criteria. The function of the pair-wise comparisons is finding the relative importance of the criteria and sub-criteria, which is rated by the nine-point scale proposed by Saaty (1980), as shown in above Table 1.1, indicating the level of relative importance from equal, moderate, strong, very strong, to extreme level by 1, 3, 5, 7, and 9, respectively. The intermediate values between two adjacent arguments were represented by 2, 4, 6, and 8.

While obtaining the pair-wise judgments for all criteria and sub-criteria, all the matrices are developed and shown in Table no. 1.6 to 1.50, the next step is the computation of local weights of elements (criteria along the rows of the matrices) in the matrix which is done by following the steps mentioned below:

a) Calculate the total of each column.
b) Divide the elements of that column by the total of the column. This is the normalized matrix.
c) Finally, add the elements in each resulting row and divide this sum by the number of elements in the row to get the average. These are the local weights corresponding to that criteria/sub-criteria.

The Consistency ratio (C.R.) is also calculated for checking the consistency of answers given by a group of 5 purchase experts. The consistency ratio (C.R.) determines the acceptance of the priority weighting. The consistency test is one of the essential features of the AHP method, which aims to eliminate the possible inconsistency revealed in the criteria weights through the computation of consistency level of each matrix. The consistency ratio (CR) was used to determine and justify the inconsistency in the pair-wise comparison made by the group of 5 purchase experts in our case study. Based on Saaty's (1980) empirical suggestion that a C.R. = 0.10 is acceptable, it is concluded that the foregoing pair-wise comparisons to obtain attribute weights are reasonably consistent. If the CR value is lower than the acceptable value, the weight results are valid and consistent. In contrast, if the CR value is larger than the acceptable value, the matrix results are inconsistent and are surveyed again or exempted for the further analysis. Estimating the consistency ratio is as follows:

a) Multiply the weights of criteria with the column elements and obtain a sum of each row elements.
b) Divide all the elements of the weighted sum of row by their respective priority.
c) Compute the average of these values to get lambda max (λ max)
d) Obtain Consistency Index (CI) by using CI= (λmax–n)/(n–1)
e) Select appropriate value of random consistency ratio, RI, for a given matrix size by using below as given in Table 1.2.
f) Then calculate the consistency ratio, CR, as follows: CR= CI/RI
g) If the value of CR is less than 0.1, the judgments are acceptable else pair-wise comparison needs to be done again

Table 1.2: Random index

Size of matrix	1	2	3	4	5	6	7	8	9	10
Random consistency	0	0	0.58	0.9	1.12	1.24	1.32	1.41	1.45	1.49

1.7. VENDOR SELECTION CRITERIA

After reviewing the literature, the list of 32 criteria applicable for outsourcing of critical Mechanical jobs for XYZ enterprise in the present scenario was listed below in Table 1.3.

Table 1.3: List of criteria and sub-criteria.

Sl. no.	*Criteria*	*Sub-criteria*
1	Quality	Product durability (lifespan of job work is as per design)
2		Product reliability (consistency over a range of past job works)
3		Quality management systems (control on procedures)
4		Percent rejections
5		Adherence to quality tools
6		Reputation and position in the market
7	Price	Competitive pricing
8		Understanding and willingness to follow financial security clause
9		Payment terms
10		Payment procedure understanding
11	Service	Attitude towards handling of complaints
12		Ability to maintain after sales service
13		Ability and willingness to provide technical support and training if req.

Table 1.3: (*Contd...*)

Table 1.3: (*Contd...*)

Sl. no.	*Criteria*	*Sub-criteria*
14		Flexibility (order volumes, mix of products, payments, freight, price reduction, order frequency and amount)
15	Business overall performance	Financial stability (sustainability)
16		Quality performance (ISO /as 9100 accreditation)
17		Knowledge of the market
18		Use of information systems (communication)
19		Management capability (includes management commitment)
20		Performance history (vendors reputation for performance)
21	Technical capability	Offering technical support when required
22		Technical know-how (vendor has the required skill set and possess good understanding of technology)
23		Vendor experience in related class of jobs
24		Responsiveness to change in quantity and due dates
25		Use of current technologies
26		Personnel technical abilities
27	On time delivery	Delivery lead time (speed)
28		Spare capacity to meet the requirements
29		Upcoming delivery commitments
30		Ability and willingness to expedite an order (continuation of cooperation)
31		Safety and security components
32		Suitable geographical location

QUESTIONAIRE PART 1

Sex	:	Male	☐	Female	☐		
Age	:	20–34	☐	35–49	☐	50–60	☐
Designation	:	Top Mgmt	☐	Mid Mgmt	☐	Lower Mgmt	☐
Specialisation	:	Engineering	☐	Commerce	☐	Others	☐
Purchase Experience	:	>5 Years	☐	<5 Years	☐		

Note: Kindly read before proceeding further

While answering the questionnaire the survey data by 10 officers is given in Table 1.5, considering that the indenter's requirement is to outsource job-work for highly accurate intricate machining job and the part requires high dimensional accuracy as well. The criticality of the part in view of all outsourced parts in XYZ enterprises. is high and can be termed as "A" class among all the machining jobs that are outsourced. The scaling criteria is given in Table 1.4.

Table 1.4: Scaling of criteria.

Item	*Unimportant*	*Little important*	*Moderately important*	*Important*	*Very important*
Scale	1	2	3	4	5

Table 1.5: Survey data by 10 officers.

Sl. no.	*Criteria*	*Sub-criteria*	*a*	*b*	*C*	*d*	*e*	*f*	*g*	*h*	*i*	*j*	*RI*	*RANK*
1	Quality	1	5	5	4	4	5	5	4	5	5	4	0.92	2
2		2	5	5	5	5	4	5	4	5	5	4	0.94	1
3		3	3	4	3	5	4	5	3	5	5	3	0.8	9
4		4	3	3	4	4	3	5	4	4	4	5	0.78	12
5		5	2	4	3	2	2	5	4	5	4	3	0.68	20
6		6	2	2	1	4	1	5	4	3	4	4	0.6	30
7	Price	1	5	5	5	1	5	4	4	4	4	4	0.82	5
8		2	2	4	3	3	4	5	3	4	4	4	0.72	18
9		3	3	3	1	3	4	3	4	4	3	1	0.58	32
10		4	2	2	2	4	3	3	4	5	4	2	0.62	29
11	Service	1	5	3	4	3	5	4	5	4	4	5	0.84	4
12		2	2	3	3	4	4	2	3	4	4	4	0.66	26
13		3	4	3	3	2	4	4	5	3	4	2	0.68	20
14		4	3	3	5	3	2	4	4	4	4	5	0.74	15
15	Business overall performance	1	5	3	3	3	5	5	4	4	4	5	0.82	5
16		2	5	4	4	4	2	3	4	3	4	1	0.68	20
17		3	5	4	3	4	3	2	5	4	3	1	0.68	20
18		4	5	3	2	3	4	4	5	4	3	4	0.74	15
19		5	5	3	3	3	4	5	4	5	4	4	0.8	9
20		6	4	3	3	4	4	4	5	4	4	5	0.8	9
21	Technical capability	1	3	3	3	3	4	4	3	4	4	2	0.66	26
22		2	4	4	2	4	5	5	4	4	4	5	0.82	5
23		3	4	4	3	4	3	4	4	4	4	5	0.78	12
24		4	4	4	3	3	2	4	3	3	3	1	0.6	30
25		5	5	4	2	4	2	4	4	4	4	1	0.68	20
26		6	3	4	3	3	3	4	4	3	3	2	0.64	28
27	On time delivery	1	5	4	4	3	5	5	5	3	5	4	0.86	3
28		2	4	4	2	4	4	4	4	4	3	4	0.74	15
29		3	5	4	5	3	3	5	4	4	4	1	0.76	14
30		4	5	4	3	3	4	4	4	5	4	5	0.82	5
31		5	3	4	5	2	3	5	3	4	4	1	0.68	20
32		6	2	3	3	4	4	4	4	5	4	3	0.72	18

In this way by following the above method we obtained the local weights for all the 6 criteria and 19 sub-criteria. The global weights are calculated by multiplying the local weights of 6 criteria with local weights of 19 sub-criteria. This method converts individual preferences into ratio scale weights that can be combined into a linear additive weight for each alternative. The resultant can be used to compare and rank the alternatives and, hence, assist the decision maker in making a choice.

Then this relative index is used to rank the criteria and by having a discussion among 5 purchasing experts we set a cut-off Relative Importance Index value of 0.7 to select all the criteria above 0.7. In this way, the final 19 parameters or criteria were finalized for further application of AHP.

1.8. BENEFITS OF THE ANALYTICAL HIERARCHY PROCESS

The advantages of AHP over other multi-criteria methods are its flexibility, intuitive appeal to the decision makers and its ability to check inconsistencies. Generally, users find the pair-wise comparison form of data input straightforward and convenient.

Additionally, the AHP method has the distinct advantage that it decomposes a decision problem into its constituent parts and build hierarchies of criteria. Here, the importance of each element (criterion) become clear.

AHP helps to capture both subjective and objective evaluation measures. While providing a useful mechanism for checking the consistency of the evaluation measures and alternatives, AHP reduces bias in decision-making.

The AHP method supports group decision-making through consensus by calculating the geometric mean of the individual pair-wise comparison.

1.9. LIMITATION OF AHP

Although the Analytic hierarchy process has been the subject of the many research papers and the consensus is that the technique is both technically valid and practically useful, there are general critics of the method.

1. Since there is no theoretical basis for constructing hierarchies, AHP users can construct different hierarchies for identical decision situation, possibly producing a different solution.

2. AHP ranking is claimed to be arbitrary because they are based on subjective opinion using the ratio scale.
3. The results from the AHP analysis totally depend upon the rating given by experts which may vary person to person so there is a risk of variation in results due to personal errors. It depends upon discussion between assessor and division manager it is very lengthy work to compare each factor in the network.
4. It is very difficult to control the consistency ratio CR below 0.1.
5. Experts are required to do the discussion for a long time to give the ratings for the whole network which increase the cost of model development
6. Many time in the AHP network development the independent factors are correlated which make the analysis unnecessarily complicated.
7. The process has no sound underlying statically theory.

Table 1.6: Pair-wise comparison of criteria.

Criteria	*Quality*	*Price*	*Service*	*Business overall performance*	*Technical capability*	*On time delivery*
Quality	1.00	9.00	7.00	7.00	1.00	3.00
Price	0.11	1.00	0.33	0.33	0.14	0.14
Service	0.14	3.00	1.00	3.00	0.20	0.20
Business overall performance	0.14	3.00	0.33	1.00	0.20	0.20
Technical capability	1.00	7.00	5.00	5.00	1.00	1.00
On time delivery	0.33	7.00	5.00	5.00	1.00	1.00
Total	2.73	30.00	18.67	21.33	3.54	5.54

Table 1.7: Normalized matrix of criteria.

Criteria	*Quality*	*Price*	*Service*	*Business overall performance*	*Technical capability*	*On time delivery*	*Criteria local weights*
Quality	0.37	0.30	0.38	0.33	0.28	0.54	0.37
Price	0.04	0.03	0.02	0.02	0.04	0.03	0.03
Service	0.05	0.10	0.05	0.14	0.06	0.04	0.07
Business overall performance	0.05	0.10	0.02	0.05	0.06	0.04	0.05
Technical capability	0.37	0.23	0.27	0.23	0.28	0.18	0.26
On time delivery	0.12	0.23	0.27	0.23	0.28	0.18	0.22

Table 1.8: Consistency check for criteria.

Criteria	*Quality*	*Price*	*Service*	*Business overall performance*	*Technical capability*	*On time delivery*	*λ value*
Quality	0.37	0.26	0.51	0.36	0.26	0.66	6.62
Price	0.04	0.03	0.02	0.02	0.04	0.03	6.21
Service	0.05	0.09	0.07	0.15	0.05	0.04	6.33
Business overall performance	0.05	0.09	0.02	0.05	0.05	0.04	6.03
Technical capability	0.37	0.20	0.37	0.26	0.26	0.22	6.41
On time delivery	0.12	0.20	0.37	0.26	0.26	0.22	6.49
						λ_{max}	6.35
				CI = 0.07	RI = 1.24	CR =0.05	

$$CI = \frac{\lambda_{max} - N}{N - 1} = \frac{6.35 - 6}{5} = 0.07$$

For 6*6 matrix value for $RI = 1.24$. So, $CR = \frac{CI}{RI} = \frac{0.07}{1.24} = 0.05$

Value of CR is less than 0.1 so judgments are acceptable.

In quality, there are four sub-criteria first is product durability, second is product reliability, third is quality management and fourth is percent rejection.

Table 1.9: Pair-wise comparison sub-criteria quality.

Quality	*Product durability (job work is as per design)*	*Product reliability (consistency over a range of past job works)*	*Quality management (control on procedures)*	*Percent rejections*
Product durability (job work is as per design)	1.00	3.00	5.00	5.00
Product reliability (consistency over a range of past job works)	0.33	1.00	5.00	3.00
Quality management systems (control on procedures)	0.20	0.20	1.00	1.00
Percent rejections	0.20	0.33	1.00	1.00

Table 1.10: Normalized matrix for sub-criteria quality.

Quality	*Product durability*	*Product reliability*	*Quality management systems*	*Percent rejections*	*Local weights*
Product durability	0.58	0.66	0.42	0.50	0.54
Product reliability	0.19	0.22	0.42	0.30	0.28
Quality management systems	0.12	0.04	0.08	0.10	0.09
Percent rejections	0.12	0.07	0.08	0.10	0.09

Table 1.11: Consistency check for quality sub-criteria.

Quality	*Product durability*	*Product reliability*	*Quality management systems*	*Percent rejections*	*Local weights*
Product durability	0.54	0.85	0.43	0.47	4.23
Product reliability	0.18	0.28	0.43	0.28	4.14
Quality management systems	0.11	0.06	0.09	0.09	4.00
Percent rejections	0.11	0.09	0.09	0.09	4.09
Product durability	0.54	0.85	0.43	0.47	4.23
				λ_{MAX}	4.12
			CI = 0.04	RI =0.9	CR =0.04

$$\text{CI} = \frac{\lambda_{\max} - N}{N-1} = \frac{4.12-4}{4-1} = \frac{0.12}{3} = 0.04$$

$$\text{CR} = \frac{CI}{RI} = \frac{0.04}{0.9} = 0.045$$

Value of CR is less than 0.1 so, judgments is acceptable which is taken for quality sub-criteria.

Table 1.12: Pair-wise comparison of sub-criteria of price.

Price	*Competitive pricing*	*Understanding and willingness to follow financial security clause*
Competitive pricing	1.00	0.20
Understanding and willingness to follow financial security clause	5.00	1.00

Table 1.13: Normalized matrix for price sub-criteria.

Price	*Competitive pricing*	*Understanding and willingness to follow financial security clause*	*Local weights*
Competitive pricing	0.17	0.17	0.17
Understanding and willingness to follow financial security clause	0.83	0.83	0.83

Table 1.14: Consistency check of price sub-criteria.

Price	*Competitive pricing*	*Understanding and willingness to follow financial security clause*	*λ value*
Competitive pricing	0.17	0.17	2.00
Understanding and willingness to follow financial security clause	0.83	0.83	2.00
		λ_{MAX}	2.00

Here sub-criteria of price are two which is competitive price and understanding and willingness to follow financial security clause, but average lambda value is two and is equal to no. of sub-criteria. So that consistency ratio is not determined.

There are two service sub-criteria. First is the attitude towards handling complaints and second is flexibility.

Table 1.15: Pair-wise comparison of service sub-criteria.

Service	*Attitude towards handling of complaints*	*Flexibility*
Attitude towards handling of complaints	1.00	5.00
Flexibility	0.20	1.00
Total	1.20	6.00

Table 1.16: Normalized matrix of service.

Service	*Attitude towards handling of complaints*	*Flexibility*	*Local weights*
Competitive pricing	0.17	0.17	2.00
Attitude towards handling of complaints	0.83	0.83	0.83
Flexibility	0.17	0.17	0.17

Table 1.17: Service consistency check.

Service consistency check	*Attitude towards handling of complaints*	*Flexibility*	*ë Value*
Attitude towards handling of complaints	0.83	0.83	2.00
Flexibility	0.17	0.17	2.00
		λ_{MAX}	2.00

Inconsistency check of service, we get max lambda value is equal to the number of sub-criteria. So that we cannot determine consistency ratio for this case.

In business overall performance there are four sub-criteria. First is financial stability, second is the use of information system, third is management capability (includes management commitment) and the last one is performance history as shown in Figs. 1.7 to 1.12.

Table 1.18: Pair-wise comparison sub-criteria - business overall performance.

Business overall performance	*Financial stability*	*Use of information systems*	*Management capability*	*Performance history*
Financial stability	1.00	5.00	0.33	0.20
Use of information systems	0.20	1.00	0.20	0.14
Management capability	3.00	5.00	1.00	1.00
Performance history	5.00	7.00	1.00	1.00
Total	9.20	18.00	2.53	2.34

Table 1.19: Normalized matrix of business overall performance.

Business overall performance	*Financial stability*	*Use of information systems*	*Management capability*	*Performance history*
Financial stability	0.15	0.27	0.12	0.09
Use of information systems	0.03	0.05	0.07	0.06
management capability	0.45	0.27	0.36	0.44
Performance history	0.75	0.38	0.36	0.44

$$\text{CI} = \frac{\lambda_{\max} - N}{N-1} = \frac{4.21-4}{4-1} = \frac{0.21}{3} = 0.07$$

For 4*4 matrix value of RI = 0.90, so CR = $\frac{CI}{RI} = \frac{0.07}{0.90} = 0.08$

Table 1.20: Consistency matrix of business overall performance.

Business overall performance	*Financial stability*	*Use of information systems*	*Management capability*	*Performance history*	*λ value*
Financial stability	0.15	0.27	0.12	0.09	4.17
Use of information systems	0.03	0.05	0.07	0.06	4.02
Management capability	0.45	0.27	0.36	0.44	4.26
Performance history	0.75	0.38	0.36	0.44	4.40
				λ_{MAX}	4.21
				CI	0.07
				RI	0.90
				CR	0.08

Value of Consistency ratio of business overall performance is 0.08 which is less than 0.1 so that judgments is acceptable.

In technical ability, there are two sub-criteria first is Technical know (the vendor has required a skill set and possess a good understanding of technology) and second is vendors experience in the related class of job.

Table 1.21: Pair-wise comparison of sub-criteria of technical.

Technical capability	*Technical know-how*	*Vendor experience in related class of jobs*
Technical know-how	1.00	7.00
Vendor experience in related class of jobs	0.14	1.00
Total	1.14	8.00

Table 1.22: Normalized matrix of technical ability.

Technical capability	*Technical know-how*	*Vendor experience in related class of jobs*	*Local weights*
Technical know-how	0.88	0.88	0.88
Vendor experience in related class of jobs	0.13	0.13	0.13

Table 1.23: Consistency check for technical ability.

Technical capability	*Technical know-how*	*Vendor experience in related class of jobs*	*λ value*
Technical know-how	0.88	0.88	2.00
Vendor experience in related class of jobs	0.13	0.13	2.00
		λ_{MAX}	2.00

Value of lambda max is equal to the no of sub-criteria so that consistency ratio is not evaluated.

There are five sub-criteria of on-time delivery. First is delivery lead time, second is the spare capacity to meet the requirement, third is an upcoming delivery commitment, fourth is the ability and willingness to expedite an order and last one is suitable geographical location.

Table 1.24: Pair-wise comparison of sub-criteria of on time delivery.

On time delivery	*Delivery lead time (speed)*	*Spare capacity to meet the requirements*	*Upcoming delivery commit-ments*	*Ability and willing-ness*	*Suitable geograp-hical location*
Delivery lead time (speed)	1.00	5.00	5.00	0.33	7.00
Spare capacity to meet the requirements	0.20	1.00	3.00	0.20	5.00
Upcoming delivery commitments	0.20	0.33	1.00	0.14	3.00
Ability and willingness	3.00	5.00	7.00	1.00	9.00
Suitable geographical location	0.14	0.20	0.33	0.11	1

Table 1.25: Normalized matrix of on time delivery sub-criteria.

On time delivery	*Delivery lead time (speed)*	*Spare capacity to meet the requirements*	*Upcoming delivery commit-ments*	*Ability and willing-ness*	*Suitable geograp-hical location*	*Local weight*
Delivery lead time (speed)	0.29	0.63	0.33	0.16	0.24	0.29
Spare capacity to meet the requirements	0.06	0.13	0.20	0.10	0.17	0.06
Upcoming delivery commitments	0.06	0.04	0.07	0.07	0.10	0.06
Ability and willingness	0.86	0.63	0.47	0.49	0.31	0.86
Suitable geographical location	0.29	0.63	0.33	0.16	0.24	0.29

$$\text{CI} = \frac{\lambda_{max} - N}{N - 1} = \frac{5.36 - 5}{5 - 1} = \frac{0.36}{4} = 0.09$$

$$\text{CR} = \frac{CI}{RI} = \frac{0.09}{1.12} = 0.08$$

Table 1.26: Consistency check for on time delivery sub-criteria.

On time delivery	*Delivery lead time (speed)*	*Spare capacity to meet the requirements*	*Upcoming delivery commit-ments*	*Ability and willing-ness*	*Suitable geograp-hical location*	*Local weight*
Delivery lead time (speed)	0.29	0.63	0.33	0.16	0.24	5.78
Spare capacity to meet the requirements	0.06	0.13	0.20	0.10	0.17	5.20
Upcoming delivery commitments	0.06	0.04	0.07	0.07	0.10	5.06
Ability and willingness	0.86	0.63	0.47	0.49	0.31	5.62
Suitable geographical location	0.04	0.03	0.02	0.05	0.03	5.15
					λ_{MAX}	5.36
				CI = 0.09	RI = 1.12	CR = 0.08

Value of consistency ratio is 0.08 which is less than 0.1 so that judgment is acceptable.

Table 1.27: Pair-wise comparison of vendors on quality sub-criteria 1.

Product durabi-lity	*Ven-dor 1*	*Ven-dor 2*	*Ven-dor 3*	*Ven-dor 4*	*Ven-dor 5*	*Ven-dor 6*	*Ven-dor 7*	*Ven-dor 8*	*Ven-dor 9*	*Ven-dor 10*	*Local weig-hts*
Vendor 1	1.00	3.00	1.00	3.00	1.00	3.00	0.33	0.33	0.33	3.00	0.091
Vendor 2	0.33	1.00	0.20	1.00	0.20	0.33	0.33	0.20	0.20	0.20	0.028
Vendor 3	1.00	5.00	1.00	3.00	0.33	0.33	0.33	0.20	0.33	3.00	0.071
Vendor 4	0.33	1.00	0.33	1.00	0.33	0.33	0.33	0.20	0.33	0.33	0.032
Vendor 5	1.00	5.00	3.00	3.00	1.00	1.00	1.00	1.00	1.00	3.00	0.130
Vendor 6	0.33	3.00	3.00	3.00	1.00	1.00	0.20	0.20	0.33	3.00	0.080
Vendor 7	3.00	3.00	3.00	3.00	1.00	5.00	1.00	1.00	3.00	3.00	0.180
Vendor 8	3.00	5.00	5.00	5.00	1.00	5.00	1.00	1.00	3.00	3.00	0.203
Vendor 9	3.00	5.00	3.00	3.00	1.00	3.00	0.33	0.33	1.00	3.00	0.128
Vendor 10	0.33	5.00	0.33	3.00	0.33	0.33	0.33	0.33	0.33	1.00	0.056
Total	13.33	36.00	19.87	28.00	7.20	19.33	5.20	4.80	9.87	22.53	

Table 1.28: Pair-wise comparison of vendors on quality sub-criteria 2.

Product reliability (consistency over a range of past job works	*Vendor 1*	*Vendor 2*	*Vendor 3*	*Vendor 4*	*Vendor 5*	*Vendor 6*	*Vendor 7*	*Vendor 8*	*Vendor 9*	*Vendor 10*	*Local weights*
Vendor 1	1.00	0.33	3.00	0.33	1.00	0.20	0.20	0.20	0.33	3.00	0.044
Vendor 2	3.00	1.00	3.00	1.00	3.00	0.33	0.20	0.20	1.00	5.00	0.078
Vendor 3	0.33	0.33	1.00	0.33	0.33	0.20	0.20	0.20	0.33	1.00	0.026
Vendor 4	3.00	1.00	3.00	1.00	3.00	0.20	1.00	0.33	0.33	3.00	0.088
Vendor 5	1.00	0.33	3.00	0.33	1.00	0.20	0.20	0.20	0.33	3.00	0.044
Vendor 6	5.00	3.00	5.00	5.00	5.00	1.00	0.33	0.33	5.00	5.00	0.175
Vendor 7	5.00	5.00	5.00	1.00	5.00	3.00	1.00	1.00	5.00	5.00	0.216
Vendor 8	5.00	5.00	5.00	3.00	5.00	3.00	1.00	1.00	3.00	5.00	0.217
Vendor 9	3.00	1.00	3.00	3.00	3.00	0.20	0.20	0.33	1.00	3.00	3.00
Vendor 10	0.33	0.20	1.00	0.33	0.33	0.20	0.20	0.20	0.33	1.00	0.33
Total	26.67	17.20	32.00	15.33	26.67	8.53	4.53	4.00	16.67	34.00	

Table 1.29: Pair-wise comparison of vendors on quality sub-criteria 3.

Quality managem-ent systems (control on procedure	*Vendor 1*	*Vendor 2*	*Vendor 3*	*Vendor 4*	*Vendor 5*	*Vendor 6*	*Vendor 7*	*Vendor 8*	*Vendor 9*	*Vendor 10*	*Local weig-hts*
Vendor 1	1.00	3.00	3.00	3.00	1.00	0.33	0.20	0.33	1.00	1.00	0.079
Vendor 2	0.33	1.00	3.00	1.00	1.00	0.20	0.20	0.33	3.00	1.00	0.061
Vendor 3	0.33	0.33	1.00	0.33	1.00	0.20	0.20	0.33	1.00	1.00	0.038
Vendor 4	0.33	1.00	3.00	1.00	1.00	0.33	0.33	0.33	1.00	0.33	0.055
Vendor 5	1.00	1.00	1.00	1.00	1.00	0.33	0.33	0.33	3.00	3.00	0.073
Vendor 6	3.00	5.00	5.00	3.00	3.00	1.00	1.00	1.00	3.00	1.00	0.172
Vendor 7	5.00	5.00	5.00	3.00	3.00	1.00	1.00	3.00	5.00	5.00	0.240
Vendor 8	3.00	3.00	3.00	3.00	3.00	1.00	0.33	1.00	5.00	5.00	0.166
Vendor 9	1.00	0.33	1.00	1.00	0.33	0.33	0.20	0.20	1.00	3.00	0.051
Vendor 10	1.00	1.00	1.00	3.00	0.33	1.00	0.20	0.20	0.33	1.00	0.064
Total	16.0	20.6	26.0	19.3	14.6	5.73	4.00	7.07	23.3	21.33	

Table 1.30: Pair-wise comparison of vendors on quality sub-criteria 4.

Percent rejections	*Ven-dor 1*	*Ven-dor 2*	*Ven-dor 3*	*Ven-dor 4*	*Ven-dor 5*	*Ven-dor 6*	*Ven-dor 7*	*Ven-dor 8*	*Ven-dor 9*	*Ven-dor 10*	*Local weig-hts*
Vendor 1	1.00	1.00	3.00	1.0	3.00	3.00	0.33	1.00	1.00	5.00	0.126
Vendor 2	1.00	1.00	0.33	1.0	0.33	1.00	0.33	0.33	0.33	1.00	0.048
Vendor 3	0.33	3.00	1.00	5.0	1.00	5.00	0.33	0.33	1.00	3.00	0.108
Vendor 4	1.00	1.00	0.20	1.0	0.33	3.00	0.33	0.20	0.33	3.00	0.059
Vendor 5	0.33	3.00	1.00	3.0	1.00	3.00	0.33	1.00	1.00	3.00	0.098
Vendor 6	0.33	1.00	0.20	0.33	0.33	1.00	0.20	0.33	0.33	3.00	0.040
Vendor 7	3.00	3.00	3.00	3.0	3.00	5.00	1.00	3.00	3.00	5.00	0.234
Vendor 8	1.00	3.00	3.00	5.0	1.00	3.00	0.33	1.00	0.33	5.00	0.128
Vendor 9	1.00	3.00	1.00	3.0	1.00	3.00	0.33	3.00	1.00	5.00	0.131
Vendor 10	0.20	1.00	0.33	0.33	0.33	0.33	0.20	0.20	0.20	1.00	0.028
Total	9.2	20.0	13.0	22.6	11.3	27.3	3.7	10.4	8.5	34.0	

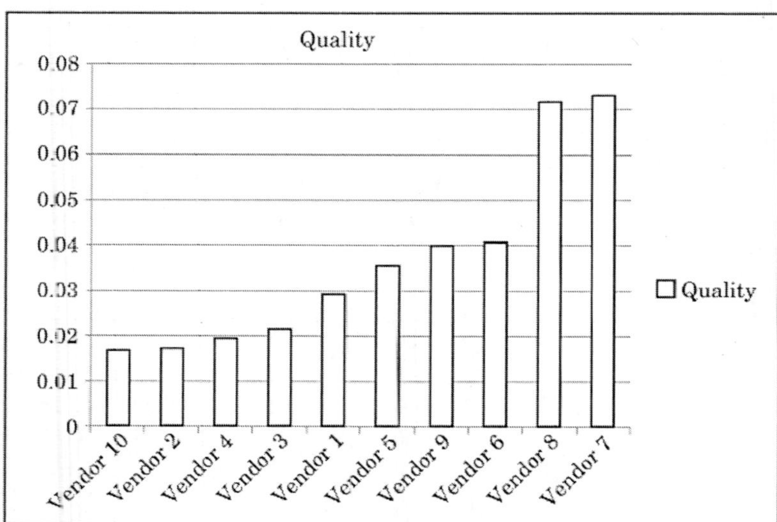

Fig. 1.6: Analysis of vendor performance w.r.t quality.

Table 1.31: Ranking of vendor according to quality.

Rank	*Vendor name*	*Global weight*	*Rank*	*Vendor name*	*Global weight*
1	Vendor 7	0.0166	2	Vendor 8	0.0171
3	Vendor 6	0.0192	4	Vendor 9	0.0215
5	Vendor 5	0.0292	6	Vendor 1	0.0357
7	Vendor 3	0.0403	8	Vendor 4	0.0406
9	Vendor 2	0.0719	10	Vendor 10	0.0733

Table 1.32: Pair-wise comparison of vendors on price sub-criteria.

Competitive pricing	*Vendor 1*	*Vendor 2*	*Vendor 3*	*Vendor 4*	*Vendor 5*	*Vendor 6*	*Vendor 7*	*Vendor 8*	*Vendor 9*	*Vendor 10*	*Local weights*
Vendor 1	1.00	3.00	0.33	3.00	0.20	1.00	0.20	0.14	0.14	0.20	0.037
Vendor 2	0.33	1.00	0.20	1.00	0.14	1.00	0.20	0.14	0.14	0.20	0.023
Vendor 3	3.00	5.00	1.00	5.00	0.33	3.00	0.33	0.20	0.20	0.33	0.070
Vendor 4	0.33	1.00	0.20	1.00	0.20	0.33	0.33	0.20	0.20	0.33	0.026
Vendor 5	5.00	7.00	3.00	5.00	1.00	3.00	0.33	0.33	1.00	3.00	0.136
Vendor 6	1.00	1.00	0.33	3.00	0.33	1.00	0.33	0.20	0.20	0.33	0.038
Vendor 7	5.00	5.00	3.00	3.00	3.00	3.00	1.00	0.33	0.33	0.33	0.119
Vendor 8	7.00	7.00	5.00	5.00	3.00	5.00	3.00	1.00	3.00	3.00	0.254
Vendor 9	7.00	7.00	5.00	5.00	1.00	5.00	3.00	0.33	1.00	3.00	0.182
Vendor 10	5.00	5.00	3.00	3.00	0.33	3.00	3.00	0.33	0.33	1.00	0.114
Total	34.67	42.0	21.07	34.0	9.54	25.33	11.73	3.22	6.55	11.73	

Table 1.33: Pair-wise comparison of vendor on price sub- criteria 2.

Understanding and willingness to follow financial security clause

Understanding and willingness to follow financial security clause	*Vendor 1*	*Vendor 2*	*Vendor 3*	*Vendor 4*	*Vendor 5*	*Vendor 6*	*Vendor 7*	*Vendor 8*	*Vendor 9*	*Vendor 10*	*Local weights*
Vendor 1	1.00	0.20	3.00	0.20	1.00	0.33	1.00	0.14	3.00	5.00	0.070
Vendor 2	5.00	1.00	5.00	1.00	3.00	1.00	1.00	1.00	3.00	3.00	0.151
Vendor 3	0.33	0.20	1.00	0.33	0.20	0.20	0.20	0.20	0.33	1.00	0.027
Vendor 4	5.00	1.00	3.00	1.00	3.00	1.00	3.00	1.00	3.00	3.00	0.157
Vendor 5	1.00	0.33	5.00	0.33	1.00	0.33	1.00	0.20	3.00	5.00	0.081
Vendor 6	3.00	1.00	5.00	1.00	3.00	1.00	3.00	1.00	3.00	3.00	0.154
Vendor 7	1.00	1.00	5.00	0.33	1.00	0.33	1.00	0.20	1.00	1.00	0.068
Vendor 8	7.00	1.00	5.00	1.00	5.00	1.00	5.00	1.00	3.00	5.00	0.200
Vendor 9	0.33	0.33	3.00	0.33	0.33	0.33	1.00	0.33	1.00	3.00	0.055
Vendor 10	0.20	0.33	1.00	0.33	0.20	0.33	1.00	0.20	0.33	1.00	0.036
Total	23.87	6.40	36.00	5.87	17.73	5.87	17.20	5.28	20.67	30.00	

Table 1.34: Vendors ranking according to price.

Rank	*Vendor name*	*Global weight*	*Rank*	*Vendor name*	*Global weight*
1	Vendor 8	0.006	2	Vendor 4	0.0039
3	Vendor 6	0.0039	4	Vendor 2	0.0037
5	Vendor 5	0.0027	6	Vendor 7	0.0023
7	Vendor 9	0.0022	8	Vendor 1	0.0019
9	Vendor 10	0.0015	10	Vendor 3	0.001

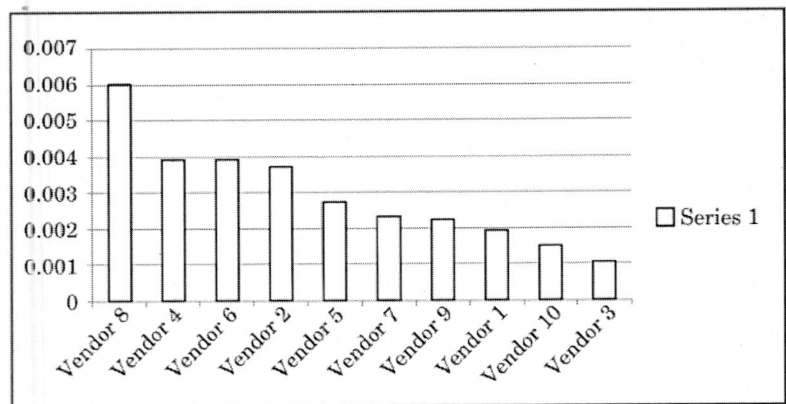

Fig. 1.7: Analysis on vendor performances on price sub-criteria.

Table 1.35: Pair-wise comparison of vendors on service sub-criteria.

Attitude towards handling of complaints	*Vendor 1*	*Vendor 2*	*Vendor 3*	*Vendor 4*	*Vendor 5*	*Vendor 6*	*Vendor 7*	*Vendor 8*	*Vendor 9*	*Vendor 10*	*Local weights*
Vendor 1	1	0.14	3.00	0.14	1.00	0.14	0.33	0.14	0.33	0.20	0.025
Vendor 2	7	1.00	7.00	5.00	7.00	3.00	5.00	1.00	5.00	5.00	0.253
Vendor 3	0.33	0.14	1.00	0.14	1.00	0.14	0.33	0.14	0.20	0.14	0.018
Vendor 4	7	0.20	7.00	1.00	7.00	1.00	5.00	1.00	5.00	3.00	0.149
Vendor 5	1	0.14	1.00	0.14	1.00	0.14	0.20	0.14	0.20	0.33	0.020
Vendor 6	7	0.33	7.00	1.00	7.00	1.00	5.00	1.00	5.00	3.00	0.153
Vendor 7	3	0.20	3.00	0.20	5.00	0.20	1.00	0.20	0.33	0.20	0.045
Vendor 8	7	1.00	7.00	1.00	7.00	1.00	5.00	1.00	7.00	7.00	0.199
Vendor 9	3	0.20	5.00	0.20	5.00	0.20	3.00	0.14	1.00	1.00	0.061
Vendor 10	5.0	0.20	7.00	0.33	3.00	0.33	5.00	0.14	1.00	1.00	0.076
Total	41.33	3.56	48.00	9.16	44.00	7.16	29.87	4.91	25.07	20.88	

Table 1.36: Pair-wise comparison of vendors on service sub-criteria 2.

Flexibility (order volume	*Vendor 1*	*Vendor 2*	*Vendor 3*	*Vendor 4*	*Vendor 5*	*Vendor 6*	*Vendor 7*	*Vendor 8*	*Vendor 9*	*Vendor 10*	*Local weights*
Vendor 1	1.00	0.20	3.00	0.20	0.33	0.14	0.33	0.14	0.33	0.20	0.027
Vendor 2	5.00	1.00	5.00	1.00	5.00	1.00	3.00	0.20	3.00	5.00	0.132
Vendor 3	0.33	0.20	1.00	0.20	0.33	0.20	0.33	0.14	0.20	1.00	0.023
Vendor 4	5.00	1.00	5.00	1.00	5.00	1.00	3.00	0.20	3.00	3.00	0.124
Vendor 5	3.00	0.20	3.00	0.20	1.00	0.20	0.33	0.14	0.33	0.33	0.035
Vendor 6	7.00	1.00	5.00	1.00	5.00	1.00	5.00	0.33	3.00	5.00	0.151
Vendor 7	3.00	0.33	3.00	0.33	3.00	0.20	1.00	0.14	0.33	3.00	0.058
Vendor 8	7.00	5.00	7.00	5.00	7.00	3.00	7.00	1.00	7.00	5.00	0.328
Vendor 9	3.00	0.33	5.00	0.33	3.00	0.33	3.00	0.14	1.00	1.00	0.069
Vendor 10	5.00	0.20	1.00	0.33	3.00	0.20	0.33	0.20	1.00	1.00	0.051
Total	39.3/3	9.47	38.00	9.60	32.67	7.28	23.33	2.65	19.20	24.53	

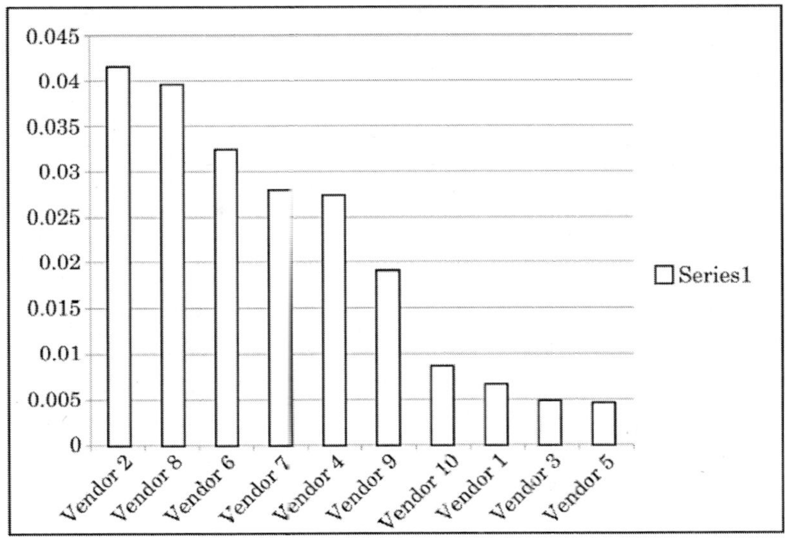

Fig. 1.8: Analysis of vendor performance on service sub-criteria.

Table 1.37: Vendor ranking according to service.

Rank	*Vendor name*	*Global weight*	*Rank*	*Vendor name*	*Global weight*
1	Vendor 2	0.017	2	Vendor 8	0.0161
3	Vendor 6	0.0111	4	Vendor 4	0.0106
5	Vendor 10	0.0052	6	Vendor 9	0.0045
7	Vendor 7	0.0035	8	Vendor 10	0.0018
9	Vendor 5	0.0016	10	Vendor 3	0.0014

Table 1.38: Pair-wise comparison of vendors on business performance sub criteria 1.

Use of information systems (communication)	*Vendor 1*	*Vendor 2*	*Vendor 3*	*Vendor 4*	*Vendor 5*	*Vendor 6*	*Vendor 7*	*Vendor 8*	*Vendor 9*	*Vendor 10*	*Local weights*
Vendor 1	1.00	0.33	1.00	0.20	0.33	0.20	1.00	0.20	1.00	0.20	0.033
Vendor 2	3.00	1.00	3.00	1.00	3.00	1.00	5.00	1.00	3.00	3.00	0.154
Vendor 3	1.00	0.33	1.00	0.14	0.14	0.20	0.33	0.14	0.33	0.20	0.025
Vendor 4	5.00	1.00	7.00	1.00	7.00	1.00	5.00	1.00	3.00	1.00	0.163
Vendor 5	3.00	0.33	7.00	0.14	1.00	0.14	0.33	0.20	0.33	0.20	0.048
Vendor 6	5.00	1.00	5.00	1.00	7.00	1.00	5.00	1.00	5.00	1.00	0.168
Vendor 7	1.00	0.20	3.00	0.20	3.00	0.20	1.00	0.14	0.33	0.20	0.039
Vendor 8	5.00	1.00	7.00	1.00	5.00	1.00	7.00	1.00	5.00	3.00	0.192
Vendor 9	1.00	0.33	3.00	0.33	3.00	0.20	3.00	0.20	1.00	1.00	0.061
Vendor 10	5.00	0.33	5.00	1.00	5.00	1.00	5.00	0.33	1.00	1.00	0.118
Total	30.00	5.87	42.00	6.02	34.48	5.94	32.67	5.22	20.00	10.80	

Table 1.39: Pair-wise comparison of vendors on business overall performance sub-criteria 2

Performance history	*Vendor 1*	*Vendor 2*	*Vendor 3*	*Vendor 4*	*Vendor 5*	*Vendor 6*	*Vendor 7*	*Vendor 8*	*Vendor 9*	*Vendor 10*	*Local weights*
Vendor 1	1.00	0.33	1.00	0.20	0.20	0.20	0.20	0.14	0.33	0.33	0.023
Vendor 2	3.00	1.00	3.00	1.00	0.33	1.00	0.20	0.33	3.00	3.00	0.091
Vendor 3	1.00	0.33	1.00	0.33	0.20	0.33	0.20	0.33	0.33	0.33	0.031
Vendor 4	5.00	1.00	3.00	1.00	1.00	1.00	0.33	1.00	3.00	5.00	0.122
Vendor 5	5.00	3.00	5.00	1.00	1.00	0.20	0.33	0.33	1.00	3.00	0.097
Vendor 6	5.00	1.00	3.00	1.00	5.00	1.00	3.00	1.00	3.00	5.00	0.179
Vendor 7	5.00	5.00	5.00	3.00	3.00	0.33	1.00	0.33	3.00	5.00	0.173
Vendor 8	7.00	3.00	3.00	1.00	3.00	1.00	3.00	1.00	3.00	5.00	0.184
Vendor 9	3.00	0.33	3.00	0.33	1.00	0.33	0.33	0.33	1.00	1.00	0.056
Vendor 10	3.00	0.33	3.00	0.20	0.33	0.20	0.20	0.20	1.00	1.00	0.043
Total	38.00	15.33	30.00	9.07	15.07	5.60	8.80	5.01	18.67	28.67	

Table 1.40: Pair-wise comparison of vendors on business overall performance sub-criteria 3

Management capability	*Vendor 1*	*Vendor 2*	*Vendor 3*	*Vendor 4*	*Vendor 5*	*Vendor 6*	*Vendor 7*	*Vendor 8*	*Vendor 9*	*Vendor 10*	*Local weights*
Vendor 1	1.00	0.20	1.00	0.14	0.33	0.14	0.33	0.14	1.00	0.33	0.027
Vendor 2	5.00	1.00	5.00	1.00	0.33	1.00	0.20	0.33	1.00	3.00	0.090
Vendor 3	1.00	0.20	1.00	0.33	0.20	0.33	0.20	0.33	0.33	0.33	0.029
Vendor 4	7.00	1.00	3.00	1.00	1.00	1.00	0.33	1.00	3.00	5.00	0.126
Vendor 5	3.00	3.00	5.00	1.00	1.00	0.20	1.00	1.00	1.00	3.00	0.110
Vendor 6	7.00	1.00	3.00	1.00	5.00	1.00	3.00	1.00	3.00	5.00	0.190
Vendor 7	3.00	5.00	5.00	3.00	1.00	0.33	1.00	0.33	3.00	5.00	0.152
Vendor 8	7.00	3.00	3.00	1.00	1.00	1.00	3.00	1.00	5.00	5.00	0.177
Vendor 9	1.00	1.00	3.00	0.33	1.00	0.33	0.33	0.20	1.00	3.00	0.059
Vendor 10	3.00	0.33	3.00	0.20	0.33	0.20	0.20	0.20	0.33	1.00	0.039

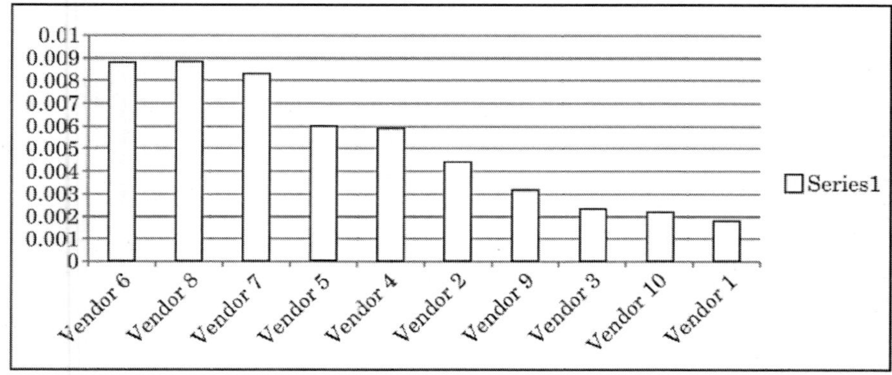

Fig. 1.9: Analysis of vendor performance on the business criteria.

Table 1.41: Ranking of vendors on the business criteria.

Vendor 6	0.0088
Vendor 8	0.0088
Vendor 7	0.0083
Vendor 5	0.006
Vendor 4	0.0059
Vendor 2	0.0044
Vendor 9	0.0032
Vendor 3	0.0023
Vendor 10	0.0022
Vendor 1	0.0018

Table 1.42: Pair-wise comparison of vendors according to technical capability sub-criteria 1

Technical know	*Vendor 1*	*Vendor 2*	*Vendor 3*	*Vendor 4*	*Vendor 5*	*Vendor 6*	*Vendor 7*	*Vendor 8*	*Vendor 9*	*Vendor 10*	*Local weights*
Vendor 1	1.00	3.00	0.33	3.00	0.33	1.00	0.33	0.33	1.00	3.00	0.070
Vendor 2	0.33	1.00	0.33	1.00	0.20	0.20	0.14	0.20	1.00	1.00	0.031
Vendor 3	3.00	3.00	1.00	3.00	0.33	3.00	0.33	0.20	0.33	3.00	0.093
Vendor 4	0.33	1.00	0.33	1.00	0.14	0.33	0.14	0.20	0.33	1.00	0.027
Vendor 5	3.00	5.00	3.00	7.00	1.00	3.00	1.00	3.00	5.00	5.00	0.217
Vendor 6	1.00	5.00	0.33	3.00	0.33	1.00	0.20	0.33	3.00	3.00	0.082
Vendor 7	3.00	7.00	3.00	7.00	1.00	5.00	1.00	3.00	5.00	5.00	0.235
Vendor 8	3.00	5.00	5.00	5.00	0.33	3.00	0.33	1.00	3.00	1.00	0.142
Vendor 9	1.00	1.00	3.00	3.00	0.20	0.33	0.20	0.33	1.00	1.00	0.061
Vendor 10	0.33	1.00	0.33	1.00	0.20	0.33	0.20	1.00	1.00	1.00	0.042
Total	16.00	32.00	16.67	34.00	4.08	17.20	3.89	9.60	20.67	24.00	

Table 1.43: Pair-wise comparison of vendors on technical capability sub-criteria 2.

Vendor experience in related class of jobs	*Vendor 1*	*Vendor 2*	*Vendor 3*	*Vendor 4*	*Vendor 5*	*Vendor 6*	*Vendor 7*	*Vendor 8*	*Vendor 9*	*Vendor 10*	*Local weights*
Vendor 1	1.00	0.33	0.33	1.00	0.20	0.33	0.14	0.20	0.14	0.33	0.024
Vendor 2	3.00	1.00	3.00	5.00	0.20	1.00	0.20	0.33	0.20	1.00	0.071
Vendor 3	3.00	0.33	1.00	3.00	0.33	1.00	0.20	0.33	0.33	0.33	0.052
Vendor 4	1.00	0.20	0.33	1.00	0.33	0.33	0.33	1.00	0.33	1.00	0.043
Vendor 5	5.00	5.00	3.00	3.00	1.00	5.00	1.00	3.00	1.00	5.00	0.196
Vendor 6	3.00	1.00	1.00	3.00	0.20	1.00	0.20	1.00	0.33	1.00	0.061
Vendor 7	7.00	5.00	5.00	3.00	1.00	5.00	1.00	5.00	3.00	5.00	0.251
Vendor 8	5.00	3.00	3.00	1.00	0.33	1.00	0.20	1.00	1.00	3.00	0.097
Vendor 9	7.00	5.00	3.00	3.00	1.00	3.00	0.33	1.00	1.00	3.00	0.148
Vendor 10	3.00	1.00	3.00	1.00	0.20	1.00	0.20	0.33	0.33	1.00	0.056
Total	38.00	21.87	22.67	24.00	4.80	18.67	3.81	13.20	7.68	20.67	

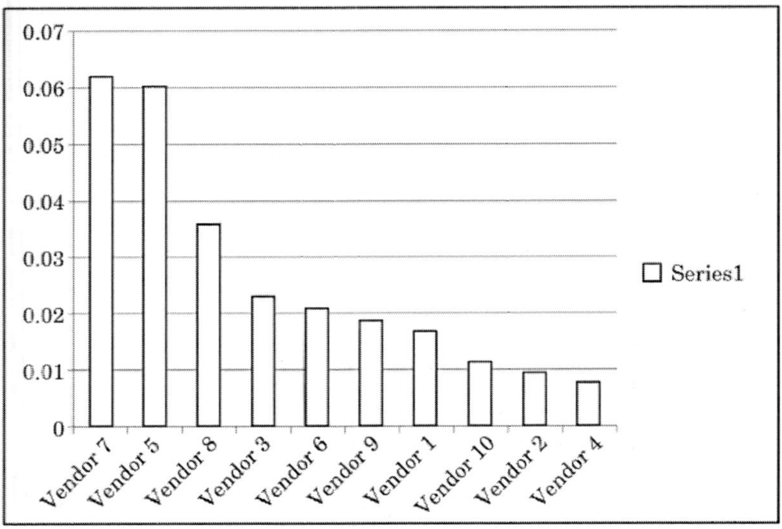

Fig. 1.10: Analysis of vendor for technical ability.

Table 1.44: Ranking of vendor according to technical ability.

Vendor 7	0.0618
Vendor 5	0.0599
Vendor 8	0.0357
Vendor 3	0.023
Vendor 6	0.0208
Vendor 9	0.0186
Vendor 1	0.0167
Vendor 10	0.0113
Vendor 2	0.0094
Vendor 4	0.0076

Table 1.45: Pair-wise comparison of vendors on time delivery sub-criteria 1.

Delivery lead time (speed)	***Vendor 1***	***Vendor 2***	***Vendor 3***	***Vendor 4***	***Vendor 5***	***Vendor 6***	***Vendor 7***	***Vendor 8***	***Vendor 9***	***Vendor 10***	***Local weights***
Vendor 1	1.00	0.20	1.00	0.14	1.00	0.14	0.33	0.14	0.33	0.33	0.024
Vendor 2	5.00	1.00	5.00	3.00	5.00	3.00	3.00	1.00	5.00	5.00	0.228
Vendor 3	1.00	0.20	1.00	0.14	1.00	0.14	0.20	0.14	0.20	0.33	0.022
Vendor 4	7.00	0.33	7.00	1.00	5.00	1.00	3.00	1.00	1.00	5.00	0.141
Vendor 5	1.00	0.20	1.00	0.20	1.00	0.14	0.20	0.14	0.14	0.33	0.023
Vendor 6	7.00	0.33	7.00	1.00	7.00	1.00	5.00	1.00	3.00	3.00	0.162
Vendor 7	3.00	0.33	5.00	0.33	5.00	0.20	1.00	0.33	0.20	3.00	0.071
Vendor 8	7.00	1.00	7.00	1.00	7.00	1.00	3.00	1.00	3.00	5.00	0.176
Vendor 9	3.00	0.20	5.00	1.00	7.00	0.33	5.00	0.33	1.00	3.00	0.107
Vendor 10	3.00	0.20	3.00	0.20	3.00	0.33	0.33	0.20	0.33	1.00	0.046

Table 1.46: Pair-wise comparison on time delivery sub-criteria 2.

Ability and willingness	*Vendor 1*	*Vendor 2*	*Vendor 3*	*Vendor 4*	*Vendor 5*	*Vendor 6*	*Vendor 7*	*Vendor 8*	*Vendor 9*	*Vendor 10*	*Local weights*
Vendor 1	1.00	0.20	3.00	0.20	3.00	0.14	0.20	0.14	0.33	1.00	0.035
Vendor 2	5.00	1.00	5.00	3.00	5.00	1.00	1.00	1.00	5.00	7.00	0.173
Vendor 3	0.33	0.20	1.00	0.20	1.00	0.20	0.20	0.14	0.33	0.33	0.023
Vendor 4	5.00	0.33	5.00	1.00	5.00	3.00	0.33	0.33	0.33	3.00	0.111
Vendor 5	0.33	0.20	1.00	0.20	1.00	0.14	0.14	0.14	0.20	0.33	0.020
Vendor 6	7.00	1.00	5.00	0.33	7.00	1.00	1.00	1.00	3.00	5.00	0.148
Vendor 7	5.00	1.00	5.00	3.00	7.00	1.00	1.00	3.00	3.00	7.00	0.195
Vendor 8	7.00	1.00	7.00	3.00	7.00	1.00	0.33	1.00	5.00	7.00	0.174
Vendor 9	3.00	0.20	3.00	3.00	5.00	0.33	0.33	0.20	1.00	5.00	0.086
Vendor 10	1.00	0.14	3.00	0.33	3.00	0.20	0.14	0.14	0.20	1.00	0.034

Table 1.47: Pair-wise comparison of vendors on time delivery sub-criteria 3.

Geographical location	*Vendor 1*	*Vendor 2*	*Vendor 3*	*Vendor 4*	*Vendor 5*	*Vendor 6*	*Vendor 7*	*Vendor 8*	*Vendor 9*	*Vendor 10*	*Local weights*
Vendor 1	1.00	1.00	1.00	5.00	1.00	1.00	1.00	1.00	5.00	5.00	0.131
Vendor 2	1.00	1.00	1.00	5.00	1.00	1.00	1.00	1.00	5.00	5.00	0.131
Vendor 3	1.00	1.00	1.00	5.00	1.00	1.00	1.00	1.00	5.00	5.00	0.131
Vendor 4	0.20	0.20	0.20	1.00	0.20	0.20	0.20	0.20	1.00	3.00	0.031
Vendor 5	1.00	1.00	1.00	5.00	1.00	1.00	1.00	1.00	5.00	5.00	0.131
Vendor 6	1.00	1.00	1.00	5.00	1.00	1.00	1.00	1.00	5.00	5.00	0.131
Vendor 7	1.00	1.00	1.00	5.00	1.00	1.00	1.00	1.00	5.00	5.00	0.131
Vendor 8	1.00	1.00	1.00	5.00	1.00	1.00	1.00	1.00	5.00	5.00	0.131
Vendor 9	0.20	0.20	0.20	1.00	0.20	0.20	0.20	0.20	1.00	3.00	0.031
Vendor 10	0.20	0.20	0.20	0.33	0.20	0.20	0.20	0.20	0.33	1.00	0.023

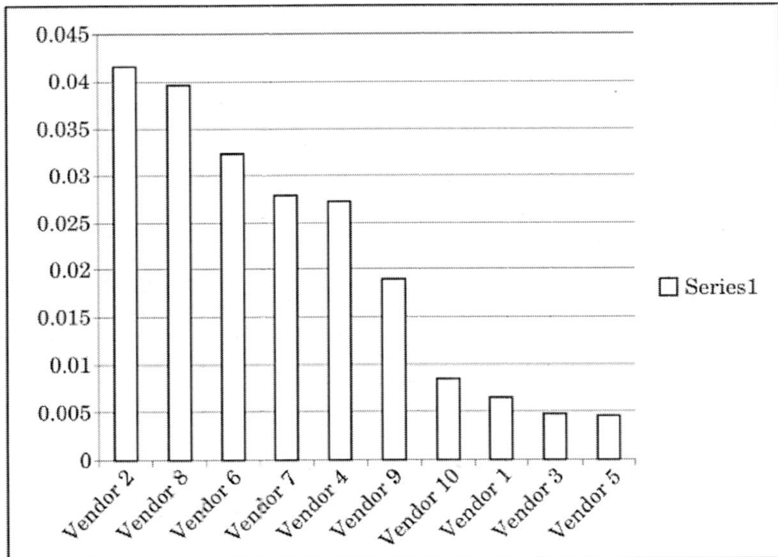

Fig. 1.11: Analysis of vendor performance w.r.t. on time delivery.

Table 1.48: Ranking of vendor's w.r.t. on time delivery.

Vendor 2	0.0415
Vendor 8	0.0395
Vendor 6	0.0324
Vendor 7	0.0279
Vendor 4	0.0273
Vendor 9	0.019
Vendor 10	0.0085
Vendor 1	0.0066
Vendor 3	0.0048
Vendor 5	0.0046

Table 1.49 (A): Global weight of vendor

Sl. no.	*Criteria*	*LW*	*Sub-criteria*	*GW*	*GW sub-criteria*	*V1 LW*	*V1 GW*	*V2 LW*	*V2 GW*	*V3 LW*	*V3 GW*
1	Quality	0.37	1	0.54	0.2	0.09	0.0179	0.03	0.0055	0.07	0.0139
2			2	0.28	0.1	0.04	0.0045	0.08	0.0081	0.03	0.0027
3			3	0.09	0.03	0.08	0.0025	0.06	0.0019	0.04	0.0012
4			4	0.09	0.03	0.13	0.0043	0.05	0.0016	0.11	0.0037
5	Price	0.03	1	0.17	0	0.04	0.0002	0.02	0.0001	0.07	0.0003
6			2	0.83	0.02	0.07	0.0017	0.15	0.0036	0.03	0.0007
7	Service	0.07	1	0.83	0.06	0.02	0.0015	0.25	0.0154	0.02	0.0011
8			2	0.17	0.01	0.03	0.0003	0.13	0.0016	0.02	0.0003
9	Business	0.05	2	0.05	0	0.03	0.0001	0.15	0.0004	0.02	0.0001
10			3	0.36	0.02	0.03	0.0005	0.09	0.0017	0.03	0.0005
11			4	0.44	0.02	0.02	0.0005	0.09	0.0021	0.03	0.0007
12	Technical	0.26	1	0.88	0.23	0.07	0.0159	0.03	0.0071	0.09	0.0213
13			2	0.13	0.03	0.02	0.0008	0.07	0.0023	0.05	0.0017
14	On time delivery	0.22	1	0.29	0.06	0.02	0.0015	0.23	0.0143	0.02	0.0014
15			2	0.07	0.01	0.04	0.0006	0.25	0.0037	0.02	0.0004
16			3	0.49	0.11	0.04	0.0038	0.17	0.0186	0.02	0.0024
17			4	0.03	0.01	0.13	0.001	0.13	0.001	0.13	0.001

Table 1.49 (B): Global weight of vendor.

Sl. no.	*Criteria*	*LW*	*Sub-criteria*	*GW*	*GW sub-criteria*	*V4 LW*	*V4 GW*	*V5 LW*	*V5 GW*	*V6 LW*	*V6 GW*
1	Quality	0.37	1	0.54	0.20	0.03	0.0064	0.13	0.0256	0.08	0.0158
2			2	0.28	0.10	0.09	0.0091	0.04	0.0045	0.18	0.0181
3			3	0.09	0.03	0.06	0.0017	0.07	0.0023	0.17	0.0054
4			4	0.09	0.03	0.06	0.0020	0.10	0.0033	0.04	0.0013
5	Price	0.03	1	0.17	0.00	0.03	0.0001	0.14	0.0007	0.04	0.0002
6			2	0.83	0.02	0.16	0.0038	0.08	0.0020	0.15	0.0037
7	Service	0.07	1	0.83	0.06	0.15	0.0091	0.02	0.0012	0.15	0.0093
8			2	0.17	0.01	0.12	0.0015	0.04	0.0004	0.15	0.0018
9	Business	0.05	0.05	0.00	0.16	0.0005	0.05	0.0001	0.17	0.0005	0.05
10			0.36	0.02	0.13	0.0023	0.11	0.0020	0.19	0.0035	0.36
11			0.44	0.02	0.12	0.0028	0.10	0.0022	0.18	0.0041	0.44
12	Technical ability	0.26	0.88	0.23	0.03	0.0062	0.22	0.0495	0.08	0.0188	0.88
13			0.13	0.03	0.04	0.0014	0.20	0.0064	0.06	0.0020	0.13
14	On time delivery	0.22	0.07	0.01	0.14	0.0020	0.02	0.0003	0.14	0.0020	0.07
15			0.49	0.11	0.11	0.0120	0.02	0.0021	0.15	0.0159	0.49
16			0.03	0.01	0.03	0.0002	0.13	0.0010	0.13	0.0010	0.03
17			0.03	0.01	0.03	0.0002	0.13	0.0010	0.13	0.0010	0.03

Table 1.49 (C): Global weight of vendor.

Sl. no.	*Criteria*	*LW*	*Sub-criteria*	*GW*	*GW sub-criteria*	*V7 LW*	*V7 GW*	*V8 LW*	*V8 GW*	*V9 LW*	*V9 GW*	*V10 LW*	*V10 GW*
1	Quality	0.37	1	0.54	0.20	0.18	0.0355	0.20	0.040	0.13	0.0253	0.06	0.011
2			2	0.28	0.10	0.22	0.0223	0.22	0.0224	0.09	0.0090	0.03	0.0026
3			3	0.09	0.03	0.24	0.0075	0.17	0.0052	0.05	0.0016	0.06	0.0020
4			4	0.09	0.03	0.23	0.0080	0.13	0.0043	0.13	0.0044	0.03	0.0009
5	Price	0.03	1	0.17	0.00	0.12	0.0006	0.25	0.0012	0.18	0.0009	0.11	0.0006
6			2	0.83	0.02	0.07	0.0017	0.20	0.0048	0.06	0.0013	0.04	0.0009
7	Service	0.07	1	0.83	0.06	0.05	0.0028	0.20	0.0121	0.06	0.0037	0.08	0.0046
8			2	0.17	0.01	0.06	0.0007	0.33	0.0040	0.07	0.0008	0.05	0.0006
9	Business	0.05	2	0.05	0.00	0.04	0.0001	0.19	0.0005	0.06	0.0002	0.12	0.0003
10			3	0.36	0.02	0.15	0.0028	0.18	0.0033	0.06	0.0011	0.04	0.0007
11			4	0.44	0.02	0.17	0.0039	0.18	0.0042	0.06	0.0013	0.04	0.0010
12	Technical	0.26	1	0.88	0.23	0.23	0.0536	0.14	0.0325	0.06	0.0138	0.04	0.0095
13			2	0.13	0.03	0.25	0.0082	0.10	0.0032	0.15	0.0048	0.06	0.0018
14	On time delivery	0.22	1	0.29	0.06	0.07	0.0045	0.18	0.0111	0.11	0.0067	0.05	0.0029
15			3	0.07	0.01	0.08	0.0012	0.21	0.0031	0.07	0.0010	0.02	0.0003
16			4	0.49	0.11	0.19	0.0209	0.17	0.0187	0.09	0.0092	0.03	0.0037
17			5	0.03	0.01	0.13	0.0010	0.13	0.0010	0.03	0.0002	0.02	0.0002

Table 1.50: Ranking of vendor on overall all six criteria.

Vendor name	*Gross weight from AHP*	*Rank*
Vendor 7	0.175218	1
Vendor 8	0.171681	2
Vendor 6	0.113674	3
Vendor 5	0.105245	4
Vendor 2	0.088963	5
Vendor 9	0.085487	6
Vendor 4	0.069785	7
Vendor 1	0.057615	8
Vendor 3	0.053376	9
Vendor 10	0.043628	10

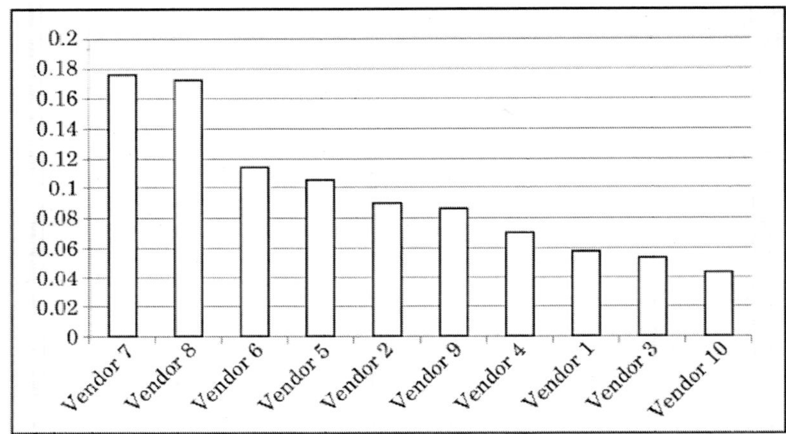

Fig. 1.12: Analysis of global weight of vendor.

1.10. VENDOR EVALUATION & RANKING THROUGH VIKOR METHOD

The VIKOR (the Serbian name is "VIšekriterijumsko KOmpromisno Rangiranje which means multi-criteria optimization and compromise solution) method was mainly established by Zeleny and later advocated by Opricovic and Tzeng.

This method helps to solve multi-criteria decision-making problems with conflicting and non-commensurable criteria, assume that a compromise can be acceptable for conflict resolution, when the decision maker wants a solution that is the closest to the ideal solution and farthest from the negative-ideal solution, and the alternatives can be evaluated with respect to all the established criteria.

It focuses on ranking and selecting the best alternative from a set of alternatives with conflicting criteria, and on proposing the compromise solution (one or more).

The compromise solution is a feasible solution, which is the closest to the ideal solution, and a compromise means an agreement established by mutual concessions made between the alternatives (Rao, 2007).

In VIKOR method, the best alternative is preferred by maximizing utility group and minimizing regret group. This method calculates ratio of positive and negative ideal solution to take wise decision on vendor selection, a methodology is proposed by combining Analytic Hierarchy Process (AHP) and VIKOR.

The outline of the proposed methodology is shown below in Fig. 1.13.

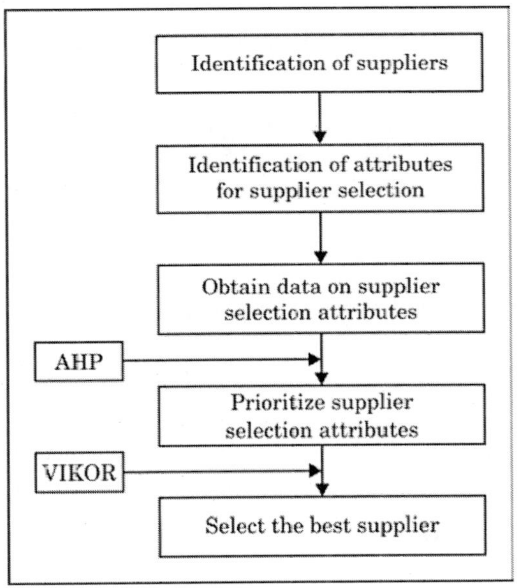

Fig. 1.13: Vendor selection using VIKOR.

In this methodology the priority structure of vendor selection attributes is obtained by using AHP. The weights of the vendor selection attributes will be reflected in determining the VIKOR index for each vendor. Based on VIKOR indices it is easier for a decision maker to identify the best vendor.

The step by step methodology is discussed below:

Step 1: Identification of vendor selection attributes the process of selection of a vendor for any firm is started with the identification vendor selection attributes. The attributes for vendor selection are usually depends on the type of firm, product, purchasing capability etc.

The top-level executives are generally involved in the identification of vendor selection attributes. There are good number of vendor selection attributes. But the most common attributes are quality, price, service, business overall performance, technical ability, on time delivery.

Step 2: Obtain data on Vendor selection attributes.

The data pertaining to Vendor attributes may be obtained through questionnaire survey. A questionnaire is developed by the management of the company to obtain the response data on Vendor selection attributes by purchase group.

Step 3: Determination of the priority structure of vendor selection attributes using AHP.

Step 4: Formulation of MCDM decision matrix:

The MCDM decision matrix has to be formed as shown below:

Where A_i = the ith alternative (i = 1, 2, 3,.......m)

Cx_j = the jth criterion (j = 1, 2, 3,........n)

X_{ij} = individual performance of the alternatives (Vendors)

	Cx_1	Cx_2	Cx_3	.	.	Cx_n
A_1	x_{11}	x_{12}	x_{13}	.	.	x_{1n}
A_2	x_{21}	x_{22}	x_{23}	.	.	x_{2n}
A_3	x_{31}	x_{32}	x_{33}	.	.	x_{3n}
.	.	.	.	.	.	.
.	.	.	.	.	.	.
A_m	x_{m1}	x_{m2}	x_{m3}	.	.	x_{mn}

Step 5: Representation of normalized decision matrix.

The normalized decision matrix can be expressed as follows:

$$F = [f_{ij}]_{m \times n} \quad (1)$$

Where, $f_{ij} = \frac{x_{ij}}{\sqrt{\sum_{i=1}^{n} x_{ij}^2}}$, i = 1,2,...,m; j = 1,2,....,n

x_{ij} is the performance of alternative A_i with respect to the *j*th criterion

Step 6: Determination of positive-ideal solution and negative-ideal solution

Determination of positive-ideal solution and negative-ideal solution

The positive ideal solution A* and the negative ideal solution A- determined as follows:

$$A^* = \max f_{ij} \mid j \in J \ or \min f_{ij} \mid j \in J \mid i = 1,2,\dots,m$$

$$= f_1^*, f_1^*, \dots . f_j^*, \dots .. f_n^*$$

$$\bar{A} = \min f_{ij} \mid j \in J \ or \ \max f_{ij} \mid j \in J \mid i = 1,2,\dots mm$$

$$= f_1^-, f_1^-, \dots . f_j^-, \dots .. f_n^-$$

Step 7: Calculation of Utility measure and Regret measure.

The Utility measure Si and Regret measure Ri for each alternative are computed using the following expressions:

$$S_i = \sum_{j=1}^{n} w_j \times \left[\frac{f_j^* - f_{ij}}{f_j^* - f_j}\right] \quad (2)$$

$$R_i = \max_j \left[w_j \times \frac{f_j^* - f_{ij}}{f_j^* - f_j}\right] \quad (3)$$

Where w_j = weight of the *j*th criterion.

Step 8: Computation of VIKOR index

The VIKOR index is calculated by using the following expression

$$Q_i = \nu \left[\frac{S_i - S^*}{\bar{S} - S^*}\right] + 1 - \boldsymbol{\nu} \left[\frac{R_i - R^*}{\bar{R} - R^*}\right] \quad (4)$$

Where Q_i represent VIKOR value, S^* represent maximum utility factor, S^- represent minimum utility value, R^* maximum regret value, R^- represent minimum utility value S_i represent utility value of alternative (Vendor) R_i represent regret value and v represent weight of maximum group utility and its value usually set 0.5.

Step 9: Rank the order of preference

The alternative which is having smallest VIKOR index value is the best solution.

Step 1: Identification of vendors and *Step 2*: Identification of attributes of vendors are already done at the initial phase of AHP and given in Table 1.51.

Step 3: Prioritizing of vendor's attribute has been done & shown in Table 1.52

Step 4: Formulation of MCDM Matrix has been summarized from Table 49A, 49B and 49C.

Table 1.51: Weight of criteria

	Quality	*Price*	*Service*	*Business*	*Tech*	*Delivery*
Weight from table 2	0.37	0.03	0.07	0.05	0.26	0.22

Table 1.52: Weight of criteria for vendors (data summarized from Tables 49A, 49B and 49C).

	Quality	*Price*	*Service*	*Business*	*Tech*	*Delivery*
Vendor 1	0.34	0.11	0.05	0.08	0.09	0.23
Vendor 2	0.22	0.17	0.39	0.34	0.10	0.78
Vendor 3	0.24	0.10	0.04	0.08	0.15	0.20
Vendor 4	0.23	0.18	0.27	0.41	0.07	0.42
Vendor 5	0.34	0.22	0.06	0.26	0.41	0.20
Vendor 6	0.47	0.19	0.30	0.54	0.14	0.58
Vendor 7	0.87	0.19	0.10	0.36	0.49	0.48
Vendor 8	0.71	0.46	0.53	0.55	0.24	0.69
Vendor 9	0.40	0.24	0.13	0.18	0.21	0.29
Vendor 10	0.17	0.15	0.13	0.20	0.10	0.12

Step 5: Normalized decision matrix has been made as per formula as shown in Table 1.53

$$F = [f_{ij}]_{m \times n}$$

Where, $f_{ij} = \frac{x_{ij}}{\sqrt{\sum_{i=1}^{n} x_{ij}^2}}$, i = 1,2,...,m; j = 1,2,....,n

x_{ij} is the performance of alternative A_i with respect to the *j*th criteria

Table 1.53: Normalized matrix of criteria

	Quality	*Price*	*Service*	*Business*	*Tech*	*Delivery*
Vendor 1	0.24	0.15	0.06	0.08	0.12	0.16
Vendor 2	0.15	0.25	0.48	0.31	0.13	0.54
Vendor 3	0.17	0.14	0.05	0.08	0.19	0.14
Vendor 4	0.16	0.26	0.34	0.38	0.09	0.29
Vendor 5	0.24	0.31	0.07	0.24	0.54	0.14
Vendor 6	0.33	0.27	0.38	0.50	0.19	0.40
Vendor 7	0.61	0.27	0.13	0.34	0.64	0.33
Vendor 8	0.50	0.65	0.66	0.51	0.31	0.48
Vendor 9	0.28	0.34	0.16	0.16	0.27	0.20
Vendor 10	0.12	0.21	0.16	0.19	0.13	0.09

Step 6: Positive-ideal solution and negative-ideal solution as per formula mentioned in above step 6 and given in Table 1.54.

Table 1.54: Most positive-ideal solution and negative-ideal solution.

	Quality	*Price*	*Service*	*Business*	*Tech*	*Delivery*
f+	0.61	0.14	0.66	0.51	0.64	0.09
f–	0.12	0.65	0.05	0.08	0.09	0.54

Step 7: Calculation of Utility measure and regret measure has been calculated as per formula mentioned in above step 7 and shown below in Table 1.55.

Step 8: Computation of VIKOR index has been measure has been calculated as per formula mentioned in above step 8 and shown below in Table 1.56.

Table 1.55: Utility measure and regret measure of vendor.

	Quality	*Price*	*Service*	*Business*	*Tech*	*Delivery*	*Utility measure* S_i	*Regret measure* R_i
Vendor 1	0.28	0.00	0.07	0.05	0.25	0.04	0.68	0.28
Vendor 2	0.35	0.01	0.02	0.02	0.24	0.22	0.86	0.35
Vendor 3	0.33	0.00	0.07	0.05	0.21	0.03	0.69	0.33
Vendor 4	0.34	0.01	0.04	0.02	0.26	0.10	0.76	0.34
Vendor 5	0.28	0.01	0.07	0.03	0.05	0.02	0.46	0.28
Vendor 6	0.21	0.01	0.03	0.00	0.21	0.15	0.62	0.21
Vendor 7	0.00	0.01	0.06	0.02	0.00	0.12	0.21	0.12
Vendor 8	0.08	0.03	0.00	0.00	0.15	0.19	0.46	0.19
Vendor 9	0.25	0.01	0.06	0.04	0.17	0.06	0.59	0.25
Vendor 10	0.37	0.00	0.06	0.04	0.24	0.00	0.71	0.37

Table 1.56: Value of Q_i of vendors.

Vendors	*Utility measure* S_i	*Regret measure* R_i	Q_i
Vendor 1	0.68	0.28	0.69198
Vendor 2	0.86	0.35	0.95615
Vendor 3	0.69	0.33	0.79896
Vendor 4	0.76	0.34	0.85825
Vendor 5	0.46	0.28	0.51433
Vendor 6	0.62	0.21	0.51148
Vendor 7	0.21	0.12	0
Vendor 8	0.46	0.19	0.33872
Vendor 9	0.59	0.25	0.55949
Vendor 10	0.71	0.37	0.88827
Maximum	0.86	0.37	
Minimum	0.21	0.12	

Table 1.57: Ranking of vendors according to VIKOR method.

Company name	*Qi*	*Rank*
Vendor 7	0	1
Vendor 8	0.338721	2
Vendor 6	0.511476	3
Vendor 5	0.514334	4
Vendor 9	0.559494	5
Vendor 1	0.691983	6
Vendor 3	0.798959	7
Vendor 4	0.858246	8
Vendor 10	0.888272	9
Vendor 2	0.956151	10

Step 9: Rank the order of preference:

Lower the Qi value, better will be ranking so ranking of all 10 vendors has been mentioned below shown in Fig. 1.4 & Table 1.57.

$$Q_i = \nu \left[\frac{S_i - S^*}{\bar{S} - S^*}\right] + 1 - \boldsymbol{\nu} \left[\frac{R_i - R^*}{\bar{R} - R^*}\right]$$

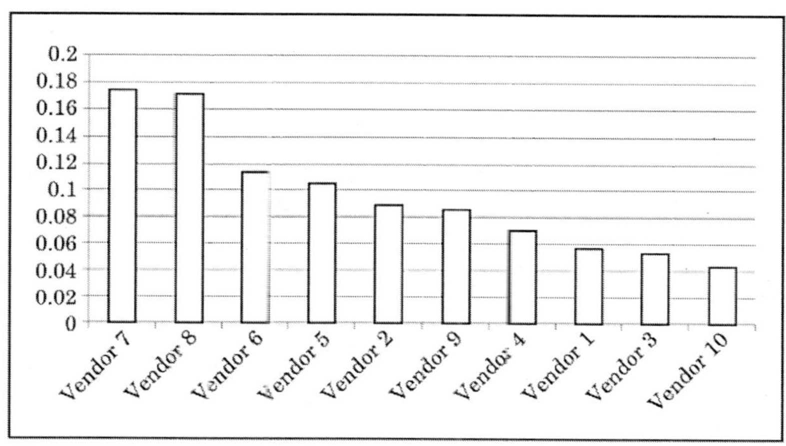

Fig. 1.14: Value of Q_i of vendor.

1.11. RESULTS AND DISCUSSION

The ultimate objective of dealing with the Vendor selection problem is to obtain a solution for prioritizing of Vendors. The best Vendor may provide faster delivery, reduced cost, good service, better business overall performance along with the improved quality in order to increase competitive advantage in the market. In the present work, AHP model and an integrated model of AHP-VIKOR for vendor selection has been developed and demonstrated the methodology through a case study conducted in XYZ manufacturing company and shown in Table 1.58.

Both models are well suited to deal with multi-criteria decisions that involve both qualitative and quantitative factors.

Our study and analysis considered 6 criteria and 32 sub-criteria to prioritize the Vendors of critical mechanical jobs. Based on survey, we considered 6 criteria and 19 sub-criteria for further analysis and remaining sub-criteria have been not considered since its weights were less than 0.7. Consistency ratio for 6 criteria and 19 sub-criteria were evaluated and found that all are less than 0.1. So every criteria and sub-criteria are meeting the AHP's philosophy.

Further for every sub-criteria, Consistency ratio for all ten vendors are evaluated and found that two sub-criteria 1. Financial stability (sustainability) and 2. Spare capacity to meet the requirements are not meeting consistency ratio limitation (< 0.1), so both sub-criteria have been eliminated further. So finally based on 6 criteria and 17 sub-criteria, prioritizing of ten Vendors has been done by both methods AHP and AHP- VIKOR for critical mechanical jobs.

We also completed the comparison analysis among AHP and VIKOR methods. Even though the ranking outcomes were the same for best 04 Vendors by AHP and AHP-VIKOR Methods and majority of ranking results of remaining Vendors are nearby as shown below:

Table 1.58: Ranking of vendor by AHP and VIKOR method.

Vendor name	*VIKOR rank*	*AHP rank*
Vendor 7	1	1
Vendor 8	2	2
Vendor 6	3	3
Vendor 5	4	4
Vendor 9	5	6
Vendor 1	6	8
Vendor 3	7	9
Vendor 4	8	7
Vendor 10	9	10
Vendor 2	10	5

The proposed methodology can be applied for Vendor selection in any manufacturing/ service sector Company.

1.12. MANAGERIAL IMPLICATIONS

In the present research AHP-VIKOR technique is effective for selection of Vendor in the supply chain. Models developed attempts to suggest how the AHP-VIKOR technique is useful for selection of Vendor. The objective of this research was to develop criteria or its sub-criteria that would help to selection of Vendor in the supply chain of industry. This technique is very useful for material selection and chosen of service.

1.13. CONCLUSIONS

Vendor play important role to boost upstream supply chain. Vendor helps to make good market reputation of the organization. Vendor selections depend on quantitative and qualitative criteria. In Vendor selection AHP-VIKOR technique has been used. Both techniques are multi decision criteria method. There are six criteria and seventeen sub-criteria have been implemented in the Vendor selection model. Vendor 7 and Vendor 8 got first and second position in Vendor selection priority.

2

Fuzzy-AHP Based Approach for Supplier Evaluation and Selection

2.1. INTRODUCTION

In this chapter a decision support system for supplier selection based on Fuzzy Analytical Hierarchy Process (FAHP) model design and implemented. The following sections of this chapter are organized as follows. Analytical hierarchy process (AHP) and Fuzzy AHP, Application Fuzzy AHP methodology is demonstrated. Today, many organizations are facing rapid changes stimulated by technological innovations and changing customer demands. These organizations realize that the effort to obtain products at the right cost, in the right quantity, with the right quality at the right time from the right source is crucial for their survival. Therefore; an efficient supplier selection process needs to be in place and of paramount importance for successful supply chain management. It begins with the realization of the need for a good supplier; determination and formulation of decision criteria; pre-qualification (initial screening and drawing up a shortlist of potential suppliers from a large list); final supplier selection; and the monitoring of the suppliers Selected (*i.e.*, continuous evaluation and assessment). The utilization of the FAHP presented in this chapter brings together good numbers of advantageous aspects of group decision-making in a fuzzy environment. That is, while many of these aspects are present in other techniques, they are most salient in this FAHP method.

Evaluation and selection of suppliers are a typical multiple criteria decision making (MCDM) problem involving multiple criteria that can be both tangible and intangible. The analysis of criteria for selecting and measuring the performance of supplier has been the focus of many researchers and purchasing practitioners as to provide a comprehensive view of the important criteria in the supplier selection decision. Supplier selection process requires a formal, systematic and rational selection model. Choices are made today in even more unpredictable situations. The fuzzy AHP gives a planned technique for assessment and weighting of the different criteria and options. Numerous possible fuzzy AHP systems exist in literature. Fuzzy AHP model was proposed to measure supplier fulfillment of catering running organization by Cebeci (2003). Failure of AHP to manage the imprecision and subjectiveness in the pair-wise comparison process has been enhanced in fuzzy AHP. According to Kuswandari (2004) rather than a fresh worth, fuzzy AHP attain quality to consolidate the choice creator's doubt. In this study, a very comprehensive application of Fuzzy Analytic Hierarchy Process (FAHP) for a case is presented to choose the best supplier. In this study, an attempt is made to further the understanding of the FAHP method introduced in Chang (1996) and developed by Zhu *et al.* (1999) which includes the utilization of the Extent Analysis method and the use of group decision making in the FAHP. As there are many different applications of Fuzzy Analytical Hierarchy Process (FAHP). In Table 2.1 below are examples of available FAHP suggested by different application.

Table 2.1: Illustrates the application of FAHP.

Sl. no	*Author name*	*Year*	*Applications*
1	Cheng Tang *et al.*,	2005	Capital investment study
2	Chi-Tai Lien *et al.*,	2007	ERP system by applying fuzzy AHP
3	Kevin P. Hwang *et al.*,	2007	Possible Strategic Alliances of Taiwanese Telecom Operators
4	Erdal Cakir *et al.*,	2009	Selecting third party logistic service provider
5	Buyukozkan *et al.*,	2011	Strategic analysis of healthcare service quality
6	Yu-Cheng Tang *et al.*,	2011	Application of the fuzzy analytic hierarchy process to the lead-free equipment selection decision
7	Golam Kabir *et al.*,	2011	Modified fuzzy analytical hierarchy process for multiple criteria inventory classification
8	Mohammad, A.H.	2010	A hybrid fuzzy MCDM Approach to Thesis Subject Selection
9	Maysam and Ashrafzadeh	2012	The application of fuzzy analytic hierarchy process approach for the selection of warehouse location: A case study
10	P. Muralidhar *et al.*,	2012	Evaluation of green supply chain management strategies
11	S. Cigdem *et al.*,	2012	Application of grey relational analysis
12	Meysam Shaverdi *et al.*,	2012	Economic cocoon traits improvement in silkworm breeding

Table 2.1: *(Contd...)*

Table 2.1: *(Contd...)*

Sl. no	*Author name*	*Year*	*Applications*
13	Anvary Rostamy and Ali Asghar	2013	Green supply chain management evaluation in publishing industry based
14	Toni Lupo *et al.*,	2013	Strategic analysis of transit service quality
15	Mohammad *et al.*,	2013	Developing a new model using Fuzzy AHP and TOPSIS methods in supplier selection problem xin supply chain management - A case study of SADRA Company in IRAN
16	K.A. Ramadan *et al.*,	2013	Decision making and evaluation system for employee recruitment
17	Ali Sorayaei *et al.*,	2014	Assessment and prioritization of business processes with the ability to outsource municipal: Case Study: Amir kola Municipality
18	Faisal Masood *et al.*,	2014	A novel model of strategically aligned portfolio optimization project prioritization
19	Farzad Tahrir *et al.*,	2014	Supplier assessment and selection using fuzzy analytic hierarchy process in a steel manufacturing company
20	Odeyale *et al.*,	2014	Evaluation and selection of an effective green supply chain management strategy: A case study.

2.2. ANALYTICAL HIERARCHY PROCESS

The analytic hierarchy process (AHP) was first introduced by Saaty in 1971 to solve the scarce resources allocation and planning needs for the military (Saaty, 1980). Since its introduction, the AHP has become one of the most widely used multiple-criteria decision-making (MCDM) methods, and has been used to solve unstructured problems in different areas of human needs and interests, such as political, economic, social and management sciences. The AHP is based on the innate human ability to make sound judgments about small problems. It facilitates decision making by organizing perceptions, feelings, judgments, and memories into a framework that exhibits the forces that influence a decision. The AHP is implemented in the software of Expert Choice and it has been applied in a variety of decisions and planning projects in nearly 20 countries (Saaty, 2001).

In AHP a problem is structured as a hierarchy. Once the hierarchy has been constructed, the decision-maker begins the prioritization procedure to determine the relative importance of the elements in each level. Prioritization involves eliciting judgments in response to questions about the dominance of one element over another with respect to a property. The scale used for comparisons in AHP enables the decision-maker to incorporate experience and knowledge intuitively and indicate how many times an element dominates another with respect to the criterion (Millet, 1997). The decision maker can express his preference between each pair of elements verbally as equally important, moderately more important, strongly more important, very strongly more important, and extremely more important. These descriptive preferences would then be translated into numerical values 1, 3, 5, 7, 9 respectively with 2, 4, 6, and 8 as intermediate values for comparisons between two successive qualitative judgments. Reciprocals of these values are used for the corresponding transposed judgments. The Table 2.2 below shows the comparison scale used by AHP.

Table 2.2: Illustrates Thomas Saaty's nine-point scale.

Intensity of importance	***Definition***	***Explanations***
1	Equal importance	Two activities contribute equally to the objective
3	Weak importance of one over another	Experience and judgment slightly favor one activity over another
5	Essential or strong important	Experience and judgment strongly favor one activity over another
7	Demonstrated importance	An activity is favored very strongly over another its dominance demonstrated in practice
9	Absolute importance	The evidence favoring one activity over another is of the highest possible order of affirmation
2, 4, 6, 8	Intermediate values between the two-adjacent judgment	When compromise is needed
Reciprocals of above non-zero	If activity I has one of the above non-zero numbers assigned to it when compared with activity j, then has the reciprocal value when compared with activity I.	A reasonable assumption

Finally, all the comparisons are synthesized to rank the alternatives. The output of AHP is a prioritized ranking of the decision alternatives based on the overall preferences expressed by the decision maker.

2.3. FUZZY ANALYTICAL HIERARCHY PROCESS (FAHP)

AHP is a standout amongst the most generally utilized MADM techniques. Dealing with the multicriteria choice analysis, the fuzzy set hypothesis could be the very well-known strategy in managing uncertain problems. AHP make the most of "pair-wise relationship" and priority weights criteria. According to Saaty selection is made by developing weights and assessing criteria (Masood, 2014). According to Zadeh (1968) the techniques are logical methodologies to possible choice and by using ideas of fuzzy set hypothesis as well as progressive structure analysis. This study utilizes one of multi-criteria decision making (MCDM) method, fuzzy analytic hierarchy process (FAHP) first appeared in Van Laarhoven and Pedrycz (1983). Previous studies have evaluated FAHP as applied to the overall issue of selection (*e.g.*, facility, vendor or building (Bozda, *et al.*, 2003; Kahraman *et al.*, 2003, 2007) supplier selection (Chan and Kumar, 2007) and project selection (Huang *et al.*, 2008) etc. These different selection processes have all benefited from FAHP and have a common characteristic: the degree of fuzziness in human decision making is fixed. They do not take into account the fact that the degree of fuzziness can vary depending on the criteria being considered.

In conventional AHP, the pair-wise comparison is established using a nine-point scale which converts the human preferences between available alternatives as equally, moderately, strongly, very strongly or extremely preferred. Even though the discrete scale of AHP has the advantages of simplicity and ease of use, it is not sufficient to consider the uncertainty associated with the mapping of one's perception to a number. Therefore, fuzzy logic is introduced into the pair-wise comparison to deal with the deficiency in the traditional AHP. This is referred to as fuzzy AHP.

By the help of FAHP, we can efficiently handle the fuzziness of the data involved in the decision of selecting best supplier. It is easier to under-stand and it can effectively handle both qualitative and quantitative data in the multi-criteria decision-making problems. In this approach triangular fuzzy numbers are used for the preferences of one criterion over another and then by using the extent analysis method, the synthetic extent value of the pair-wise comparison is calculated. Based on this approach, the weight vectors are decided and normalized, thus the normalized weight vectors will be determined. As a result, based on the different weights of criteria and criteria's the final priority weights of the alternative suppliers are decided. The highest priority would be given to the supplier with highest weight (Chan and Kumar, 2005).

2.4. FUZZY SET THEORY

In 1965, Lotfi A. Zadeh proposed a new approach to a rigorous, precise theory of approximation and vagueness based on generalization of standard set theory to fuzzy sets. Fuzzy sets and fuzzy logic are powerful mathematical tools for modeling: uncertain systems in industry, nature and humanity; and facilitators for common-sense reasoning in decision making in the absence of complete and precise information. Their role is significant when applied to complex phenomena not easily described by traditional mathematical methods, especially when the goal is to find a good approximate solution. Some basic definitions of fuzzy sets, fuzzy numbers and linguistic variables are reviewed from Zadeh (1975), Buckley (1985), Kaufmann and Gupta (1991). The basic definitions and notations below will be used throughout this thesis until otherwise stated.

2.4.1. Fuzzy Sets

Definition 1. A fuzzy set A in a universe of discourse *X* is characterized by a membership Function $U_A(x)$ which associates with each element x in X a real number in the interval[0,1]. The function value $U_A(x)$ is termed the grade of membership of x in A (Kaufmann and Gupta, 1991).

$$U_{\tilde{A}}(x) = \begin{bmatrix} 1 & for\ x \in A \\ 0 & for\ x \notin A \end{bmatrix} \quad \text{.......................(2.1)}$$

Definition 2. A fuzzy set A in a universe of discourse *X* is convex if and only if

$$U_A(\lambda_{x1}+(1-\lambda)x_2) \geq \min\ (U_A(x_1),\ U_A(x_2)\ ) \quad \text{..............(2.2)}$$

For all x_1, x_2 in X and all $\lambda\varepsilon[0,1]$, where min denotes the minimum operator (Klir and Yuan, 1995).

Definition 3. The height of a fuzzy set is the largest membership grade attained by any element in that set. A fuzzy set A in the universe of discourse X is called normalized when the height of A is equal to 1 (Klir and Yuan, 1995).

2.5. TRIANGULAR FUZZY NUMBERS

According to the principle of incompatibility (Zadeh, 1973), when facing a complex decision, human beings have difficulty in making a precise decision. Consequently, the data of human subjective judgment are usually fuzzy and imprecise in nature. Fuzzy data can be expressed in linguistic terms or in fuzzy numbers (Chen and Hwang, 1992). Thus, we consider the value of fuzzy measures as a linguistic value and then convert linguistic terms to fuzzy numbers. Zhang (1992a,) developed the elementary concepts and theorems of a fuzzy number-valued fuzzy measure ((z) fuzzy measure) and a fuzzy number-valued fuzzy integral on the fuzzy set.

In some situation, when decision-makers are asked to express their opinions, it is very difficult to solve vague and uncertain problem by traditional linguistic method, but it can convert into fuzzy numbers. A fuzzy number is a fuzzy subset in real numbers which have two properties, the convexity and the normality. Fuzzy numbers become more meaningful to quantify a subjective measurement into a range rather than in an exact value (Chan *et al.*, 2000). The TFNs and Tr FNs are a special case of fuzzy numbers. The weights of decisive criteria and the ratings of qualitative criteria which are assessed by the TFNs and Tr FNs in linguistic variables can solve lots of uncertain problems and practical applications (Chen *et al.*, 2006).

Fuzzy numbers are the special classes of fuzzy quantities. A fuzzy number is a fuzzy quantity M that represents a generalization of a real number r intuitively; M(x) should be a measure of how well M(x) "approximates" r. A fuzzy number M is a convex normalized fuzzy set. A fuzzy number is characterized by a given interval of real numbers, each with a grade of membership between 0 and 1 (Deng, 1999). Fuzzy Sets can be defined as follows.

2.5.1. Definitions of Fuzzy Numbers

Definition 1. A fuzzy number is a fuzzy subset in the universe of discourse X that is both convex and normal. Fig. 2.1 show a fuzzy number n in the universe of discourse X that conforms to this definition (Kaufmann and Gupta, 1991).

Definition 2. Suppose, a positive triangular fuzzy number (PTFN) is$\tilde{A}$ and that can be defined as (l, m, u) shown in Fig. 2.2. The membership function (x) m n is defined as: A triangular fuzzy number (TFN), M is shown in Figure. Triangular fuzzy numbers (Table 2.3) are defined by three real numbers, expressed as (l, m, u). The parameters l, m, and u respectively, indicate the smallest possible value, the most promising value, and the largest possible value that describe a fuzzy event. Their membership functions are described as.

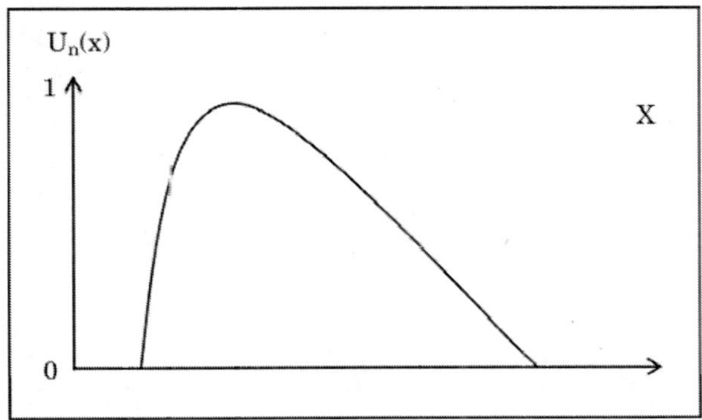

Fig. 2.1: A fuzzy number n

$$U_{\bar{A}}(x) = \begin{cases} 0 & x \leq 1 \\ \dfrac{x-l}{m-l} & l < x \leq m \\ \dfrac{u-x}{u-m} & m < x \leq u \end{cases} \tag{2.3}$$

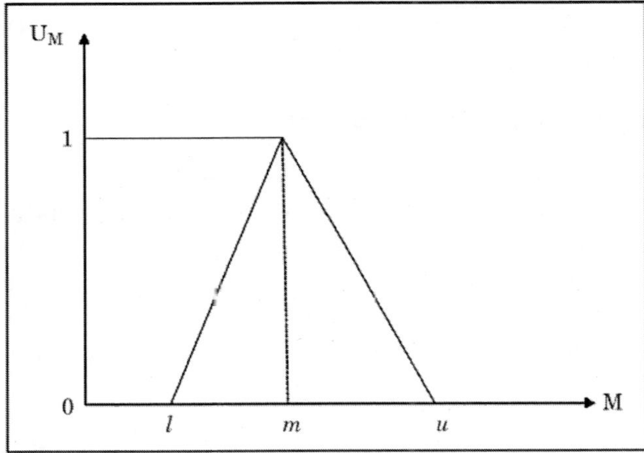

Fig. 2.2: A triangular fuzzy numbers

In applications it is convenient to work with TFNs because of their computational simplicity, and they are useful in promoting representation and information processing in a fuzzy environment. In this study TFNs in the FAHP is adopted.

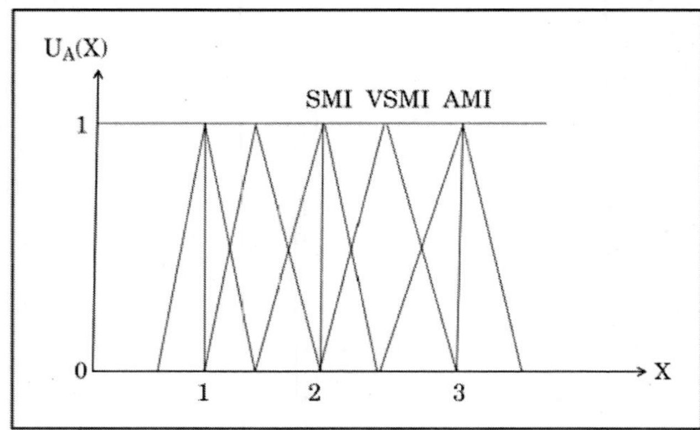

Fig. 2.3: Linguistic scale for relative importance

Table 2.3: Triangular fuzzy numbers.

Sl. no.	*Linguistic variables*	*Positive triangular fuzzy numbers*	*Positive reciprocal triangular fuzzy numbers*
1	Just Equal (JE)	(1, 1, 1)	(1, 1, 1)
2	Equally Important (EI)	(1/2, 1, 3/2)	(2/3, 1, 2)
3	Weakly More Important (WMI)	(1, 3/2, 2)	(1/2, 2/3, 1)
4	Strongly More Important (SMI)	(3/2, 2, 5/2)	(2/5, 1/2, 2/3)
5	Very Strongly More Important (VSMI)	(2, 5/2, 3)	(1/3, 2/5, 1/2)
6	Absolutely More Important (AMI)	(5/2, 3, 7/2)	(2/7, 1/3, 2/5)

2.5.2. Algebraic Operations on TFNs

Although we are familiar with algebraic operations with crisp numbers, when we want to use fuzzy sets in applications, then dealing with fuzzy numbers is must. Author have defined various operations on TFNs. But in this section, three important operations used in this study are illustrated (Tang and Beynon, 2005). If we define, two TFNs A and B by the triplets $A = (l_1,m_1,u_1)$ and $B = (l_2,m_2,u_2)$ Then

Addition:

$$A+B = (l_1,m_1,u_1)+(l_1,m_2,u_2)=(l_1+l_2)+(m_1+m_2)+(u_1+u_2) \quad (2.4)$$

Multiplication:

$$A.B= (l_1,m_1,u_1)+(l_1,m_2,u_2)=(l_1.l_2,\ m_1.m_2,\ u_1.u_2) \quad (2.5)$$

Inverse:

$$(l_1,m_1,u_1)–1= (1/u_1,\ 1/m_1,\ 1/l_1) \quad (2.6)$$

2.6. LINGUISTIC VARIABLE

A linguistic variable is the variable whose values are not expressed in numbers but words or sentences in a natural or artificial language (Zadeh, 1975). The concept of a linguistic variable is very useful in dealing with situations, which are too complex or not well defined to be reasonably described in conventional quantitative expressions (Zimmermann, 1991). Linguistic variables represent crisp information in a form and precision appropriate for the problem. The linguistic assessment of human feelings and judgments are vague, and it is not reasonable to represent it in terms of precise numbers. To give interval judgments than fixed value judgments is more confident for decision makers. So, triangular fuzzy numbers are used to decide the priority of one decision variable over other in fuzzy AHP (Chan and Kumar, 2005). For example, 'weight' is a linguistic variable whose values are 'very low', 'low', 'medium', 'high', 'very high', etc. Fuzzy numbers can also represent these linguistic values.

2.7. ALGORITHM OF FAHP METHOD

In this study the extent FAHP is utilized, which was originally introduced by Chang (1996). Let X = $\{x_1, x_2, x_3,......., x_n\}$ an object set, and G = $\{g_1, g_2, g_3,......., g_n\}$ be a goal set. According to the method of Chang's extent analysis, each object is taken and extent analysis for each goal performed respectively. Therefore, m extent analysis values for each object can be obtained, with the following signs:

$$M_{1gi}, M_{2gi},............M_{mgi}, \qquad i = 1, 2,....,n, \tag{2.7}$$

Where M_{jgi} (j = 1, 2, ..., m) all are TFNs. The steps of Chang's extent analysis can be given as in the following

Let A = $(a_{ij})_{max}$ be a fuzzy pair-wise comparison matrix, where a_{ij} = (l_{ij}, m_{ij}, u_{ij}). The steps used for the Chang method are as follows. Initially, pair-wise comparison is made using fuzzy numbers.

Step 1: The value of fuzzy synthetic extent with respect to the ith object is defined as

$$\sum_{j=1}^{m} M_{gi}^{J} = (l_{i1}, m_{i1}, u_{i1}) \oplus (l_{i2}, m_{i2}, u_{i2}) \oplus (l_{im}, m_{im}, u_{im}) \tag{2.8}$$

$$S S_i = \sum_{j=1}^{m} M_{gi}^{J} \otimes \left[\sum_{i=1}^{m} \sum_{j=1}^{m} M_{gi}^{J} \right]^{-1} \tag{2.9}$$

Step 2: As $M_1(l_1, m_2, u_3)$ and $M_2(l_2, m_2, u_3)$ are two triangular fuzzy numbers, the degree of possibility of $M_2(l_2, m_2, u_2) \geq M_1(l_1, m_1, u_1)$ defined as:

$$V(S_i \geq S_k) = \underset{x \geq y}{SUP}\left(\min\{U_{M_1}(x), U_{M_2}(x)\}\right)$$

$$V\ (S_i \geq S_j) = \text{height}\ (S_i \cap S_j) = U_{M2}(d)$$

$$V(S2 \geq S1) = \begin{cases} 1 & \text{if } b2 \geq b1 \\ 0 & \text{if } a1 \geq c2 \\ \dfrac{a1 - c2}{(b2 - c2) - (b1 - c1)}, & \text{Otherwise} \end{cases} \quad (2.11)$$

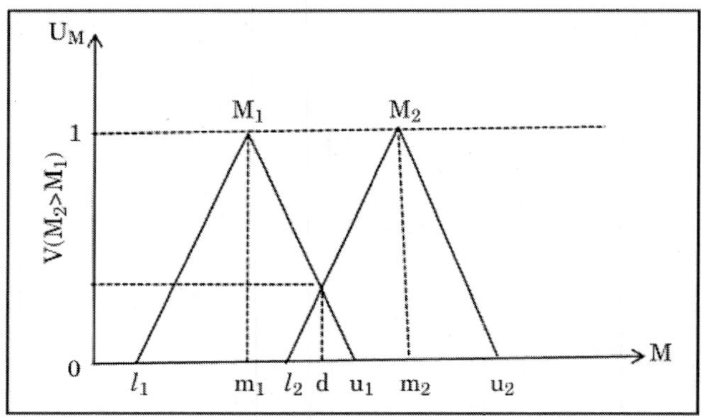

Fig. 2.4: Intersection point between M_2 and M_2.

Step 3: The degree possibility for a convex fuzzy number to be greater than k convex fuzzy $M_i(I = 1, 2, k)$ numbers can be defined by

$$V(MM_1M_2 \ldots M_k) = V(M_2 \geq M_1) \text{ and } (M_1 \geq M_2) \text{ and} \ldots (Mi \geq Mk) \quad (2.12)$$

$$= \min V(M > Mi)\ i = 1,2,3,\ldots.k \quad (2.13)$$

Assume that $(d(A_i) = \min (S_i \geq S_j)$

$$V\ (S \geq S1, S2, S3, \ldots, Sk,) \qquad \text{for } i = 1,2,3,\ldots,k$$

$$= V\ (S \geq S1) \text{ and } (S \geq S2) \text{ and} \ldots (S \geq Sk) \quad (2.14)$$

$$V = \min V\ (S \geq Si\ ) \qquad \text{for } i = 1,2,3,..,.k, \quad (2.15)$$

Then the weight vector is given by

Assume that (d' Ai) = minV (S>Si), for i = 1,2,3,..,.k (2.16)

Then the weight vector is defined as

$$W' = ((d(A1).d(A2)......d(An))^T \quad (2.17)$$

Where Ai (i = 1, 2,3,...,n)are the n elements.

Finally, the weight vectors are then normalized as follows. And to give crisp weight vector represented by

$$W = (W'/ \Sigma W') \quad (2.18)$$

$$W = d\ (A1),\ d(A2),...,d\ (An)^T \quad (2.19)$$

Where W is a non-fuzzy number, and this gives the importance weights of one indicator over other.

2.8. APPLICATION OF METHODOLOGY ADOPTED

Researchers have done a variety of application of FAHP to various fields, the ability to rapidly incorporate feedback and a possibility of simple comparison to actual results makes it very powerful method in multi criteria decision making in various manufacturing areas. In addition to the wide application of FAHP in manufacturing areas, recent research and industrial activities of applying FAHP on other selection problems are also quite active.

The study not only provides the evidence that the FAHP proved much better in MCDM, but also aids the decision makers and researchers in applying FAHP effectively. It has the advantage of mathematically representing uncertainty and vagueness, and of providing formalized tools for dealing with the imprecision intrinsic to many problems (Ming Lang Tseng and Yuan Hsu Lin, 2008). A schematic representation of the methodology is given in Fig. 2.5.

2.9. WORKING STEPS OF CALCULATIONS AND APPLICATIONS FOR FAHP IN SUPPLIER EVALUATION AND SELECTION

The following steps have been considered to set the Supplier selection model.

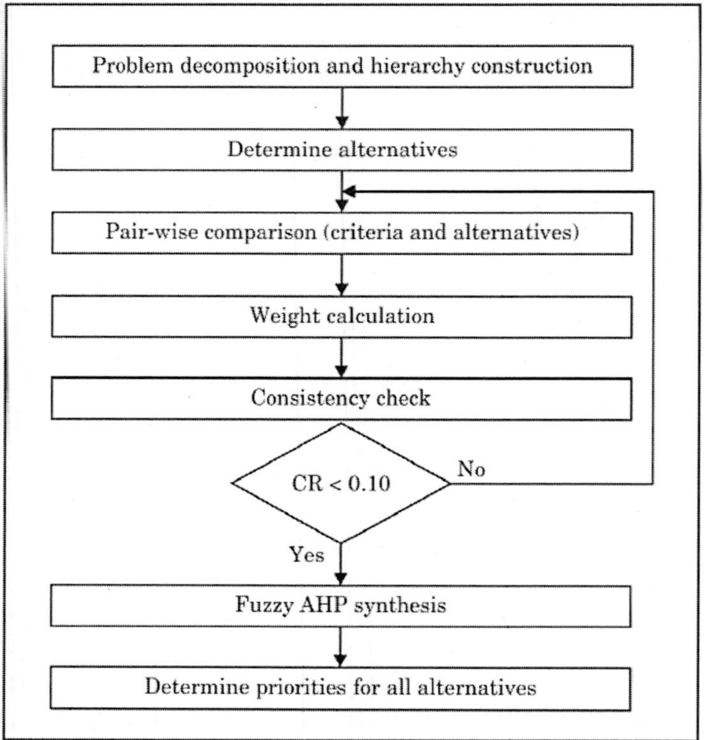

Fig. 2.5: Schematic representation of the methodology.

2.9.1. Establishing a Group of Decision-Makers

In the first step, a group of decision-makers is formed. The members of the group are experienced experts and managers in industry. The decision-makers must determine the relative weights of each output indicator.

2.9.2. Model Development and Problem Formulation

In this step, the decision problem is structured into its important components. The relevant criteria and alternatives are chosen from the review of literature and discussion with few both from industry and academia. The relevant criteria and alternatives are structured in the form of a control hierarchy where the criteria at the top level in the model have the highest strategic value. The Goal in this model is Supplier evaluation and selection.

In this FAHP model, each of the five dimensions has some enablers, which help to achieve that dimension. For example, the dimension Cost is supported by the enablers Material Cost, Transportation Cost, Quantity Discount and Order and Invoicing Cost. These enablers also have some interdependency on one another. The degree of interdependency may vary from case to case and would be captured in later steps. The developed approach can adequately handle the inherent uncertainty and imprecision of the human decision-making process and can provide the flexibility and robustness needed for the decision maker to understand the decision priorities (Ming and Yuan, 2008). The objective of this hierarchy is to select the best possible alternative that will best meet the goals of Supplier selection in automotive component industry. The FAHP model so developed is presented in Fig. 2.4. The alternatives that the decision maker wishes to evaluate are shown at the bottom of the model. The opinion of the material manager of the company was sought in the comparisons of the relative importance of the criteria and the formation of pair-wise comparison matrices to be used in the FAHP model. The results of determinants would be used in the calculation of Supplier selection.

2.9.3. Determining the Linguistic Variables and Fuzzy Conversion Scale

Weights are determined by using a pair-wise comparison of the importance or preference between each pair of indicators. The comparison of one indicator over another can be done with the help of questionnaire. In this case TFNs are used to represent subjective pair-wise comparisons of decision-makers namely "just equal", "equally important", "weakly more important", "strongly more important", "very strongly more important" and "absolutely more important". The triangular fuzzy conversion scales and linguistic scales, which is proposed by Kahraman *et al*. (1991), is used to convert such linguistic values into fuzzy scales is demonstrated in Fig. 2.3 and Table 2.3.

2.9.4. Constructing the Fuzzy Comparison Matrix

Let us consider a problem with n indicators, where the relative importance of indicator *i* to *j* is represented by fuzzy triangular numbers $a_{ij}(l_{ij}, m_{ij}, u_{ij})$. As in the traditional AHP, the comparison matrix $A(a_{ij})$ max can be constructed, such that

$$A(a_{ij})_{max} = \begin{bmatrix} 1 & a_{12} & \ldots\ldots & a_{1n} \\ a_{21} & 1 & \ldots\ldots & a_2n \\ \ldots & \ldots\ldots & \ldots & \ldots\ldots \\ a_{1n} & a_{2n} & \ldots\ldots & 1 \end{bmatrix} = \begin{bmatrix} 1 & a_{12} & \ldots\ldots & a_{1n} \\ 1/a_{12} & 1 & \ldots\ldots & a_{2n} \\ \ldots\ldots & \ldots\ldots & \ldots & \ldots\ldots \\ 1/a_{1n} & 1/a_{2n} & \ldots\ldots & 1 \end{bmatrix}$$

2.9.5. Calculating the Consistency Index and Consistency Ratio of Fuzzy Comparison Matrix

To assure a certain quality level of a decision, the consistency of an evaluation has to be analyzed. Saaty proposed a consistency index to measure consistency. This index can be used to indicate how consistent the pair-wise comparison matrices are. To investigate the consistency, the fuzzy comparison matrices need to be converted into crisp matrices. There are some defuzzifications methods are for obtaining a crisp number from the triangular Fuzzy number. We select the fuzzy mean and spread method to defuzzify the fuzzy numbers. A triangular fuzzy number denoted as a = (l, m, u) can be defuzzified to a crisp number as follows.

$$a\_crisp = (l+m+u)/3 \tag{2.20}$$

The consistence index, CI, for a comparison matrix can be computed with the use of following equation.

$$CI = \frac{\lambda \max - n}{n-1}$$

Where, λ_{max} is the largest eigen value of the comparison matrix, n is the dimension of the matrix.

2.9.5.1. *Consistency ratio*

$$CR = CI/RCI \tag{2.22}$$

Where CI = Consistency index, CR = Consistency ratio, RCI = Random consistency index, n = Number of elements

2.9.5.2. *Random consistency index (RCI)*

Where random consistency index (RCI) varies depending upon the order of matrix. Table 2.4 shows the value of the Random Consistency Index (RCI) for matrices of order 1 to 10 obtained by approximating random indices using a sample size of 500.

Table 2.4: Random consistency index (RCI).

N	3	4	5	6	7	8	9	10	11	12	13	14	15
RCI	0.58	0.9	1.12	1.24	1.32	1.41	1.45	1.51	1.52	1.54	1.56	1.6	1.59

The acceptable CR range varies according to the size of matrix *i.e.*, 0.05 for a 3 by 3 matrix, 0.08 for a 4 by 4 matrix and 0.1 for all larger matrices, $m \geq 5$. If the value of CR is equal to, or less than that value, it implies that the evaluation within the matrix is acceptable or indicates a good level of consistency in the comparative judgments represented in that matrix. In contrast, if CR is more than the acceptable value, inconsistency of judgments within that matrix has occurred and the evaluation process should therefore be reviewed, reconsidered and improved.

2.9.6. Calculating the Weights

When the consistency in the comparison matrix is accepted, the extended analysis fuzzy AHP method is then employed to identify the weights of output indicators. These weights are used to determine the assurance regions for each output indicator in the supplier selection model.

2.9.7. Defining the Supplier Selection Constraints

The Supplier Selection constraints are done by defining upper and lower bounds for each weight indicator. These bounds are now ranges for preference weights for each of the indicators as defined by the decision-makers.

2.10. HIERARCHY MODEL FOR SUPPLIER SELECTION

Based on the challenge stated, a practical example is defined, and a mathematical model is generated. Early supplier involvement approach recognizes that qualified suppliers can offer a firm more than just manufacturing according to given specifications. Such supplier can provide early insights on how to produce efficiently and effectively given their capabilities, how to simplify a product's design, thus affecting quality and cost levels. The Supplier selection criteria are structured as a hierarchy shown in Fig. 2.6 below. The hierarchy structure includes goal, criteria, sub-criteria, rating levels, and desired result for the Supplier selection problem. The goal for the best Supplier selection is put on the highest level of the hierarchy and the criteria on the second level. The hierarchy is easily extended to more detailed levels by breaking down the criteria into sub-criteria. The sub-criterion on the third level and the fourth and last level is desire result.

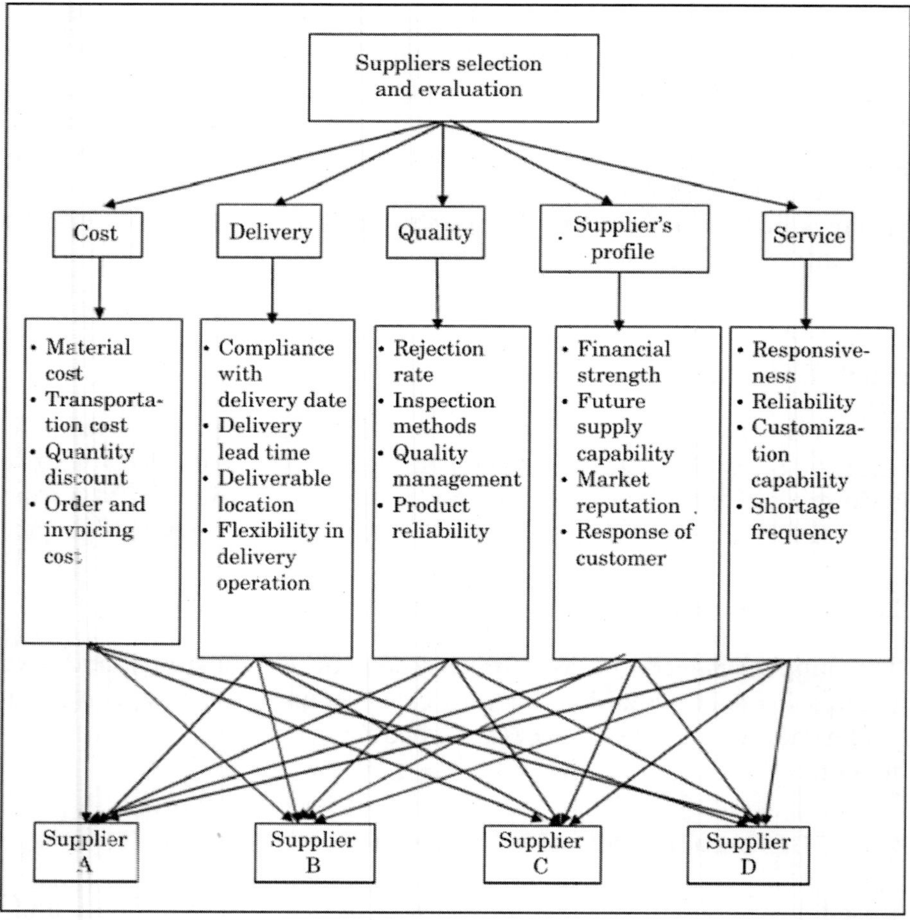

Fig. 2.6: Hierarchy model for supplier selection.

2.10.1. Description of the Model

A thorough analysis of the problem is required along with the identification of the important criteria involved. The selections of criteria have been determined through literature survey and discussions held with experts during industrial visits. The criteria and sub-criteria used in the FAHP model for Material supplier selection in supply chain are given in Table 2.5.

Table 2.5: Criteria and sub-criteria used in supplier selection.

Major criteria	*Sub-criteria*	*Major criteria*	*Sub-criteria*
Cost	Material cost Transportation cost Quantity discount Ordering and invoicing cost	Delivery	Compliance with delivery date Delivery lead time Deliverable location Flexibility in delivery operation
Quality	Rejection rate Inspection method Quality management Product reliability	Supplier's profile	Financial strength Future supply capability Market reputation Response of customers
Service	Responsiveness Reliability Customization capability Shortage frequency		

2.10.2. Data Input and Analysis Using Fuzzy AHP

Fuzzy AHP analysis has been explained step-wise and presented in the Table no. 2.6 to 2.58.

Table 2.6: Pair-wise comparison matrix for the criteria of goal.

	Cost	*Delivery*	*Quality*	*Customer's profile*	*Service*
Cost	1, 1, 1	1/2, 1, 3/2	1, 1, 1	1, 3/2, 2	1/2, 1, 3/2
Delivery	2/3, 1, 2	1, 1, 1	2/3, 1, 2	1, 3/2, 2	1, 3/2, 2
Quality	1, 1, 1	1/2, 1, 3/2	1, 1, 1	3/2, 2, 5/2	3/2, 2, 5/2
Customer's profile	1/2, 2/3, 1	1/2, 2/3, 1	2/5, 1/2, 2/3	1, 1, 1	1/2, 1, 3/2
Service	2/3, 1, 2	1/2, 2/3, 1	2/5, 1/2, 2/3	2/3, 1, 2	1, 1, 1
CR = 0.022					

Synthetic extent normalized row sum. Row sum value of row in table using equation (4.8)

RS_1 = (4.00, 5.50, 7.00) RS_2 = (4.33, 6.00, 9.00), RS_3 = (5.50, 7.00, 8.50), RS_4 = (2.90, 3.83, 5.17), RS_5 = (2.90, 3.83, 5.17)

The synthetic extent normalized row sums obtained from using equation (4.9)

S1 = (0.110, 0.210, 0.350), S2 = (0.120, 0.230, 0.450), S3 = (0.150, 0.260, 0.430), S4 = (0.080, 0.140, 0.260), S5 = (0.090, 0.160, 0.330)

Table 2.7: Degree of possibility for criteria of goal.

V(Si ≥ Sj)	*Degree of possibility (Calculated by using equation 2.11)*	*Value*	*d'(Ai)- minV (Si ≥ Sj)*
S1 ≥ S2	$\frac{(0.120 - 0.350)}{(0.207 - 0.350) - (0.230 - 0.120)}$	0.909	d'(A1)- minV(S1 > Sj)
S1 ≥ S3	$\frac{(0.150 - 0.350)}{(0.207 - 0.350) - (0.260 - 0.150)}$	0.769	= 0.769
S1 ≥ S4	b1 ≥ b4	1.00	
S1 ≥ S5	b1 ≥ b5	1.00	
S2 ≥ S1	b2 ≥ b1	1.00	d'(A2)- minV(S2 > Sj)
S2 ≥ S3	$\frac{(0.150 - 0.450)}{(0.230 - 0.450) - (0.260 - 0.150)}$	0.890	= 0.890
S2 ≥ S4	b2 ≥ b4	1.00	
S2 ≥ S5	b2 ≥ b5	1.00	
S3 ≥ S1	b2 ≥ b1	1.00	d'(A3)- minV(S2 > Sj)
S3 ≥ S2	b3 ≥ b2	1.00	= 1.00
S3 ≥ S4	b3 ≥ b4	1.00	
S3 ≥ S5	b3 ≥ b5	1.00	
S4 ≥ S1	$\frac{(0.110 - 0.260)}{(0.140 - 0.260) - (0.207 - 0.110)}$	0.691	d'(A4)- minV(S4 ≥ Sj)
S4 ≥ S2	$\frac{(0.120 - 0.260)}{(0.140 - 0.260) - (0.230 - 0.120)}$	0.608	= 0.478
S4 ≥ S3	$\frac{(0.150 - 0.260}{(0.140 - 0.260) - (0.260 - 0.150)}$	0.478	
S4 ≥ S5	$\frac{(0.09 - 0.260}{(0.140 - 0.260) - (0.160 - 0.09)}$	0.894	
S5 ≥ S1	$\frac{(0.110 - 0.330)}{(0.160 - 0.330) - (0.207 - 0.110)}$	0.823	d'(A5)- minV(S5 ≥ Sj)
S5 ≥ S4	$\frac{(0.120 - 0.330)}{(0.160 - 0.330) - (0.230 - 0.120)}$	0.760	= 0.630
S5 ≥ S4	$\frac{(0.08 - 0.330)}{(0.160 - 0.330) - (0.140 - 0.08)}$	0.630	
S5 ≥ S4	b4 ≥ b5	1.00	

Determination of Weight Vectors

Weight vector {d'(Ai), d'(A2) ----------- d(An)}^{T}

$$W' = (0.769, 0.890, 1.00, 0.478, 0.630)$$

The above vector is normalized to give crisp weight vector

$$W = (0.204, 0.236, 0.265, 0.127, 0.167)^T$$

Table 2.8: Pair-wise comparison matrix for the sub-criteria of "Cost".

Criteria's	*Material cost*	*Transportation cost*	*Quantity discount*	*Overhead cost*
Material cost	1, 1, 1	1, 3/2, 2	1/2, 1, 3/2	3/2, 2, 5/2
Transportation cost	1/2, 2/3, 1	1, 1, 1	2/3, 1, 2	1, 3/2, 2
Quantity discount	2/3, 1, 2	1/2, 1, 3/2	1, 1, 1	3/2, 2, 5/2
Overhead cost	2/5, 1/2, 2/3	1/2, 2/3, 1	2/5, 1/2, 2/3	1, 1, 1
		CR = 0.006		

Synthetic Extent Normalized Row Sum. Row sum value of row in table using equation (2.8)

RS1 = (4.00, 5.50, 7.00) RS2 = (3.16, 4.16, 6.00)

RS3 = (3.66, 5.00' 7.00) RS4 = (2.30, 2.66, 3.33)

The synthetic extent normalized row sums obtained from above using equation (2.9)

S1 = (0.171, 0.317, 0.527) S2 = (0.135, 0.240, 0.452)

S3 = (0.156, 0.288, 0.527) S4 = (0.0985, 0.153, 0.251)

Table 2.9: Degree of possibility for main criteria 'Cost'.

V(Si ≥Sj)	*Degree of possibility (Calculated by using equation 4.11)*	*Value*	*d'(Ai)- minV (Si ≥Sj)*
S1 ≥ S2	b1 ≥ b2	1.00	d'(A1)- minV(S1 ≥ Sj)
S1 ≥ S3	b1 ≥ b3	1.00	= 1.00
S1 ≥ S4	b1 ≥ b4	1.00	
S2 ≥ S1	$\frac{(0.171-0.452)}{(0.153-0.452)-(0.317-0.171)}$	0.784	d'(A2)- minV(S2 ≥ Sj)

Table 2.9 (*Contd...*)

Table 2.9: (*Contd...*)

V(Si ≥ Sj)	*Degree of possibility (Calculated by using equation 4.11)*	*Value*	*d'(Ai)- minV (Si ≥ Sj)*
S2 ≥ S3	$\frac{(0.156 - 0.452}{(0.240 - 0.452) - (0.288 - 0.156)}$	0.860	= 0.784
S2 ≥ S4	b2 ≥ b4	1.00	
S3 ≥ S1	$\frac{(0.171 - 0.527)}{(0.288 - 0.527) - (0.317 - 0.171)}$	0.924	d'(A3)- minV(S3 ≥ Sj)
S3 ≥ S2	b3 ≥ b2	1.00	= 0.924
S3 ≥ S4	b3 ≥ b4	1.00	
S4 > S1	$\frac{(0.171 - 0.251}{(0.153 - 0.251) - (0.3317 - 0.171)}$	0.327	d'(A3)- minV(S3 ≥ Sj)
S4 ≥ S2	$\frac{(0.135 - 0.251}{(0.135 - 0.251) - (0.240 - 0.135)}$	0.524	= 0.327
S4 ≥ S3	$\frac{(0.156 - 0.251}{(0.153 - 0.251) - (0.288 - 0.156)}$	0.413	

Determination of Weight Vectors

Weight vector {d'(Ai), d'(A2) ----------- d(An)}^{T}

$$W' = (1.00, 0.784, 0.924, 0.327)$$

The above vector is normalized to give crisp weight vector

$$W = (0.329, 0.258, 0.304, 0.1077)^T$$

Table 2.10: Pair-wise comparison matrix for the sub-criteria of "Delivery".

	Compliance with delivery date	*Lead time*	*Deliverable location*	*Flexibility in delivery operation*
Compliance with delivery date	1, 1, 1	3/2, 2, 5/3	2, 5/2, 3	1, 3/2, 2
Lead time	2/5, 1/2, 2/3	1, 1, 1	2, 5/2, 3	3/2, 5/2, 3
Deliverable location	1/3, 2/5, 1/2	1/3, 2/5, 1/2	1, 1, 1	1, 3/2, 2
Flexibility in delivery operation	1/2, 2/3, 1	1/3, 2/5, 2/3	1/2, 2/3, 1	1, 1, 1
		CR = 0.073		

Synthetic extent normalized row sum. Row sum value of row in table using equation (2.8)

RS1 = (5.50, 7.00, 8.50)   RS2 = (4.90, 6.50, 7.66)

RS3 = (2.66, 3.33, 4.00)   RS4 = (2.33, 2.73, 3.66)

The synthetic extent normalized row sums obtained from above using equation (2.9)

S1 = (0.230, 0.358, 0.552)   S2 = (0.205, 0.332, 0.497)

S3 = (0.111, 0.168, 0.259)   S4 = (0.097, 0.139, 0.237)

Table 2.11: Degree of possibility for 'Delivery' main criteria.

V(Si ≥Sj)	*Degree of possibility (Calculated by using equation 4.11)*	*Value*	*d'(Ai)- minV (Si ≥Sj)*
S1 ≥ S2	b2 ≥ b1	1.00	d'(A1) -minV(S1 ≥ Sj)
S1 ≥ S3	b1 ≥ b3	1.00	= 1.00
S1 ≥ S4	b1 ≥ b4	1.00	
S2 ≥ S1	$\frac{(0.230 - 0.497}{(0.332 - 0.497) - (0.358 - 0.230)}$	0.911	d'(A2) -minV(S2 ≥ Sj)
S2 ≥ S3	b2 ≥ b3	1.00	= 0.911
S2 ≥ S4	b2 ≥ b4	1.00	
S3 ≥ S1	$\frac{(0.230 - 0.259}{(0.168 - 0.259) - (0.258 - 0.230)}$	0.132	d'(A3) -minV(S3 ≥ Sj)
S3 ≥ S2	$\frac{(0.205 - 0.259}{(0.168 - 0.259) - (0.332 - 0.205)}$	0.247	= 0.132
S3 ≥ S4	b3 ≥ b4	1.00	
S4 ≥ S1	$\frac{(0.230 - 0.237)}{(0.139 - 0.237) - (0.358 - 0.230)}$	0.030	d'(A3) -minV(S3 ≥ Sj)
S4 ≥ S2	$\frac{(0.205 - 0.237)}{(0.139 - 0.237) - (0.332 - 0.205)}$	0.142	= 0.030
S4 ≥ S3	$\frac{(0.111 - 0.237)}{(0.139 - 0.237) - (0.168 - 0.111)}$	0.810	

Determination of Weight Vectors

Weight vector {d'(Ai), d'(A2) ----------- d(An)}^{T}

$$W = (1.00, 0.911, 0.132, 0.030)$$

The above vector is normalized to give crisp weight vector

$$W' = (0.482, 0.439, 0.063, 0.0144)$$

Table 2.12: Pair-wise comparison matrix for the sub-criteria of "Quality".

	Rejection rate	*Inspection method*	*Quality management*	*Product reliability*
Rejection rate	1, 1, 1	1, 3/2, 2	2, 5/2, 3	3/2, 2, 5/2
Inspection method	1/2, 2/3, 1	1, 1, 1	2, 5/2, 3	1, 1, 1
Quality management	1/3, 2/5, 1/2	1/3, 2/5, 1/2	1, 1, 1	1/2, 1, 3/2
Product reliability	2/5, 1/2, 2/3	111	2/3, 1, 2	1, 1, 1
		CR = 0.022		

Synthetic extent normalized row sum. Row sum value of row in table using equation (2.8)

RS1 = (5.50, 7.00, 8.5) RS2 = (4.50, 5.166, 6.00)

RS3 = (2.166, 2.80, 3.50) RS4 = (3.06, 3.50, 4.66)

The synthetic extent normalized row sums obtained from above using equation (2.9)

S1 = (0.242, 0.379, 0.558) S2 = (0.198, 0.279, 0.394)

S3 = (0.095, 0.151, 0.229) S4 = (0.135, 0.189, 0.306)

Table 2.13: Degree of possibility for 'Quality' sub-criteria.

V(Si ≥ Sj)	*Degree of possibility (Calculated by using equation 4.11)*	*Value*	*d'(Ai)- minV (Si ≥ Sj)*
S1 ≥ S2	b2 ≥ b1	1.00	d'(A1)-minV(S1 > Sj)
S1 ≥ S3	b2 ≥ b1	1.00	= 1.00
S1 ≥ S4	b2 ≥ b1	1.00	
S2 ≥ S1	$\frac{(0.242 - 0.394)}{(0.279 - 0.394) - (0.379 - 0.242)}$	0.603	d'(A2)-minV(S2 ≥ Sj)
S2 ≥ S3	b2 ≥ b3	1.00	= 0.603
S2 ≥ S4	b2 ≥ b4	1.00	
S3 ≥ S1	a1 ≥ c3	0	d'(A3)-minV(S3 ≥ Sj)
S3 ≥ S2	$\frac{(0.198 - 0.229)}{(0.157 - 0.229) - (0.279 - 0.198)}$	0.194	= 0.00
S3 ≥ S4	$\frac{(0.135 - 0.229)}{(0.151 - 0.229) - (0.189 - 0.135)}$	0.710	
S4 ≥ S1	$\frac{(0.242 - 0.306)}{(0.189 - 0.306) - (0.279 - 0.242)}$	0.251	d'(A4)-min V(S4 ≥ Sj)
S4 ≥ S2	$\frac{(0.198 - 0.306}{(0.189 - 0.306) - (0.279 - 0.198)}$	0.142	= 0.251
S4 ≥ S3	$\frac{(0.095 - 0.306)}{(0.189 - 0.306) - (0.151 - 0.095)}$	0.810	

Determination of Weight Vectors

Weight vector {d'(Ai), d'(A2) ----------- d(An)}^{T}

W'= (1.00, 0.603, 0.00, 0.251)

The above vector is normalized to give crisp weight vector

W = (0.539, 0.325, 0, 0.315)

Table 2.14: Pair-wise comparison matrix for the sub-criteria of "Supplier's Profile".

	Financial strength	*Future supply capability*	*Market reputation*	*Response of customer*
Financial strength	1, 1, 1	2/3, 1, 2	1, 1, 1	2, 5/2, 3
Future supply capability	1/2, 1, 3/2	1, 1,1	1, 1, 1	1, 1, 1
Market reputation	1, 1, 1	1, 1, 1	1, 1, 1	1/2, 1, 3/2
Response of customer	1/3, 2/5, 1/2	1, 1, 1	2/3, 1, 2	1, 1, 1
		CR = 0.034		

Synthetic extent normalized row sum. Row sum value of row in table using equation (2.8)

RS1 = (4.66, 5.5, 7.00) RS2 = (3.50, 4.00, 4.50)

RS3 = (3.50, 4.00, 4.50) RS4 = (3, 3.40, 4.50)

The synthetic extent normalized row sums obtained from above using equation (2.9)

S1= (0.227, 0.325, 0.477) S2 = (0.170, 0.236, 0.306)

S3 = (0.170, 0.236, 0.306) S4 = (0.146, 0.201, 0.306)

Table 2.15: Degree of possibility for supplier's profile sub-criteria.

V(Si ≥ Sj)	*Degree of possibility (Calculated by using equation 4.11)*	*Value*	*d'(Ai)- minV (Si ≥ Sj)*
S1 ≥ S2	b1 ≥ b2	1.00	d'(A1)-min V(S1 ≥ Sj)
S1 ≥ S3	b1 ≥ b3	1.00	= 1.00
S1 ≥ S4	b1 ≥ b4	1.00	
S2 ≥ S1	$\frac{(0.227 - 0.306}{(0.236 - 0.306) - (0.325 - 0.227)}$	0.470	d'(A2)-min V(S2 ≥ Sj)
S2 ≥ S3	b2 ≥ b3	1.00	= 0.470
S2 ≥ S4	b2 ≥ b4	1.00	

Table 2.15 (*Contd...*)

Table 2.15: *(Contd...)*

V(Si ≥ Sj)	*Degree of possibility (Calculated by using equation 4.11)*	*Value*	*d'(Ai)- minV (Si ≥ Sj)*
S3 ≥ S1	$\frac{(0.227 - 0.306)}{(0.236 - 0.306) - (0.325 - 0.227)}$	0.470	d'(A3)-min V(S3 ≥ Sj)
S3 ≥ S2	b3 ≥ b2	1.00	=0.470
S3 ≥ S4	b3 ≥ b4	1.00	
S4 ≥ S1	$\frac{(0.227 - 0.306)}{(0.201 - 0.306) - (0.325 - 0.227)}$	0.389	d'(A4)-min V(S4 ≥ Sj)
S4 ≥ S2	$\frac{(0.170 - 0.306)}{(0.201 - 0.306) - (0.236 - 0.170)}$	0.795	= 0.389
S4 ≥ S3	$\frac{(0.170 - 0.306)}{(0.201 - 0.306) - (0.236 - 0.170)}$	0.795	

Determination of Weight Vectors

Weight vector {d'(Ai), d'(A2) ....d(An)}^{T}

W = (1.00, 0.470, 0.470, 0.389)

The above vector is normalized to give crisp weight vector

W = (0.429, 0.201, 0.201, 0.167)

Table 2.16: Pair-wise comparison matrix for the sub-criteria of "Service".

	Responsiveness	*Reliability*	*Customization capability*	*Shortage frequency*
Responsiveness	1,1,1	2/5,1/2, 2/3	2/5, 1/2, 2/3	1, 3/2, 2
Reliability	3/2, 2, 5/2	1,1,1	1/2, 1, 3/2	3/2, 2, 5/2
Customization Capability	3/2, 2, 5/2	2/3, 1, 2	1, 1, 1	3/2, 2, 5/2
Shortage Frequency	1/2, 2/3, 1	2/5, 1/2, 2/3	2/5, 1/2, 2/3	1,1,1
		CR= 0.013		

Synthetic extent normalized row sum. Row sum value of row in table using equation (2.8)

RS1= (2.80, 5.00, 8.33) RS2= (4.5, 6.00, 7.50)

RS3 = (4.66, 6.00, 8.00) RS4 = (2.30, 2.66, 3.33)

The synthetic extent normalized row sums obtained from above using equation (2.9)

S1= (0.103, 0.254, 0.583) S2 = (0.165, 0.305, 0.525)

S3= (0.171, 0.305, 0.560) S4 = (0.084, 0.135, 0.233)

Table 2.17: Degree of possibility for 'Service' sub-criteria.

V(Si ≥Sj)	*Degree of possibility (Calculated by using equation 4.11)*	*Value*	*d'(Ai)- minV (Si ≥Sj)*
S1 ≥ S2	$\frac{(0.165-0.583)}{(0.254-0.583)-(0.305-0.165)}$	0.891	d'(A1)min -V(S1 ≥ Sj)
S1 ≥ S3	$\frac{0.171-0.583)}{(0.254-0.583)-(0.355-0.171)}$	1.00	= 0.889
S1 ≥ S4	b1 ≥ b4	1.00	
S2 ≥ S1	b2 ≥ b1	1.00	d'(A2)min -V(S2 ≥ Sj)
S2 ≥ S3	b2 ≥ b3	1.00	= 1.00
S2 ≥ S4	b2 ≥ b4	1.00	
S3 ≥ S1	b3 ≥ b1	1.00	d'(A3)min -V(S3 ≥ Sj)
S3 ≥ S2	b3 ≥ b2	1.00	= 1.00
S3 ≥ S4	b3 ≥ b4	1.00	
S4 ≥ S1	$\frac{(0.103-0.233)}{(0.135-0.233)-(0.254-0.103)}$	0.522	d'(A4)min -V(S4 ≥ Sj)
S4 ≥ S2	$\frac{(0.165-0.233)}{(0.135-0.233)-(0.305-0.165)}$	0.285	= 0.267
S4 ≥ S3	$\frac{(0.171-0.233)}{(0.135-0.233)-(0.305-0.171)}$	0.795	

Determination of Weight Vectors

Weight vector {d'(Ai), d'(A2) ---------- d(An)}^{T}

$$W = (0.889,\ 1.00,\ 1.00,\ 0.267)$$

The above vector is normalized to give crisp weight vector

$$W = (0.281,\ 0.316,\ 0.316,\ 0.084)$$

Table 2.18: Pair-wise comparison for the alternatives regarding to sub-criteria of "Material Cost".

Material cost	*Supplier A*	*Supplier B*	*Supplier C*	*Supplier D*
Supplier A	1,1,1	1,3/2,2	1/3,2/5,1/2	1, 3/2, 2
Supplier B	1/2,2/3,1	1,1,1	2/3, 1, 2	1/2, 1, 3/2
Supplier C	2, 5/2, 3	1/2, 1, 3/2	1, 1, 1	1, 3/2, 2
Supplier D	1/2, 2/3, 1	2/3, 1, 2	1/2, 2/3, 1	1,1,1
		CR= 0.087		

Synthetic extent normalized row sum. Row sum value of row in table using equation (2.8)

RS1 = (3.33, 4.40, 5.5)   RS2 = (2.66, 3.66, 5.50)

RS3 = (4.50, 6.00, 7.50)   RS4 = (2.66, 3.33, 5.00)

The synthetic extent normalized row sums obtained from above using equation (2.9)

S1 = (0.141, 0.253, 0.418)   S2 = (0.113, 0.210, 0.418)

S3 = (0.191, 0.345, 0.570)   S4 = (0.113, 0.191, 0.380)

Table 2.19: Degree of possibility for the alternatives regarding to sub-criteria of "Material Cost".

V(Si ≥Sj)	*Degree of possibility (Calculated by using equation 4.11)*	*Value*	*d'(Ai)min - V (Si ≥Sj)*
S1 ≥ S2	b1 ≥ b2	1	d'(A1)min V(S1 ≥ Sj)
S1 ≥ S3	$\frac{(0.191 - 0.418)}{(0.253 - 0.418) - (0.345 - 0.191)}$	0.711	= 0.711
S1 ≥ S4	b1 ≥ b4	1	
S2 ≥ S1	$\frac{(0.141 - 0.418)}{(0.210 - 0.418) - (0.253 - 0.141)}$	0.866	d'(A2)min V(S2 ≥ Sj)
S2 ≥ S3	$\frac{(0.191 - 0.418)}{(0.210 - 0.418) - (0.345 - 0.191)}$	0.627	= 0.627
S2 ≥ S4	b2 ≥ b4	1	
S3 ≥ S1	b3 ≥ b1	1	d'(A3)min V(S3 ≥ Sj)
S3 ≥ S2	b3 ≥ b2	1	= 1.00
S3 ≥ S4	b3 ≥ b4	1	
S4 ≥ S1	$\frac{(0.141 - 0.380)}{(0.191 - 0.380) - (0.253 - 0.141)}$	0.794	d'(A3)min V(S3 ≥ Sj)
S4 ≥ S2	$\frac{(0.113 - 0.380)}{(0.191 - 0.380) - (0.210 - 0.113)}$	0.934	= 0.551
S4 ≥ S3	$\frac{(0.191 - 0.380)}{(0.191 - 0.380) - (0.345 - 0.191)}$	0.551	

Determination of Weight Vectors {d'(Ai), d'(A2) ----------- d(An)}T

W' = (0.711, 0.627, 1.00, 0.551)

The above vector is normalized to give crisp weight vector

W = (0.246, 0.217, 0.346, 0.190)

Table 2.20: Pair-wise comparison for the alternatives regarding to sub-criteria of "Transportation Cost".

	Supplier A	*Supplier B*	*Supplier C*	*Supplier D*
Supplier A	1, 1, 1	1/2, 2/3, 1	1/3, 2/5, 1/2	2/5, 1/2, 2/3
Supplier B	1, 3/2,	1, 1, 1	1/2, 2/3, 1	2/3, 1, 2
Supplier C	2, 5/2, 3	1, 3/2, 2	1, 1, 1	1/2, 1, 3/2
Supplier D	3/2, 2, 5/2	1/2, 1, 3/2	2/3, 1, 2	1, 1, 1
		CR = 0.016		

Synthetic extent normalized row sum. Row sum value of row in table using equation (2.8)

RS1 = (2.23, 2.56, 3.16) RS2= (3.16, 4.16, 6.00)

RS3 = (4.50, 6.00, 7.50) RS4 = (3.66, 5.00, 7.00)

The synthetic extent normalized row sums obtained from above using equation (2.9)

S1 = (0.094, 0.144, 0.233) S2 = (0.133, 0.234, 0.442)

S3 = (0.190, 0.338, 0.553) S4 = (0.154, 0.282, 0.516)

Table 2.21: Degree of possibility for the alternatives regarding to sub-criteria of "Transportation Cost".

V(Si ≥ Sj)	*Degree of possibility (Calculated by using equation 4.11)*	*Value*	*d'(Ai) - min V (Si ≥ Sj)*
S1 ≥ S2	$\frac{(0.133 - 0.233)}{(0.144 - 0.233) - (0.234 - 0.133)}$	0.526	d'(A1)-min V(S1 ≥ Sj)
S1 ≥ S3	$\frac{(0.190 - 0.233)}{(0.144 - 0.233) - (0.338 - 0.190)}$	0.365	= 0.364
S1 ≥ S4	$\frac{(0.154 - 0.233)}{(0.144 - 0.233) - (0.282 - 0.154)}$	0.364	
S2 ≥ S1	b2 ≥ b1	1.00	d'(A2)-min V(S2 ≥ Sj)
S2 ≥ S3	$\frac{(0.190 - 0.442)}{(0.234 - 0.442) - (0.338 - 0.190)}$	0.708	= 0.708
S2 ≥ S4	$\frac{(0.154 - 0.442)}{(0.234 - 0.442) - (0.282 - 0.154)}$	0.857	
S3 ≥ S1	b3 ≥ b1	1.00	d'(A3)-min V(S3 ≥ Sj)
S3 ≥ S2	b3 ≥ b2	1.00	= 1.00
S3 ≥ S4	b3 ≥ b4	1.00	
S4 ≥ S1	b4 ≥ b1	1.00	d'(A3)-min V(S3 ≥ Sj)
S4 ≥ S2	b4 ≥ b2	1.00	= 0.853
S4 ≥ S3	$\frac{(0.190 - 0.516)}{(0.282 - 0.516) - (0.338 - 0.190)}$	0.853	

Determination of Weight Vectors {d'(Ai), d'(A2) ----------- d(An)}^{T}

$$W' = (0.364, 0.708, 1.00, 0.853)$$

The above vector is normalized to give crisp weight vector

$$W = (0.124, 0.242, 0.341, 0.291)$$

Table 2.22: Pair-wise comparison for the alternatives regarding to sub-criteria of "Quantity Discount".

	Supplier A	*Supplier B*	*Supplier C*	*Supplier D*
Supplier A	1, 1, 1	2/3, 1, 2	2/5, 1/2, 2/3	1/2, 2/3, 1
Supplier B	1/2, 1, 3/2	1,1,1	2/3, 1, 2	1, 3/2, 2
Supplier C	3/2, 2, 5/2	1/2, 1, 3/2	1, 1, 1	3/2, 2, 5/2
Supplier D	1, 3/2, 2	1/2, 2/3, 1	2/5, 1/2, 2/3	1, 1, 1
		CR = 0.054		

Synthetic extent normalized row sum. Row sum value of row in table using equation (2.8)

RS1 = (2.56, 3.16, 4.66) RS2 = (3.16, 4.50, 6.50)

RS3 = (4.50, 6.00, 7.50) RS4 = (2.90, 3.66, 4.66)

The synthetic extent normalized row sums obtained from above using equation (2.9)

S1 = (0.109, 0.182, 0.355) S2 = (0.135, 0.259, 0.495)

S3 = (0.192, 0.346, 0.571) S4 = (0.124, 0.211, 0.355)

Table 2.23: Degree of possibility for the alternatives regarding to sub-criteria of "Quantity Discount".

V(Si ≥ Sj)	*Degree of possibility (Calculated by using equation 4.11)*	*Value*	*d'(Ai) min - V (Si ≥ Sj)*
S1 ≥ S2	$\frac{(0.135 - 0.355)}{(0.182 - 0.355) - (0.259 - 0.135)}$	0.740	d'(A1)min-V(S1 ≥ Sj)
S1 ≥ S3	$\frac{(0.192 - 0.355)}{(0.182 - 0.355) - (0.346 - 0.192)}$	0.498	= 0.498
S1 ≥ S4	$\frac{(0.124 - 0.355)}{(0.182 - 0.355) - (0.211 - 0.124)}$	0.888	
S2 ≥ S1	b2 ≥ b1	1.00	d'(A2)min-V(S2 ≥ Sj)

Table 2.23 *(Contd...)*

Table 2.23: (*Contd...*)

V(Si ≥Sj)	*Degree of possibility (Calculated by using equation 4.11)*	*Value*	*d'(Ai) min - V (Si ≥Sj)*
S2 ≥ S3	$\frac{(0.192 - 0.495)}{(0.259 - 0.495) - (0.346 - 0.192)}$	0.776	= 0.776
S2 ≥ S4	b2 ≥ b4	1.00	
S3 ≥ S1	b3 ≥ b1	1.00	d'(A2)min-V(S2 ≥ Sj)
S3 ≥ S2	b3 ≥ b2	1.00	=1.00
S3 ≥ S4	b3 ≥ b4	1.00	
S4 ≥ S1	b4 ≥ b1	1.00	d'(A2)min-V(S2 ≥ Sj)
S4 ≥ S2	$\frac{(0.135 - 0.355)}{(0.211 - 0.355) - (0.259 - 0.135)}$	0.820	= 0.546
S4 ≥ S3	$\frac{(0.192 - 0.355)}{(0.211 - 0.355) - (0.346 - 0.192)}$	0.546	

Determination of Weight Vectors {d'(Ai), d'(A2) ----------- d(An)}^{T}

$$W' = (0.498, 0.776, 1.00, 0.546)$$

The above vector is normalized to give crisp weight vector

$$W = (0.176, 0.275, 0.324, 0.225)$$

Table 2.24: Pair-wise comparison for the alternatives regarding to sub-criteria of "Overhead Cost".

	Supplier A	*Supplier B*	*Supplier C*	*Supplier D*
Supplier A	1,1,1	1, 3/2,2	1,1,1	1/2, 1, 3/2
Supplier B	1/2, 2/3, 1	1,1,1	1/2, 1, 3/2	1, 3/2, 2
Supplier C	1,1,1	2/3, 1, 2	1, 1, 1	1/2, 1, 3/2
Supplier D	2/3, 1, 2	1/2, 2/3, 1	2/3, 1, 2	1,1,1
		CR = 0.023		

Synthetic extent normalized row sum. Row sum value of row in table using (2.8)

RS1 = (3.50, 4.50, 5.50) RS2 = (3.00, 4.16, 5.50)

RS3 = (3.16, 4.00, 5.50) RS4 = (2.83, 3.66, 6.00)

The synthetic extent normalized row sums obtained from above using equation (2.9)

S1 = (0.155, 0.275, 0.440) S2 = (0.133, 0.255, 0.440)

S3 = (0.140, 0.245, 0.440) S3 = (0.125, 0.224, 0.480)

Table 2.25: Degree of possibility for the alternatives regarding to sub-criteria of "Overhead Cost".

V(Si ≥Sj)	*Degree of possibility (Calculated by using equation 4.11)*	*Value*	*d'(Ai) min V (Si ≥Sj)*
S1 ≥ S2	b1 ≥ b2	1.00	d'(A1)min V(S1 ≥ Sj)
S1 ≥ S3	b1 ≥ b3	1.00	= 1.00
S1 ≥ S4	b1 ≥ b4	1.00	
S2 ≥ S1	$\frac{(0.155 - 0.440)}{(0.255 - 0.440) - (0.275 - 0.155)}$	0.934	d'(A1)min V(S1 ≥ Sj)
S2 ≥ S3	b2 ≥ b3	1.00	= 0.934
S2 ≥ S4	b2 ≥ b4	1.00	
S3 ≥ S1	$\frac{(0.155 - 0.440)}{(0.245 - 0.440) - (0.275 - 0.155)}$	0.904	d'(A1)min- V(S1 ≥ Sj)
S3 ≥ S2	$\frac{(0.133 - 0.440)}{(0.245 - 0.440) - (0.255 - 0.133)}$	0.968	= 0.904
S3 ≥ S4	$\frac{(0.125 - 0.440)}{(0.245 - 0.440) - (0.224 - 0.125)}$	0.915	
S4 ≥ S1	$\frac{(0.155 - 0.480)}{(0.224 - 0.480) - (0.275 - 0.155)}$	0.864	d'(A1)min- V(S1 ≥ Sj)
S4 ≥ S2	$\frac{(0.133 - 0.480)}{(0.224 - 0.480) - (0.255 - 0.133)}$	0.917	= 0.864
S4 ≥ S3	$\frac{(0.133 - 0.480)}{(0.224 - 0.480) - (0.255 - 0.133)}$	0.941	

Determination of Weight Vectors {d'(Ai), d'(A2) ----------- d(An)}^{T}

$$W = (1.00,\ 0.934,\ 0.904,\ 0.864)$$

The above vector is normalized to give crisp weight vector

$$W = (0.270,\ 0.252,\ 0.244,\ 0.233)$$

Table 2.26: Pair-wise comparison for the alternatives regarding to sub-criteria of "compliance with Due Date".

	Supplier A	*Supplier B*	*Supplier C*	*Supplier D*
Supplier A	1, 1, 1	3/2, 2, 5/2	1, 3/2, 2 `	2, 5/2, 3
Supplier B	2/5, 1/2, 2/3	1, 1, 1	2/3, 1, 2	1, 3/2, 2
Supplier C	1/2, 2/3, 1	1/2, 1, 3/2	1, 1, 1	1/2, 1, 3/2
Supplier C	1/3, 2/5, 1/2	1/2, 2/3, 1	2/3, 1, 2	1, 1, 1
		CR = 0.014		

Synthetic extent normalized row sum. Row sum value of row in table using (2.8)

RS1 = (5.50, 7.00, 8.50)   RS2 = (3.06, 4.00, 5.66)

RS3 = (2.50, 3.66, 5.00)   RS4 = (2.50, 3.06, 4.50)

The synthetic extent normalized row sums obtained from above using equation (2.9)

S1 = (0.232, 0.395, 0.626)   S2 = (0.129, 0.225, 0.417)

S3 = (0.105, 0.206, 0.368)   S4 = (0.105, 0.172, 0.331)

Table 2.27: Degree of possibility for the alternatives regarding to sub-criteria of "Compliance with Due Date'.

V(Si ≥ Sj)	*Degree of possibility (Calculated by using equation 4.11)*	*Value*	*d'(Ai) - min V (Si ≥ Sj)*
S1 ≥ S2	b1 ≥ b2	1.00	d'(A1)-minV(S1 ≥ Sj)
S1 ≥ S3	b1 ≥ b3	1.00	= 1.00
S1 ≥ S4	b1 ≥ b4	1.00	
S2 ≥ S1	$\frac{(0.232 - 0.417)}{(0.225 - 0.417) - (0.395 - 0.232)}$	0.521	d'(A2)-minV(S2 ≥ Sj)
S2 ≥ S3	b2 ≥ b3	1.00	= 0.521
S2 ≥ S4	b2 ≥ b4	1.00	
S3 ≥ S1	$\frac{(0.232 - 0.368}{(0.206 - 0.368) - (0.395 - 0.232)}$	0.418	d'(A3)-minV(S3 ≥ Sj)
S3 ≥ S2	$\frac{(0.129 - 0.368)}{(0.206 - 0.368) - (0.225 - 0.129)}$	0.926	= 0.418
S3 ≥ S4	b3 ≥ b4	1.00	
S4 ≥ S1	$\frac{(0.232 - 0.331)}{(0.172 - 0.331) - (0.395 - 0.232)}$	0.307	d'(A4)-minV(S4 ≥ Sj)
S4 ≥ S2	$\frac{(0.129 - 0.331)}{(0.172 - 0.331) - (0.225 - 0.129)}$	0.792	0.307
S4 ≥ S3	$\frac{(0.105 - 0.331)}{(0.172 - 0.331) - (0.206 - 0.105)}$	0.869	

Determination of Weight Vectors {d'(Ai), d'(A2) ----------- d(An)}^{T}

W' = (1.00, 0.521, 0.418, 0.307)

The above vector is normalized to give crisp weight vector

W = (0.445, 0231, 0.186, 0.138)

Table 2.28: Pair-wise comparison for the alternatives regarding to sub-criteria of "Delivery Lead Time".

	Supplier A	*Supplier B*	*Supplier C*	*Supplier D*
Supplier A	1, 1, 1	3/2, 2, 5/2	1, 3/2, 2 `	1/2, 1, 3/2
Supplier B	2/5, 1/2, 2/3	1, 1, 1	1/2, 2/3, 1	2/3, 1, 2
Supplier C	1/2, 2/3, 1	1, 3/2, 2	1, 1, 1	1, 3/2, 2
Supplier D	2/3, 1, 2	1/2, 1, 3/2	1/2, 2/3, 1	1, 1, 1
		CR = 0.035		

Synthetic Extent Normalized Row Sum. Row sum value of row in table using equation (2.8)

RS1 = (4.00, 5.50, 6.00) RS2 = (2.56, 3.166, 4.66)

RS3 = (3.50, 4.66, 6.00) RS4 = (2.66, 3.66, 5.50)

The synthetic extent normalized row sums obtained from above using equation (2.9)

S1= (0.180, 0.232, 0.471) S2 = (0.115, 0.186, 0.366)

S3 = (0.157, 0.274, 0.471) S4 = (0.120, 0.215, 0.432)

Table 2.29: Degree of possibility for the alternatives regarding to sub-criteria of "Delivery Lead Time".

V(Si ≥ Sj)	*Degree of possibility (Calculated by using equation 4.11)*	*Value*	*d'(Ai)min - V (Si ≥ Sj)*
S1 ≥ S2	b1 ≥ b2	1.00	d'(A1)min-V(S1 ≥ Sj)
S1 ≥ S3	b1 ≥ b3	1.00	= 1.00
S1 ≥ S4	b1 ≥ b4	1.00	
S2 ≥ S1	$\frac{(0.180 - 0.366)}{(0.186 - 0.366) - (0.323 - 0.180)}$	0.575	d'(A2)min-V(S2 ≥ Sj)
S2 ≥ S3	$\frac{(0.157 - 0.366)}{(0.186 - 0.366) - (0.274 - 0.157)}$	0.703	= 0.575
S2 ≥ S4	$\frac{(0.120 - 0.366)}{(0.186 - 0.366) - (0.215 - 0.120)}$	0.894	
S3 ≥ S1	$\frac{(0.180 - 0.471)}{(0.274 - 0.471) - (0.323 - 0.180)}$	0.855	d'(A2)min-V(S2 ≥ Sj)
S3 ≥ S2	b3 ≥ b2	1.00	= 0.855
S3 ≥ S4	b3 ≥ b4	1.00	
S4 ≥ S1	$\frac{(0.180 - 0.432)}{(0.215 - 0.432) - (0.323 - 0.180)}$	0.710	d'(A2)min-V(S4 ≥ Sj)
S4 ≥ S2	b4 ≥ b2	1.00	= 0.710
S4 ≥ S3	$\frac{(0.157 - 0.432)}{(0.215 - 0.432) - (0.274 - 0.157)}$	0.823	

Determination of Weight Vectors {d'(Ai), d'(A2) ----------- d(An)}T

$$W' = (1.00, 0.575, 0.855, 0.710)$$

The above vector is normalized to give crisp weight vector

$$W = (0.318, 0.183, 0.272, 0.227)$$

Table 2.30: Pair-wise comparison for the alternatives regarding to sub-criteria of "Deliverable Location".

	Supplier A	*Supplier B*	*Supplier C*	*Supplier D*
Supplier A	1,1,1	1/2, 1, 3/2	3/2, 2, 5/2	1, 3/2, 2
Supplier B	2/3, 1, 2	1,1,1	2/3, 1, 2	2/5,1/2,2/3
Supplier C	2/5,1/2,2/3	1/2, 1, 3/2	1, 1, 1	1/2, 2/3, 1
Supplier D	1/2, 2/3, 1	3/2, 2, 5/2	1, 3/2, 2	1,1,1
		CR = 0.041		

Synthetic extent normalized row sum. Row sum value of row in table using equation (2.8)

RS1 = (4.00, 5.50, 7.00) RS2 = (2.73, 3.50, 5.66)

RS3 = (2.40, 3.16, 4.166) RS4 = (4.00, 4.166, 6.50)

The synthetic extent normalized row sums obtained from above using equation (2.9)

S1 = (0.171, 0.336, 0.533) S2 = (0.117, 0.214, 0.431)

S3 = (0.102, 0.193, 0.317) S4 = (0.171, 0.255, 0.495)

Table 2.31: Degree of possibility for the alternatives regarding to sub-criteria of "Deliverable Location".

V(Si ≥ Sj)	*Degree of possibility (Calculated by using equation 4.11)*	*Value*	*d'(Ai)min - V (Si ≥ Sj)*
S1 ≥ S2	b1 ≥ b2	1.00	d'(A1)min-V(S1 ≥ Sj)
S1 ≥ S3	b1 ≥ b3	1.00	= 1.00
S1 ≥ S4	b1 ≥ b4	1.00	
S2 ≥ S1	$\frac{(0.171 - 0.431)}{(0.214 - 0.431) - (0.336 - 0.171)}$	0.680	d'(A2)min-V(S2 ≥ Sj)
S2 ≥ S3	b2 ≥ b3	1.00	= 0.680
S2 ≥ S4	$\frac{(0.171 - 0.431)}{(0.214 - 0.431) - (0.255 - 0.171)}$	0.863	

Table 2.31 *(Contd...)*

Table 2.31: *(Contd...)*

V(Si ≥ Sj)	*Degree of possibility (Calculated by using equation 4.11)*	*Value*	*d'(Ai)min - V (Si ≥ Sj)*
S3 ≥ S1	$\frac{(0.171 - 0.317)}{(0.193 - 0.317) - (0.336 - 0.171)}$	0.505	d'(A3)min-V(S3 ≥ Sj)
S3 ≥ S2	$\frac{(0.117 - 0.317)}{(0.193 - 0.317) - (0.214 - 0.117)}$	0.904	= 0.855
S3 ≥ S4	$\frac{(0.171 - 0.317}{(0.193 - 0.317) - (0.255 - 0.117)}$	0.701	
S4 ≥ S1	$\frac{(0.171 - 0.495)}{(0.255 - 0.495) - (0.336 - 0.171)}$	0.800	d'(A3)min-V(S3 ≥ Sj)
S4 ≥ S2	b4 ≥ b2	1.00	= 0.800
S4 ≥ S3	b4 ≥ b3	1.00	

Determination of Weight Vectors {d'(Ai), d'(A2) ----------- d(An)}^{T}

$$W' = (1.00,\ 0.680,\ 0.505,\ 0.800)$$

The above vector is normalized to give crisp weight vector

$$W = (0.335,\ 0.227,\ 0.170,\ 0.268)$$

Table 2.32: Pair-wise comparison for the alternatives regarding to sub-criteria of "Flexibility in Delivery Operation".

	Supplier A	*Supplier B*	*Supplier C*	*Supplier D*
Supplier A	1, 1, 1	2, 5/2, 3	1/2, 1, 3/2	3/2, 2, 5/2
Supplier B	1/3, 2/5, 3	1, 1, 1	2/3, 1, 2	1/2, 2/3, 1
Supplier C	2/3, 1, 2	1/2, 1, 3/2	1, 1, 1	1, 3/2, 2
Supplier D	2/5, 1/2, 2/3	1, 3/2, 2	1/2, 2/3, 1	1, 1, 1
		CR = 0.050		

Synthetic extent normalized row sum. Row sum value of row in table using equation (2.8)

RS1 = (5.00, 6.50, 8.00) RS2 = (2.50, 3.06, 4.50)

RS3 = (3.166, 4.50, 6.50) RS4 = (2.90, 3.66, 4.66)

The synthetic extent normalized row sums obtained from above using equation (2.9)

S1 = (0.211, 0.366, 0.592) S2 = (0.105, 0.172, 0.331)

S3 = (0.133, 0.253, 0.479) S4 = (0.122, 0.206, 0.343)

Table 2.33: Degree of possibility for the alternatives regarding to sub-criteria of "Flexibility in Delivery Operation".

V(Si ≥ Sj)	*Degree of possibility (Calculated by using equation 4.11)*	*Value*	*d'(Ai)min V (Si ≥ Sj)*
S1 ≥ S2	b1 ≥ b2	1.00	d'(A1)min V(S1 ≥ Sj)
S1 ≥ S3	b1 ≥ b3	1.00	=1.00
S1 ≥ S4	b1 ≥ b4	1.00	
S2 ≥ S1	$\frac{(0.211 - 0.331)}{(0.172 - 0.331) - (0.366 - 0.211)}$	0.382	d'(A2)min V(S2 ≥ Sj)
S2 ≥ S3	b2 ≥ b3	1.00	= 0.382
S2 ≥ S4	b2 ≥ b4	1.00	
S3 ≥ S1	$\frac{(0.211 - 0.479)}{(0.253 - 0.479) - (0.366 - 0.211)}$	0.703	d'(A3)min V(S3 ≥ Sj)
S3 ≥ S2	$\frac{(0.105 - 0.479)}{(0.253 - 0.479) - (0.172 - 0.105)}$	0.864	= 0.703
S3 ≥ S4	b3 ≥ b4	1.00	
S4 ≥ S1	$\frac{(0.211 - 0.343)}{(0.206 - 0.343) - (0.366 - 0.211)}$	0.450	d'(A3)min V(S3 ≥ Sj)
S4 ≥ S2	$\frac{(0.105 - 0.343)}{(0.206 - 0.343) - (0.172 - 0.105)}$	0.550	= 0.450
S4 ≥ S3	$\frac{(0.133 - 0.343)}{(0.205 - 0.343) - (0.253 - 0.133)}$	0.817	

Determination of Weight Vectors {d'(Ai), d'(A2) ----------- d(An)}^{T}

$$W' = (1.00, 0.382, 0.703, 0.450)$$

The above vector is normalized to give crisp weight vector

$$W = (0.394, 0.152, 0.277, 0.177)$$

Table 2.34: Pair-wise comparison for the alternatives regarding to sub-criteria of "Rejection Rate".

	Supplier A	*Supplier B*	*Supplier C*	*Supplier D*
Supplier A	1, 1, 1	1/2, 1, 3/2	1, 3/2, 2	1, 3/2, 2
Supplier B	2/3, 1, 2	1, 1, 1	1, 3/2, 2	1/2, 1, 3/2
Supplier C	1/2, 2/3, 1	1/2, 2/3, 1	1, 1, 1	1/2, 2/3, 1
Supplier D	1/2, 2/3, 1	2/3, 1, 2	1, 3/2, 2	1, 1, 1
		CR = 0.007		

Synthetic extent normalized row sum. Row sum value of row in table using (2.8)

RS1 = (3.50, 5.00, 6.50) RS2 = (3.166, 4.50, 6.50)

RS3 = (2.50, 3.00, 4.00) RS4 = (3.166, 4.166, 6.00)

The synthetic extent normalized row sums obtained from above using equation (2.9)

S1 = (0.152, 0.300, 0.527) S2 = (0.137, 0.270, 0.527)

S3 = (0.108, 0.180, 0.324) S4 = (0.137, 0.250, 0.486)

Table 2.35: Degree of possibility for the alternatives regarding to sub-criteria of "Rejection Rate".

V(Si ≥ Sj)	*Degree of possibility (Calculated by using equation 4.11)*	*Value*	*d'(Ai)min - V (Si ≥ Sj)*
S1 ≥ S2	b1 ≥ b2	1.00	d'(A1)min-V(S1 ≥ Sj)
S1 ≥ S3	b1 ≥ b3	1.00	1.00
S1 ≥ S4	b1 ≥ b4	1.00	
S2 ≥ S1	$\frac{(0.152 - 0.324)}{(0.180 - 0.324) - (0.300 - 0.152)}$	0.925	d'(A2)min-V(S2 ≥ Sj)
S2 ≥ S3	b2 ≥ b3	1.00	= 0.925
S2 ≥ S4	b2 ≥ b4	1.00	
S3 ≥ S1	$\frac{(0.152 - 0.324)}{(0.180- 0.324) - (0.300 - 0.152)}$	0.589	d'(A3)min-V(S3 ≥ Sj)
S3 ≥ S2	$\frac{(0.137 - 0.324)}{(0.180- 0.324) - (0.270 - 0.137)}$	0.675	= 0.589
S3 ≥ S4	$\frac{(0.137 - 0.324)}{(0.180- 0.324) - (0.250 - 0.137)}$	0.727	
S4 ≥ S1	$\frac{(0.152 - 0.486}{(0.250 - 0.486) - (0.300 - 0.152)}$	0.869	d'(A4)min-V(S4 ≥ Sj)
S4 ≥ S2	$\frac{(0.137 - 0.486)}{(0.250 - 0.486) - (0.270 - 0.137)}$	0.945	= 0.869
S4 ≥ S3	b4 ≥ b3	1.00	

Determination of Weight Vectors {d'(Ai), d'(A2) ----------- d(An)}^{T}

W = (1.00, 0.925, 0.589, 0.869)

The above vector is normalized to give crisp weight vector

W = (0.295, 0.273, 0.175, 0.256)

Table 2.36: Pair-wise comparison for the alternatives regarding to sub-criteria of "Inspection Method".

	Supplier A	*Supplier B*	*Supplier C*	*Supplier D*
Supplier A	1, 1, 1	1, 1, 1	1/2, 1, 3/2	1, 3/2, 2
Supplier B	1, 1, 1	1, 1, 1	1/2, 1, 3/2	1, 3/2, 2
Supplier C	2/3, 1, 2	2/3, 1, 2	1, 1, 1	1/2,, 1, 3/2
Supplier D	1/2, 2/3, 1	1/2, 2/3, 1	2/3, 1, 2	1, 1, 1
		CR = 0.007		

Synthetic extent normalized row sum. Row sum value of row in table using equation (2.8)

RS1 = (3.50, 4.50, 5.50) RS1 = (3.50, 4.50, 5.50)

RS3 = (2.83, 4.00, 5.50, 6.50) RS4 = (2.66, 4.16, 5.00)

The synthetic extent normalized row sums obtained from above using equation (2.9)

S1 = (0.155, 0.262, 0.440) S2 = (0.155, 0.262, 0.440)

S3 = (0.125, 0.233, 0.520) S4 = (0.118, 0.242, 0.400)

Table 2.37: Degree of possibility for the alternatives regarding to sub-criteria of "Inspection Method".

V(Si ≥Sj)	*Degree of possibility (Calculated by using equation 4.11)*	*Value*	*d'(Ai)min V (Si ≥Sj)*
S1 ≥ S2	b1 ≥ b2	1.00	d'(Ai)min V(Si ≥ Sj)
S1 ≥ S3	b1 ≥ b3	1.00	= 1.00
S1 ≥ S4	b1 ≥ b4	1.00	
S2 ≥ S1	b2 ≥ b1	1.00	d'(Ai)min V(Si ≥ Sj)
S2 ≥ S3	b2 ≥ b3	1.00	= 1.00
S2 ≥ S4	b2 ≥ b4	1.00	
S3 ≥ S1	$\frac{(0.155 - 0.520)}{(0.233 - 0.520) - (0.262 - 0.155)}$	0.926	d'(Ai)min V(Si ≥ Sj)
S3 > S2	$\frac{(0.155 - 0.520)}{(0.233 - 0.520) - (0.262 - 0.155)}$	0.926	= 0.926
S3 > S4	$\frac{(0.118 - 0.520)}{(0.233 - 0.520) - (0.242 - 0.118)}$	0.978	
S4 > S1	$\frac{(0.155 - 0.400)}{(0.242 - 0.400) - (0.262 - 0.155)}$	0.924	d'(Ai)min V(Si ≥ Sj)
S4 > S2	$\frac{(0.155 - 0.400)}{(0.242 - 0.400) - (0.262 - 0.155)}$	0.924	= 0.924
S4 > S3	b4 ≥ b3	1.00	

Determination of Weight Vectors {d'(Ai), d'(A2) ----------- d(An)}^{T}

W'= (1.00, 1.00, 0.926, 0.924)

The above vector is normalized to give crisp weight vector

W = (0.259, 0.259, 0.241, 0.241)

Table 2.38: Pair-wise comparison for the alternatives regarding to sub-criteria of "Quality Management".

	Supplier A	*Supplier B*	*Supplier C*	*Supplier D*
Supplier A	1, 1,1	1/2, 1,3/2	2/3, 1, 2	1, 3/2, 2
Supplier B	2/3, 1, 2	1, 1, 1	2/5, 1/2, 2/3	1/2, 2/3, 1
Supplier C	1/2, 1, 3/2	3/2, 2, 5/2	1, 1, 1	2/3, 1, 2
Supplier D	1/2, 2/3, 1	1, 3/2, 2	1/2, 1, 3/2	1, 1, 1
		CR = 0.0029		

Synthetic extent normalized row sum. Row sum value of row in table using equation (.8)

RS1 = (3.16, 4.50, 6.50) RS2 = (2.56, 3.16, 4.66)

RS3 = (3.66, 5.00, 7.00) RS4 = (3.00, 4.16, 5.50)

The synthetic extent normalized row sums obtained from above using equation (2.9)

S1 = (0.133, 0.267, 0.524) S2 = (0.108, 0.188,0 .376)

S3 = (0.154, 0.297, 0.564) S4 = (0.126, 0.247, 0.443)

Table 2.39: Degree of possibility for the alternatives regarding to sub-criteria of "Quality Management".

V(Si ≥ Sj)	*Degree of possibility (Calculated by using equation 4.11)*	*Value*	*d'(Ai)min - V (Si ≥ Sj)*
S1 ≥ S2	b1 ≥ b2	1.00	d'(A1)min-V(S1 ≥ Sj)
S1 ≥ S3	$\frac{(0.154 - 0.524)}{(0.267 - 0.524) - (0.297 - 0.154)}$	0.925	= 0.925
S1 ≥ S4	b1 ≥ b4	1.00	
S2 ≥ S1	$\frac{(0.133 - 0.376)}{(0.188 - 0.376) - (0.267 - 0.133)}$	0.754	d'(A2)min-V(S2 ≥ Sj)
S2 ≥ S3	$\frac{(0.154 - 0.376)}{(0.188 - 0.376) - (0.297 - 0.154)}$	0.670	= 0.670

Table 2.39 *(Contd...)*

Table 2.39: *(Contd...)*

V(Si ≥Sj)	*Degree of possibility (Calculated by using equation 4.11)*	*Value*	*d'(Ai)min - V (Si ≥Sj)*
S2 ≥ S4	$\frac{(0.126 - 0.376)}{(0.188 - 0.376) - (0.247 - 0.126)}$	0.809	
S3 ≥ S1	b3 ≥ b1	1.00	d'(A3)min-V(S3 ≥ Sj)
S3 ≥ S2	b3 ≥ b2	1.00	= 1.00
S3 ≥ S4	b2 ≥ b1	1.00	
S4 ≥ S1	$\frac{(0.108 - 0.443)}{(0.247 - 0.443) - (0.267 - 0.108)}$	0.943	d'(A4)min-V(S4 ≥ Sj)
S4 ≥ S2	b4 ≥ b2	1.00	= 0.852
S4 ≥ S3	$\frac{(0.154 - 0.443)}{(0.247 - 0.443) - (0.267 - 0.108)}$	0.852	

Determination of Weight Vectors {d'(Ai), d'(A2) ----------- d(An)}^{T}

$$W' = (0.925, 0.670, 1, 0.852)$$

The above vector is normalized to give crisp weight vector

$$W = (0.268, 0.194, 0.290, 0.247)$$

Table 2.40: Degree of possibility for the alternatives regarding to sub-criteria of "Quality Management".

	Supplier A	*Supplier B*	*Supplier C*	*Supplier D*
Supplier A	1, 1, 1	3/2, 2, 5/2	2, 5/2, 3	1, 3/2, 2
Supplier B	2/5, 1/2, 2/3	1, 1, 1	1/2, 1, 3/2	3/2, 2, 5/2
Supplier C	1/3, 2/5,1/2	2/3, 1, 2	1, 1, 1	2/3, 1, 2
Supplier D	1/2, 2/3, 1	2/5, 1/2, 2/3	1/2, 1, 3/2	1, 1, 1
		CR = 0.042		

Synthetic extent normalized row sum. Row sum value of row in table using equation (2.8)

RS1 = (5.50, 7.00, 8.50) RS2 = (3.40, 4.50, 5.66)

RS3 = (2.66, 3.40, 5.50) RS4 = (2.40, 3.16, 4.16)

The synthetic extent normalized row sums obtained from above using equation (2.9)

S1 = (0.230, 0.387, 0.608) S2= (0.142, 0.249, 0.405)

S3 = (0.111, 0.118, 0.393) S4= (0.100, 0.175, 0.298)

Table 2.41: Degree of possibility for the alternatives regarding to sub-criteria of "Product Reliability".

V(Si ≥ Sj)	*Degree of possibility (Calculated by using equation 4.11)*	*Value*	*d'(Ai)min - V (Si ≥ Sj)*
S1 ≥ S2	b1 ≥ b2	1.00	d'(A1)-minV(S1 ≥ Sj)
S1 ≥ S3	b1 ≥ b3	1.00	= 1.00
S1 ≥ S4	b1 ≥ b4	1.00	
S2 ≥ S1	$\frac{(0.230 - 0.405)}{(0.249 - 0.405) - (0.249 - 0.230)}$	0.559	d'(A2)min-V(S2 ≥ Sj)
S2 ≥ S3	b2 ≥ b3	1.00	= 0.559
S2 ≥ S4	b2 ≥ b4	1.00	
S3 ≥ S1	$\frac{(0.230 - 0.393)}{(0.188 - 0.393) - (0.387 - 0.230)}$	0.450	d'(A3)min-V(S3 ≥ Sj)
S3 ≥ S2	$\frac{(0.142 - 0.393)}{(0.188 - 0.393) - (0.249 - 0.142)}$	0.804	= 0.450
S3 ≥ S4	$\frac{(0.100 - 0.393)}{(0.188 - 0.393) - (0.175 - 0.100)}$	0.561	
S4 ≥ S1	$\frac{(0.230 - 0.298)}{(0.175 - 0.298) - (0.387 - 0.230)}$	0.242	d'(A4)min-V(S4 ≥ Sj)
S4 ≥ S2	$\frac{(0.142 - 0.298)}{(0.175 - 0.298) - (0.249 - 0.142)}$	0.678	= 0.242
S4 ≥ S3	b4 ≥ b3	1.00	

Determination of Weight Vectors {d'(Ai), d'(A2) ----------- d(An)}

$$W' = (1.00,\ 0.559,\ 0.450,\ 0.242)$$

The above vector is normalized to give crisp weight vector

$$W = (0.444,\ 0.249,\ 0.199,\ 0.108)$$

Table 2.42: Pair-wise comparison for the alternatives regarding to sub-criteria of "Financial Strength".

	Supplier A	*Supplier B*	*Supplier C*	*Supplier D*
Supplier A	1, 1, 1	1/2, 1, 3/2	2, 5/2, 3	1/2, 1, 3/2
Supplier B	2/3, 1, 2	1, 1, 1	3/2, 2, 5/2	1, 3/2, 2
Supplier C	1/3, 2/5, 1/2,	2/5, 1/2, 2/3	1, 1, 1	2/3, 1, 2
Supplier D	2/3, 1, 2	1/2, 2/3, 1	1/2, 1, 3/2	1, 1, 1
		CR = 0.036		

Synthetic extent normalized row sum. Row sum value of Row in table using equation (2.8)

RS1 = (4.00, 5.50, 7.00) RS2 = (2.66, 3.66, 5.50)

RS3 = (4.16, 5.50, 7.50) RS4 = (2.40, 2.90, 4.16)

The synthetic extent normalized row sums obtained from above using equation (2.9)

S1 = (0.165, 0.313, 0.529) S2 = (0.172, 0.313, 0.567)

S3 = (0.099, 0.165, 0.314) S4 = (0.110, 0.208, 0.416)

Table 2.43: Degree of possibility for the alternatives regarding to sub-criteria of "Financial Strength".

V(Si ≥ Sj)	*Degree of possibility (Calculated by using equation 4.11)*	*Value*	*d'(Ai)min - V (Si ≥ Sj)*
S1 ≥ S2	b1 ≥ b2	1.00	d'(A1)min-V(S1 ≥ Sj)
S1 ≥ S3	b1 ≥ b3	1.00	= 1.00
S1 ≥ S4	b1 ≥ b4	1.00	
S2 ≥ S1	b2 ≥ b1	1.00	d'(A1)min-V(S1 ≥ Sj
S2 ≥ S3	b2 ≥ b3	1.00	= 1.00
S2 ≥ S4	b2 ≥ b4	1.00	
S3 ≥ S1	$\frac{(0.165 - 0.314)}{(0.165 - 0.314) - (0.313 - 0.165)}$	0.501	d'(A1)min-V(S1 ≥ Sj)
S3 ≥ S2	$\frac{(0.172 - 0.314)}{(0.165 - 0.314) - (0.313 - 0.172)}$	0.489	= 0.450
S3 ≥ S4	b3 ≥ b4	1.00	
S4 ≥ S1	$\frac{(0.165 - 0.416)}{(0.208 - 0.416) - (0.313 - 0.165)}$	0.705	d'(A1)min-V(S1 ≥ Sj
S4 ≥ S2	$\frac{(0.172 - 0.416)}{(0.208 - 0.416) - (0.313 - 0.172)}$	0.699	= 0.699
S4 ≥ S3	b4 ≥ b3	1.00	

Determination of Weight Vectors {d'(Ai), d'(A2) ----------- d(An)}^{T}

W' = (1.00, 1.00, 0.489, 0.699)

The above vector is normalized to give crisp weight vector

W = (0.313, 0.313, 0.154, 0.220

Table 2.44: Pair-wise comparison for the alternatives regarding to sub-criteria of "Future Supply Capability".

	Supplier A	*Supplier B*	*Supplier C*	*Supplier D*
Supplier A	1, 1, 1	1, 1, 1	1, 3/2, 2	1, 3/2, 2
Supplier B	1, 1, 1	1, 1, 1	1, 3/2, 2	1, 3/2, 2
Supplier C	1/2, 2/3, 1	1/2, 2/3, 1	1, 1, 1	1, 1, 1
Supplier D	1/2, 2/3, 1	1/2, 2/3, 1	1, 1, 1	1, 1, 1
		CR = 0.013		

Synthetic Extent Normalized Row Sum. Row sum value of row in table using equation (2.8)

RS1 = (4.00, 5.00, 6.00) RS2 = (4.00, 5.00, 6.00)
RS3 = (3.00, 3.33, 4.00) RS4 = (3.00, 3.33, 4.00)

The synthetic extent normalized row sums obtained from above using equation (2.9)

S1 = (0.200, 0.300, 0.428) S2 = (0.200, 0.300, 0.428)
S3 = (0.150, 0.199, 0.285) S4 = (0.150, 0.199, 0.285)

Table 2.45: Degree of possibility for the alternatives regarding to sub-criteria of "Future Supply Capability".

V(Si ≥ Sj)	*Degree of possibility (Calculated by using equation 4.11)*	*Value*	*d'(Ai)min V (Si ≥ Sj)*
S1 ≥ S2	b2 ≥ b1	1.00	d'(A1)min V(S2 ≥ Sj)
S1 ≥ S3	b2 ≥ b1	1.00	= 1.00
S1 ≥ S4	b2 ≥ b1	1.00	
S2 ≥ S1	b2 ≥ b1	1.00	d'(A2)min V(S2 ≥ Sj)
S2 ≥ S3	b2 ≥ b1	1.00	= 1.00
S2 ≥ S4	b2 ≥ b1	1.00	
S3 ≥ S1	$\frac{(0.200-0.285)}{(0.199-0.285)-(0.300-0.200)}$	0.456	d'(A4)min V(S4 ≥ Sj)
S3 ≥ S2	$\frac{(0.200-0.285)}{(0.199-0.285)-(0.300-0.200)}$	0.456	= 0.456
S3 ≥ S4	b2 ≥ b1	1.00	
S4 ≥ S1	$\frac{(0.200-0.285)}{(0.199-0.285)-(0.300-0.200)}$	0.456	d'(A4)min V(S4 ≥ Sj)
S4 ≥ S2	$\frac{(0.200-0.285)}{(0.199-0.285)-(0.300-0.200)}$	0.456	= 0.456
S4 ≥ S3	b2 ≥ b1	1.00	

Determination of Weight Vectors {d'(Ai), d'(A2) ----------- d(An)}^{T}

$$W' = (1.00, 1.00, 0.456, 0.456)$$

The above vector is normalized to give crisp weight vector

$$W = (0.344, 0.344, 0.156, 0.156)$$

Table 2.46: Pair-wise comparison for the alternatives regarding to sub-criteria of "Market Reputation".

	Supplier A	*Supplier B*	*Supplier C*	*Supplier D*
Supplier A	1, 1, 1	1/2, 1, 3/2	3/2, 2, 5/2	1/2, 1, 3/2
Supplier B	2/3, 1, 2	1, 1, 1	2/5, 1/2, 2/3	2/3, 1, 2
Supplier C	2/5, 1/2, 2/3	3/2, 2, 5/2	1, 1, 1	1/2, 2/3, 1
Supplier D	2/3, 1, 2	1/2, 1, 3/2	1, 3/2, 2	1, 1, 1
		CR = 0.076		

Synthetic extent normalized row sum. Row sum value of row in table using equation (2.8)

RS1 = (3.50, 5.00, 6.50) RS2 = (2.73, 3.50, 5.66)

RS3 = (3.40, 4.16, 5.16) RS4 = (3.166, 4.50, 6.50)

The synthetic extent normalized row sums obtained from above using equation (2.9)

S1 = (0.146, 0.291, 0.507) S2 = (0.114, 0.203, 0.442)

S3 = (0.142, 0.242, 0.403) S3 = (0.132, 0.262, 0.507)

Table 2.47: Degree of possibility for the alternatives regarding to sub-criteria of "Market Reputation".

V(Si ≥Sj)	*Degree of possibility (Calculated by using equation 4.11)*	*Value*	*d'(Ai)min - V (Si ≥Sj)*
S1 > S2	b1 ≥ b2	1.00	d'(A1)min-V(S1 > Sj)
S1 > S3	b1 ≥ b3	1.00	= 1.00
S1 > S4	b1 ≥ b4	1.00	
S2 > S1	$\frac{(0.146-0.442)}{(0.203-0.442)-(0.291-0.146)}$	0.770	d'(A2)min-V(S2 > Sj)
S2 > S3	$\frac{(0.142-0.442)}{(0.203-0.442)-(0.242-0.142)}$	0.884	= 0.770
S2 > S4	$\frac{(0.132-0.442)}{(0.203-0.442)-(0.262-0.132)}$	0.840	
S3 > S1	$\frac{(0.146-0.403)}{(0.242-0.403)-(0.291-0.146)}$	0.839	d'(A3)min-V(S3 > Sj)
S3 > S2	b3 ≥ b2	1.00	= 0.839
S3 > S4	$\frac{(0.132-0.403)}{(0.242-0.403)-(0.262-0.132)}$	0.931	
S4 > S1	$\frac{(0.146-0.507)}{(0.262-0.403)-(0.291-0.146)}$	0.925	d'(A4)min-V(S4 > Sj)
S4 > S2	b4 ≥ b2	1.00	= 0.925
S4 > S3	b4 ≥ b3	1.00	

Determination of Weight Vectors {d'(Ai), d'(A2) ----------- d(An)}T

W' = (1.00, 0.770, 0.839, 0.925)

The above vector is normalized to give crisp weight vector

W = (0.283, 0.218, 0.237, 0.261)

Table 2.48: Pair-wise comparison for the alternatives regarding to sub-criteria of "Response of Customer".

	Supplier A	*Supplier B*	*Supplier C*	*Supplier D*
Supplier C	1, 1, 1	2, 5/2, 3	3/2, 2, 5/2	1/2, 1, 3/2
Supplier B	1/3, 2/5, 1/2,	1, 1, 1	1/2, 1, 3/2	1, 3/2, 2
Supplier C	2/5, 1/2, 2/3	2/3, 1, 2	1, 1, 1	1/2, 2/3, 1
Supplier D	2/3, 1, 2	1/2, 2/3, 1	1, 3/2, 2	1, 1, 1
		CR= 0.060		

Synthetic extent normalized row sum. Row sum value of row in table using equation (2.8)

RS1 = (5.00, 6.50, 8.00) RS2 = (2.83, 3.90, 5.00)

RS3 = (2.56, 3.16, 4.66) RS4 = (3.16, 4.166, 6.00)

The synthetic extent normalized row sums obtained from above using equation (2.9)

S1 = (0.211, 0.366, 0.590) S2 = (0.119, 0.220, 0.369)

S3 = (0.108, 0.178, 0.343) S4 = (0.133, 0.235, 0.442)

Table 2.49: Degree of possibility for the alternatives regarding to sub-criteria of "Response of Customer".

V(Si ≥ Sj)	*Degree of possibility (Calculated by using equation 4.11)*	*Value*	*d'(Ai)min - V (Si ≥ Sj)*
S1 > S2	b1 ≥ b2	1.00	d'(A1)-min V(S1 > Sj)
S1 > S3	b1 ≥ b3	1.00	= 1.00
S1 > S4	b1 ≥ b4	1.00	
S2 > S1	$\frac{(0.211 - 0.369)}{(0.220 - 0.369) - (0.366 - 0.211)}$	0.519	d'(A2)-min V(S2 > Sj)
S2 > S3	b2 ≥ b3	1.00	= 0.519
S2 > S4	$\frac{(0.133 - 0.369)}{(0.220 - 0.369) - (0.235 - 0.133)}$	0.940	
S3 > S1	$\frac{(0.211 - 0.343)}{(0.178 - 0.343) - (0.235 - 0.133)}$	0.412	d'(A3)-min V(S3 > Sj)
S3 > S2	$\frac{(0.119 - 0.343)}{(0.178 - 0.343) - (0.220 - 0.119)}$	0.842	= 0.412
S3 > S4	$\frac{(0.133 - 0.343)}{(0.178 - 0.343) - (0.235 - 0.211)}$	0.786	
S4 > S1	$\frac{(0.211 - 0.442)}{(0.235 - 0.442) - (0.366 - 0.211)}$	0.638	d'(A4)-min V(S4 > Sj)
S4 > S2	b4 ≥ b2	1.00	= 0.638
S4 > S3	b4 ≥ b3	1.00	

Determination of Weight Vectors {d'(Ai), d'(A2) ----------- d(An)}^{T}

$$W' = (1.00, 0.519, 0.412, 0.638)$$

The above vector is normalized to give crisp weight vector

$$W = (0.389, 0.202, 0.160, 0.249)$$

Table 2.50: Pair-wise comparison for the alternatives regarding to sub-criteria of "Responsiveness".

	Supplier A	*Supplier B*	*Supplier C*	*Supplier D*
Supplier A	1, 1,1	1/2, 1, 3/2	2, 5/2, 3	3/2, 2,5/2
Supplier B	2/3, 1, 2	1, 1,1	3/2, 2,5/2	1, 3/2, 2
Supplier C	1/3, 2/5, 1/2	2/5, 1/2, 2/3	1, 1,1	1/2, 1, 3/2
Supplier D	2/5, 1/2, 2/3	1/2, 2/3, 1	2/3, 1, 2	1, 1,1
		CR = 0.006		

Synthetic extent normalized row sum. Row sum value of row in table using equation (2.8)

RS1 = (5.00, 6.50, 8.00) RS2 = (4.16, 5.50, 7.50)

RS3 = (2.23, 2.90, 4.66) RS4 = (2.56, 3.16, 4.66)

The synthetic extent normalized row sums obtained from above using equation (2.9)

S1 = (0.209, 0.359, 0.573) S2 = (0.174, 0.304, 0.537)

S3 = (0.0936, 0.160, 0.262) S4 = (0.107, 0.175, 0.333)

Table 2.51: Degree of possibility for the alternatives regarding to sub-criteria of "Responsiveness".

V(Si ≥Sj)	*Degree of possibility (Calculated by using equation 4.11)*	*Value*	*d'(Ai)min - V (Si ≥Sj)*
S1 > S2	b1 ≥ b2	1.00	d'(A1)-minV(S1>Sj) = 1.00
S1 > S3	b1 ≥ b3	1.00	
S1 > S4	b1 ≥ b4	1.00	
S2 > S1	$\frac{(0.209 - 0.537)}{(0.304 - 0.537) - (0.359 - 0.209)}$	0.856	d'(A2)-minV(S2 > Sj) = 0.856
S2 > S3	b2 ≥ b3	1.00	
S2 > S4	b2 ≥ b4	1.00	

Table 2.51 (*Contd...*)

Table 2.51: (*Contd...*)

V(Si ≥ Sj)	*Degree of possibility (Calculated by using equation 4.11)*	*Value*	*d'(Ai)min - V (Si ≥ Sj)*
S3 > S1	$\frac{(0.209-0.262)}{(0.160-0.262)-(0.359-0.209)}$	0.210	d'(A3)-minV(S3 > Sj)
S3 > S2	$\frac{(0.174-0.262)}{(0.160-0.262)-(0.304-0.174)}$	0.379	= 0.210
S3 > S4	$\frac{(0.107-0.262)}{(0.160-0.262)-(0.175-0.107)}$	0.911	
S4 > S1	$\frac{(0.209-0.333)}{(0.175-0.333)-(0.359-0.209)}$	0.402	d'(A4)-minV(S4 > Sj)
S4 > S2	$\frac{(0.209-0.333)}{(0.175-0.333)-(0.359-0.209)}$	0.402	= 0.402
S4 > S3	b4 ≥ b3	1.00	

Determination of Weight Vectors {d'(Ai), d'(A2) ----------- d(An)}^{T}

$$W' = (1.00, 0.856, 0.210, 0.402)$$

The above vector is normalized to give crisp weight vector

$$W = (0.405, 0.347, 0.085, 0.163)$$

Table 2.52: Pair-wise comparison for the alternatives regarding to sub-criteria of "Reliability".

	Supplier A	*Supplier B*	*Supplier C*	*Supplier D*
Supplier A	1, 1, 1	1, 3/2, 2	3/2, 2, 5/2	1, 1, 1
Supplier B	1/2, 2/3, 1	1, 1, 1	2/3, 1, 2	1/2, 2/3, 1
Supplier C	2/5, 1/2, 2/3	1/2, 1, 3/2	1, 1, 1	1, 3/2, 2
Supplier D	1, 1, 1	1, 3/2, 2	1/2, 2/3, 1	1, 1, 1
		CR= 0.048		

Synthetic extent normalized row sum. Row sum value of row in table using equation (2.8)

RS1 = (4.50, 5.50, 6.50) RS2 = (2.66, 3.33, 5.00)

RS3 = (2.90, 4.00, 5.66) RS4 = (3.50, 4.16, 5.00)

The synthetic extent normalized row sums obtained from above using equation (2.9)

S1 = (0.207, 0.323, 0.479) S2 = (0.122, 0.195, 0.338)

S3 = (0.133, 0.235, 0.380) S4 = (0.161, 0.245, 0.368)

Table 2.53: Degree of possibility for the alternatives regarding to sub-criteria of "Reliability".

V(Si ≥Sj)	*Degree of possibility (Calculated by using equation 4.11)*	*Value*	*d'(Ai)min V (Si ≥Sj)*
S1 > S2	b1 ≥ b2	1.00	d'(A1)-min V(S1 > Sj)
S1 > S3	b1 ≥ b3	1.00	= 1.00
S1 > S4	b1 ≥ b4	1.00	
S2 > S1	$\frac{(0.207 - 0.368)}{(0.195 - 0.368) - (0.323 - 0.207)}$	0.557	d'(A2)-min V(S2 > Sj)
S2 > S3	$\frac{(0.133 - 0.368)}{(0.195 - 0.368) - (0.235 - 0.133)}$	0.854	= 0.557
S2 > S4	$\frac{(0.161 - 0.368)}{(0.195 - 0.368) - (0.245 - 0.161)}$	0.805	
S3 > S1	$\frac{(0.207 - 0.380)}{(0.235 - 0.380) - (0.323 - 0.207)}$	0.662	d'(A3)-min V(S3 > Sj)
S3 > S2	b3 ≥ b2	1.00	= 0.662
S3 > S4	$\frac{(0.161 - 0.380)}{(0.235 - 0.380) - (0.245 - 0.161)}$	0.956	
S4 > S1	$\frac{(0.207 - 0.368)}{(0.245 - 0.363) - (0.323 - 0.207)}$	0.673	d'(A4)-min V(S4 > Sj)
S4 > S2	b4 ≥ b2	1.00	= 0.402
S4 > S3	b4 ≥ b3	1.00	

Determination of Weight Vectors {d'(Ai), d'(A2) ----------- d(An)}^{T}

W' = (1.00, 0.557, 0.662, 0.673)

The above vector is normalized to give crisp weight vector

W = (0.346, 0.192, 0.229, 0.233

Table 2.54: Pair-wise comparison for the alternatives regarding to sub-criteria's of "Customization Capacity".

	Supplier A	*Supplier B*	*Supplier C*	*Supplier D*
Supplier A	1, 1, 1	1, 3/2, 2	3/2, 2,5/2	1/2, 1, 3/2
Supplier B	1/2, 2/3, 1	1, 1, 1	2/3, 1, 2	1/2, 2/3, 1
Supplier C	2/5, 1/2, 2/3	1/2, 1, 3/2	1, 1, 1	2/5, 1/2, 2/3
Supplier D	2/3, 1, 2	1, 3/2, 2	3/2, 2,5/2	1, 1, 1
		CR = 0.001		

Synthetic Extent Normalized Row Sum. Row sum value of Row in table using equation (2.8)

RS1 = (4.00, 5.50, 7.00) RS2 = (2.66, 3.33, 5.00)

RS3 = (2.33, 3.00, 3.83) RS4 = (4.16, 5.50, 7.50)

The synthetic extent normalized row sums obtained from above using equation (2.9)

S1 = (0.171, 0.317, 0.533) S2 = (0.114, 0.192, 0.380)

S3 = (0.0985, 0.173, 0.291) S4 = (0.178, 0.317, 0.571)

Table 2.55: Degree of possibility for the alternatives regarding to sub-criteria of "Customization Capacity".

V(Si ≥ Sj)	*Degree of possibility (Calculated by using equation 4.11)*	*Value*	*d'(Ai)min V (Si ≥ Sj)*
S1 > S2	b1 ≥ b2	1.00	d'(A1)-minV(S1 > Sj)
S1 > S3	b1 ≥ b3	1.00	= 1.00
S1 > S4	b1 ≥ b4	1.00	
S2 > S1	$\frac{(0.171 - 0.380)}{(0.192 - 0.380) - (0.317 - 0.171)}$	0.625	d'(A2)-minV(S2 > Sj)
S2 > S3	b2 ≥ b 3	1.00	= 0.625
S2 > S4	$\frac{(0.178 - 0.380)}{(0.192 - 0.380) - (0.317 - 0.178)}$	0.626	
S3 > S1	$\frac{(0.171 - 0.291)}{(0.173 - 0.291) - (0.317 - 0.171)}$	0.454	d'(A3)-minV(S3 > Sj)
S3 > S2	$\frac{(0.114 - 0.291)}{(0.173 - 0.291) - (0.192 - 0.114)}$	0.903	= 0.454
S3 > S4	$\frac{(0.178 - 0.291)}{(0.173 - 0.291) - (0.317 - 0.178)}$	0.459	
S4 >S1	b4 ≥ b1	1.00	d'(A4)-minV(S4 > Sj)
S4 >S2	4 ≥ b2	1.00	= 1.00
S4 >S3	b4 ≥ b3	1.00	

Determination of Weight Vectors {d'(Ai), d'(A2) ----------- d(An)}^{T}

W' = (1.00, 0.625, 0.454, 1.00)

The above vector is normalized to give crisp weight vector

W = (0.329, 0.203, 0.147, 0.326)

Table 2.56: Pair-wise comparison for the alternatives regarding to sub-criteria of "Shortage Frequency".

	Supplier A	*Supplier B*	*Supplier C*	*Supplier D*
Supplier A	1, 1, 1	1/2, 1, 3/2	2, 5/2, 3	1, 3/2, 2
Supplier B	2/3, 1, 2	1, 1, 1	1/2, 2/3, 1	2/5, 1/2, 2/3
Supplier C	1/3, 2/5,1/2	1, 3/2, 2	1, 1, 1	2/3, 1, 2
Supplier D	1/2, 2/3, 1	3/2, 2,5/2	1/2, 1, 3/2	1, 1, 1
		CR= 0.084		

Synthetic extent normalized row sum. Row sum value of row in table using equation (4.8)

RS1 = (4.50, 6.00, 7.50) RS2 = (2.56,3.16, 4.66)

RS3 = (3.00, 3.90, 5.50) RS4 = (3.50, 4.66, 6.00)

The synthetic extent normalized row sums obtained from above using equation (2.9)

S1 = (0.190, 0.338, 0.553) S2 = (0.108, 0.178, 0.343)

S3 = (0.126, 0.220, 0.405) S4 = (0.147, 0.262, 0.442)

Table 2.57: Degree of possibility for the alternatives regarding to sub-criteria of "Shortage Frequency".

V(Si ≥Sj)	*Degree of possibility (Calculated by using equation 4.11)*	*Value*	*d'(Ai)min V (Si ≥Sj)*
S1 > S2	b1 ≥ b2	1.00	d'(A1)-min V(S1 > Sj)
S1 > S3	b1 ≥ b3	1.00	= 1.00
S1 > S4	b1 ≥ b4	1.00	
S2 > S1	$\frac{(0.190 - 0.343)}{(0.178- 0.343) - (0.338 - 0.190)}$	0.488	d'(A2)-min V(S2 > Sj)
S2 > S3	$\frac{(0.126 - 0.343)}{(0.178- 0.343) - (0.220 - 0.126)}$	0.370	= 0.370
S2 > S4	$\frac{(0.147 - 0.343)}{(0.178- 0.343) - (0.262 - 0.147)}$	0.700	
S3 >S1	$\frac{(0.190 - 0.405)}{(0.220- 0.405) - (0.338 - 0.190)}$	0.645	d'(A3)-min V(S3 > Sj)
S3 >S2	b3 ≥ b2	1.00	= 0.645
S3 >S4	$\frac{(0.147 - 0.405)}{(0.220- 0.405) - (0.262 - 0.147)}$	0.860	
S4 > S1	$\frac{(0.190 - 0.442)}{(0.220- 0.442) - (0.338 - 0.190)}$	0.768	d'(A4)-min V(S4 > Sj)
S4 > S2	b4 ≥ b2	1.00	= 0.768
S4 > S3	b4 ≥ b3	1.00	

Determination of Weight Vectors {d'(Ai), d'(A2) ----------- d(An)}^{T}

W' = (1.00, 0.370, 0.645, 0.768)

The above vector is normalized to give crisp weight vector

W = (0.360, 0.133, 0.231, 0.276)

2.11. DISCUSSION AND MANAGERIAL IMPLICATIONS

In this chapter methodology of the FAHP has been implemented to solve the problem of best supplier selection. The present work proposes a FAHP approach for the selection of suppliers in a supply chain. The major advantages of this research are that it can be used for both qualitative and quantitative criteria and consider fuzziness and vagueness. Pair-wise comparison used in this work reduces the dependency of the model on human judgment. A systematic approach using FAHP approach with linguistic variable has been applied for supplier's selection.

According to the results as given in Table 2.58 found in the supplier evaluation and selection, Supplier A appears to be the best choice of all 4 suppliers based on its highest total score. In above table based on the comparison of suppliers and the method applied and observed that Supplier-A is the best supplier. This chapter has presented an integrated supplier selection and order allocation problem for a supply chain management over a planning horizon with multiple time periods when customer flexibility exists. A Fuzzy AHP model has been developed to maximize the manufacturer's profit by determining the production quantity of each product variant and by selecting the most suitable suppliers based on the selection criteria and their capacity and splitting the orders among these suppliers. A preferred, since to has the highest weight of (0.31085) among four suppliers, suppliers B is at the second choice (0.24365), supplier D is at third choice (0.2275) and supplier C the last choice (0.2180).

It is pertinent here to discuss the priority values of the determinants, which influence this decision. From Table 2.58, it is observed that quality (26.50 percent) is the most important determinant in the selection of suppliers. It is followed by delivery (23.6 percent), cost (20.4 percent), service (16.7 percent), and supplier's profile (12.70 percent). According to these results, Supplier selection ought to improve their operational and financial performance firstly providing advanced technology offerings around strategy, planning, collaboration, data management, decision support, integration, and flexibility.

Table 2.58: Priority weight of main criteria and sub-criteria and alternatives.

Main	*Main*	*Sub-Criteria*	*Sub-Criteria*	*Supplier A*	*Supplier B*	*Supplier C*	*Supplier D*	*Weight A*	*Weight B*	*Weight C*	*Weight D*
Cost	0.204	Material cost	0.329	0.246	0.217	0.346	0.190	0.0165	0.0145	0.02321	0.01274
		Transportation cost	0.258	0.124	0.242	0.341	0.291	0.00652	0.01270	0.0179	0.0153
		Quantity discount	0.304	0.176	0.275	0.324	0.225	0.0109	0.01705	0.0200	0.01395
		Overhead cost	0.107	0.270	0.252	0.244	0.233	0.00588	0.00549	0.005319	0.005079
Delivery	0.236	Compliance with delivery date	0.482	0.445	0.231	0.186	0.138	0.05061	0.02626	0.02114	0.01569
		Lead time	0.439	0.318	0.183	0.272	0.227	0.0329	0.01895	0.02817	0.02351
		Deliverable location	0.063	0.335	0.227	0.170	0.268	0.00498	0.003373	0.002526	0.003982
		Flexibility in delivery operation	0.0144	0.394	0.152	0.277	0.177	0.001338	0.000516	0.000941	0.000601
Quality	0.265	Rejection rate	0.539	0.295	0.273	0.175	0.256	0.04213	0.03898	0.02499	0.03655
		Inspection methods	0.325	0.259	0.259	0.241	0.241	0.0223	0.0223	0.02085	0.02075
		Quality management	0.00	0.268	0.194	0.290	0.247	0.000	0.000	0.000	0.000
		Product reliability	0.135	0.444	0.249	0.199	0.108	0.01585	0.00888	0.007104	0.00385
Supplier's profile	0.127	Financial strength	0.429	0.313	0.313	0.154	0.220	0.01702	0.01702	0.008377	0.01196
		Future supply capacity	0.201	0.344	0.344	0.156	0.156	0.008781	0.008781	0.003978	0.01078
		Market reputation	0.201	0.283	0.218	0.237	0.261	0.007216	0.00559	0.006043	0.006655
Service	0.167	Responsiveness	0.281	0.405	0.347	0.085	0.163	0.01900	0.01628	0.003988	0.007647
		Reliability	0.316	0.346	0.192	0.229	0.233	0.01825	0.01013	0.01208	0.01229
		Customization capacity	0.316	0.329	0.203	0.147	0.326	0.01736	0.01071	0.007751	0.017203
		Shortage frequency	0.084	0.360	0.133	0.231	0.276	0.00505	0.00186	0.003238	0.003869
		Total weight						0.31085	0.24365	0.2180	0.2275

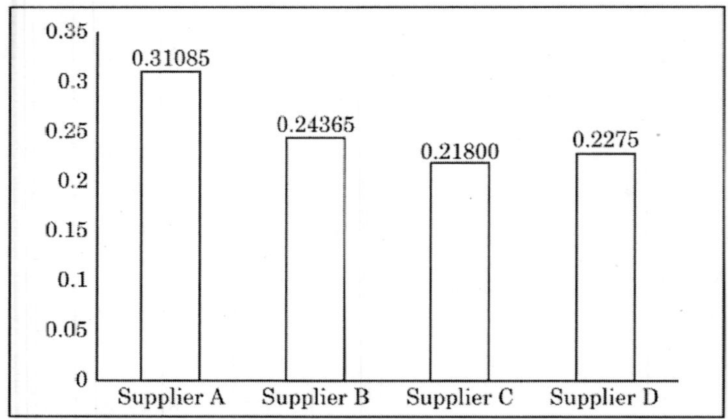

Fig. 2.7: Overall comparison of supplier rating using FAHP.

Fuzzy Sets and AHP to take advantage of their individual abilities to integrate group opinions, ability to handle ambiguity and address multi criteria decision making problems. Weights of the sub criteria adjudged low in significance have become zero. The results of the model reflect the present preferences of the company decision makers. When shared with the suppliers, this allows identification of individual areas for improvement and builds up business prospects with the company. The overall predictability brought about in supplier selection reduces uncertainty in supply. This will lead to significant cutting down on ordering and operating costs thus enhancing the supply chain surplus. Based on the analysis of FAHP results, Supplier A would be the appropriate choice for the case company. Selecting supplier-A meets the goal the case company and its product specification as illustrated in Fig. 2.7. In addition to low price, Supplier A provides a high level of trustworthy and a lower level of expected risks.

3

Analytic Network Process (ANP) Approach for Green Supplier Selection in Supply Chain Management

3.1. INTRODUCTION

The purchasing function has become vital in determining the profitability and survival of business organization; it has been receiving considerable attention. Supplier-manufacturer relationships based solely on price are no longer acceptable. The increasing importance of supplier selection decisions is forcing organizations to re-think their purchasing and evaluation strategies because a successful purchasing decision directly depends on selecting the "right" supplier.

Recent advances in multi criteria decision making process generated economic and strategic changes in many aspects of modern business, *e.g.*, manufacturing, logistics, finance and marketing. Analytic network process (ANP) is a general theory of relative measurement used to derive composite priority ratio scales from individual ratio scales that represent relative measurements of the influence of elements that interacts within the sub criteria. The ANP is a general form of the AHP. The basic structure is an influence network of clusters and nodes contained within the clusters. Priorities are established in the same way they are in the AHP using Pair-wise comparisons and judgment. Whereas analytic hierarchical process (AHP) models a decision-making frame work that assumes a unidirectional hierarchical relationship among decision levels, ANP allows for more complex interrelationship among decision levels.

Many decision problems cannot be building as hierarchical because of dependencies (inner/outer), influences between and within clusters (criteria's, alternatives). ANP is very useful to solve this kind of problems.

ANP provides a general framework to deal with decisions without making assumptions about the independence of higher-level elements from lower level elements and about the independence of the elements within a level. In fact, ANP uses a network without the need to specify levels as in a hierarchy. Not only does the importance of the criteria determine the importance of the alternatives as in a hierarchy, but also the importance of the alternatives themselves determines the importance of the criteria. Feedback enables us to factor the future into the present to determine what we must do to attain a desired future.

ANP is a multi-attribute, decision-making approach based on the reasoning, knowledge, and experience of the experts in the field. ANP can act as a valuable aid for decision making involving both tangible as well as intangible attributes that are associated with the model under study. ANP relies on the process of eliciting managerial inputs, thus allowing for a structured communication among decision makers. Thus, it can act as a qualitative tool for strategic decision-making problems.

3.2. OBJECTIVES AND ISSUES

In modern era of industrialization and globalization, every organization must ensure that their product must meet the international Environmental standards and quality requirements to remain competent in this rapidly changing industrial environment. To achieve this, it is a necessary requirement that supplier from which the organization is getting the supply of raw material or any other kind of necessary inputs should be selected correctly as per the Green requirements. This multi criteria decision making problem is now a part of day to day affair of all the organizations.

3.3. ANALYTIC NETWORK PROCESS

Analytic network process (ANP) is a general theory of relative measurement used to derive composite priority ratio scales from individual ratio scales that represent relative measurements of the influence of elements that interacts within the sub criteria. The ANP is a general form of the AHP (Saaty, 1980). The basic structure is an influence network of clusters and nodes contained within the clusters. Priorities are established in the same way they are in the AHP using pair-wise comparisons and judgment. Whereas analytic hierarchical process (AHP) models a decision-making frame work that assumes a unidirectional hierarchical relationship among decision levels, ANP allows for more complex interrelationship among decision levels.

3.4. STEPS TO ANALYTIC NETWORK PROCESS

Saaty 1991, proposed the following four main steps of the ANP as enlisted below and explained in Fig. 3.2.

Step 1: Model development and problem formulation.

Step 2: Pair-wise comparison matrices and priority vectors.

Step 3: Super-matrix formation:

Step 4: Selection of the best alternative.

3.4.1. Model Construction and Problem Structuring

The problem should be clearly stated and decomposed into a rational system, such as a network. The framework can be determined based on the decision maker's opinion through brainstorming or other appropriate methods.

3.4.2. Pair-wise Comparison Matrices and Priority Vectors

Like the AHP, pairs of the decision elements at each cluster are compared with respect to their importance towards their control criteria. The clusters themselves are also compared pair-wise with respect to their contribution to the objective. The decision maker is asked to respond to a series of pair-wise comparisons of two elements or two clusters to be evaluated in terms of their contribution to their upper level criteria. In addition, interdependencies among the elements of a cluster must also be examined pair-wise where the influence of each element on other elements can be represented by an eigenvector. The relative importance values are determined with Saaty's 1–9 scale (Table 3.1) where a score of 1 represents equal importance between the two elements and a score of 9 indicates the extreme importance of one element (row cluster in the matrix) compared to the other one (column cluster in the matrix). A reciprocal value is assigned to the inverse comparison, *i.e.*, aij = 1/aji, where aij denotes the importance of ith element over jth one. Like with the AHP, pair-wise comparison in the ANP is also performed in the framework of a matrix (decision matrix). The local priority vector can be derived as an estimate of the relative importance associated with the elements (or clusters) being compared by solving the given equations.

3.4.2.1. *The decision matrix (comparative judgments)*

A set of comparison matrices of all elements in a level of the hierarchy with respect to an element of the immediately higher level are constructed to prioritize and convert individual comparative judgments into ratio scale measurements. The preferences are quantified by using a nine-point scale. The meaning of each scale measurement is explained in Table 3.1. The pair-wise comparisons are given in terms of how much more element A is important than element B. As the ANP approach is a subjective methodology, information and the priority weights of elements may be obtained from a decision maker of the company using direct questioning or a questionnaire method.

Table 3.1: Thomas Saaty's nine-point scale.

Intensity of importance	*Definition*	*Explanations*
1	Equal importance	Two activities contribute equally to the objective
3	Weak importance of one over another	Experience and judgment slightly favor one activity over another
5	Essential or strong important	Experience and judgment strongly favor one activity over another
7	Demonstrated importance	An activity is favored very strongly over another; its dominance demonstrated in practice
9	Absolute importance	The evidence favoring one activity over another is of the highest possible order of affirmation
2,4,6,8	Intermediate values between the two-adjacent judgment	When compromise is needed
Reciprocals of above nonzero	If activity I has one of the above non-zero numbers assigned to it when compared with activity j, then has the reciprocal value when compared with I	A reasonable assumption

3.4.2.3. *Synthesis of priorities and the measurement of consistency*

The pair-wise comparisons of the criteria of green supplier selection problem generate a matrix of relative rankings for each level of the network. The number of matrices depends on the number of elements at each level. The number of elements at each level decides the order of every matrix of the next higher level. After all matrices are developed, eigenvectors or the relative weights (the degree of relative importance amongst the elements) and the maximum Eigen-value (? max) for each matrix are calculated. The ? max value is an important validating parameter in ANP. It is used for calculating the consistency ratio CR of the estimated vector to validate whether the pair-wise comparison matrix provides a completely consistent evaluation. The consistency ratio is calculated as per the following steps:-

Step 1: Calculate the Eigen-vector or the relative weights and ? max for each matrix of order m

AW = ? max W .... (1)

Step 2: Compute the consistency index for each matrix of order n by the formulae:-

CI = (? max -m)/ (m-1)................................ (2)

Step 3: The consistency ratio is then calculated using the formulae:-

CR = CI/RCI... (3)

Where CI = Consistency index

CR = Consistency ratio

RCI = Random consistency index

M = Number of elements

3.4.2.4. *Random consistency index (RCI)*

Where random consistency index (RCI) varies depending upon the order of matrix. Table 3.2 shows the value of the Random Consistency Index (RCI) for matrices of order 1 to 10 obtained by approximating random indices using a sample size of 500.

Table 3.2: Random consistency index (RCI).

N	3	4	5	6	7	8	9	10	11	12	13	14	15
RCI	0.58	0.9	1.12	1.24	1.32	1.41	1.45	1.51	1.52	1.54	1.56	1.57	1.59

The acceptable CR range varies according to the size of matrix *i.e.*, 0.05 for a 3 by 3 matrix, 0.08 for a 4 by 4 matrix and 0.1 for all larger matrices, m > = 5. If the value of CR is equal to, or less than that value, it implies that the evaluation within the matrix is acceptable or indicates a good level of consistency in the comparative judgments represented in that matrix. In contrast, if CR is more than the acceptable value, inconsistency of judgments within that matrix has occurred and the evaluation process should therefore be reviewed, reconsidered and improved. An acceptable consistency ratio helps to ensure decision-maker reliability in determining the priorities of a set of criteria.

3.4.3. Super-matrix Formation

3.4.3.1. *Formation of the initial super-matrix*

Elements in ANP are the entities in the system that interacts with each other. The determination of relative weights mentioned above is based on pair-wise comparisons as in the standard AHP. The weights are then put into the super-matrix that represents the interrelationships of elements in the system. The general form of the super-matrix is described here below where CN denotes the Nth cluster, eNn denotes the nth element in the Nth cluster, and Wij is a block matrix consisting of priority weight vectors (w) of the influence of the elements in the ith cluster with respect to the jth cluster. The Super Matrix is shown below in Fig. 3.1.

		C_1				C_2				…	C_N			
		e_{11}	e_{22}	…	e_{1n}	e_{21}	e_{22}	…	e_{2n}	…	e_{N1}	e_{N2}	…	e_N
C_1	e_{11}	W_{11}				W_{12}					W_{1N}			
	e_{22}													
	…													
	e_{1n}													
…	…	…				…				…	…			
C_N	e_{N1}	W_{N1}				W_{N2}					W_{NN}			
	e_{N2}													
	…													
	e_N													

Fig. 3.1: Initial super-matrix

3.4.3.2. *Formation of weighted super-matrix*

The initial or "un-weighted" super-matrix consists of several eigenvectors each of which sums to one. The clusters in the initial super-matrix must be weighted and transformed to a matrix in which each of its columns sums to unity.

3.4.3.3. *Calculation of global priority vectors and weights*

In the final step, the weighted super-matrix is raised to limiting power to get the global priority vectors as in equation.

$$W^* = \lim_{k-\infty} W^k$$

Steps of Analytical Network Process

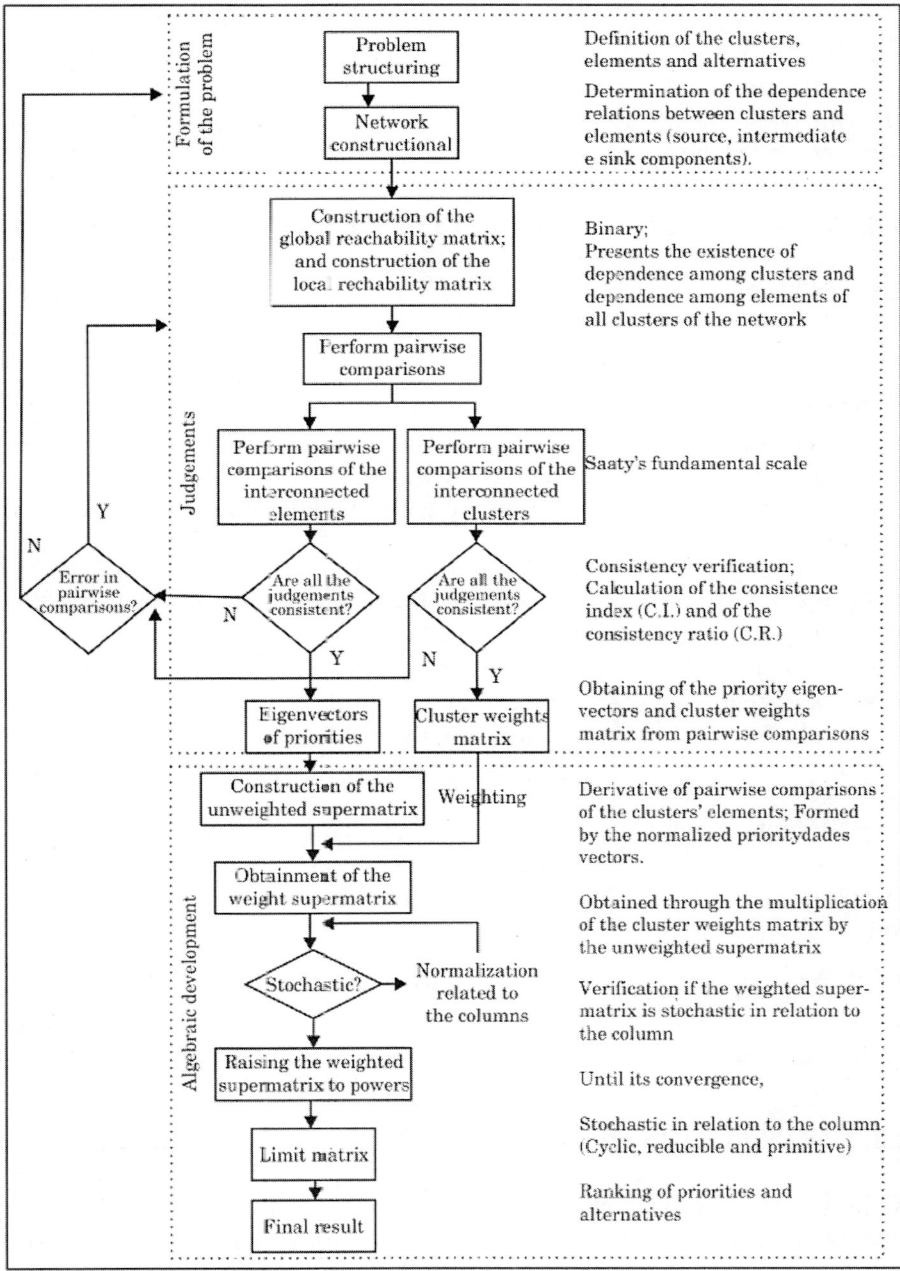

Fig. 3.2: Steps of analytical network process.

3.4.4. Selection of the Best Alternative

If the super-matrix formed in step 3 covers the whole network, the priority weights of the alternatives can be found in the column of alternatives in the normalized super-matrix. On the other hand, if the super-matrix only comprises clusters that are interrelated, additional calculations must be made to obtain the overall priorities of the alternatives. The alternative with the highest overall priority should be selected as the best choice.

3.5. BENEFITS OF THE ANALYTIC NETWORK PROCESS

ANP is a comprehensive technique that allows for the inclusion of all the relevant criteria; tangible as well as intangible, which have some bearing on decision-making process.

AHP models a decision-making framework that assumes unidirectional hierarchical relationship among decision levels, whereas ANP allows for more complex relationship among the decision levels and attributes as it does not require a strict hierarchical structure.

In decision-making problems, it is very important to consider the interdependent relationship among criteria because of the characteristics of interdependence that exists in real life problems. The ANP methodology allows for the consideration of interdependencies among and between levels of criteria and thus is an attractive multi-criteria decision-making tool. This feature makes it superior from AHP which fails to capture interdependencies among different enablers, criteria, and sub-criteria.

ANP methodology is beneficial in considering both qualitative as well as quantitative characteristics which need to be considered, as well as taking non-linear interdependent relationship among the attributes into consideration.

ANP is unique in the sense that it provides synthetic scores, which is an indicator of the relative ranking of different alternatives available to the decision maker.

3.6. LIMITATIONS OF ANALYTIC NETWORK PROCESS

Although the analytic Network process has been the subject of many research papers and the consensus is that the technique is both technically valid and practically useful, there are critics of the method.

Their criticism has included

Identifying the relevant attributes of the problem and determining their relative importance in decision making process requires extensive discussion and brainstorming sessions. Also, data acquisition is a very time intensive process for ANP methodology.

ANP requires more calculations and formation of additional pair-wise comparison matrices as compared to the AHP process. Thus, a careful track of matrices and pair-wise comparisons of attributes is necessary.

The pair-wise comparison of attributes under consideration can only be subjectively performed, and hence their accuracy of the results depends on the user's expertise knowledge in the area concerned.

3.7. METHOD APPLICATION

Researchers have done a variety of application of ANP to various fields, the ability to rapidly incorporate feedback and a possibility of simple comparison to actual results makes it very powerful method in multi

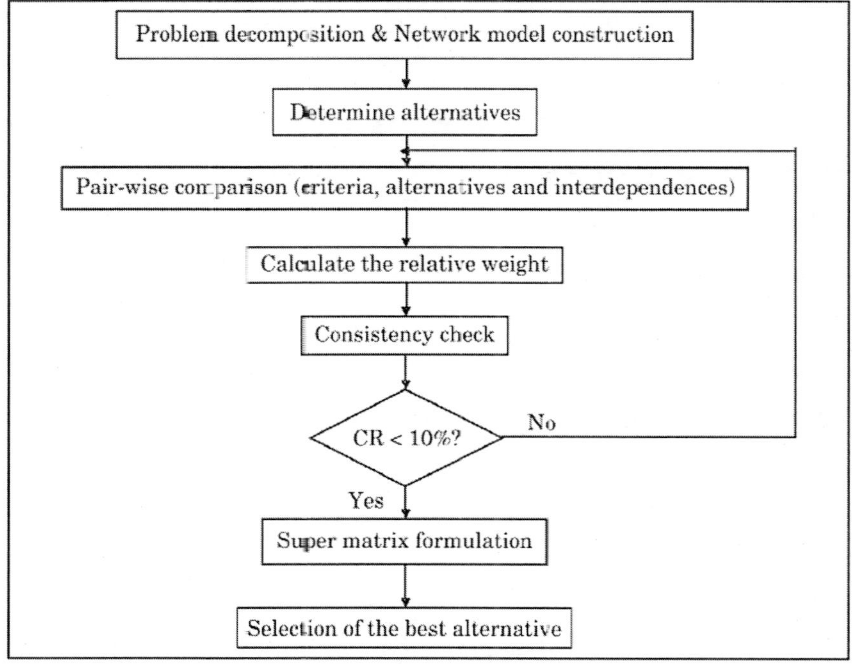

Fig. 3.3: Schematic representation of the methodology.

criteria decision making in various manufacturing areas. In addition to the wide application of ANP in manufacturing areas, recent research and industrial activities of applying ANP on other selection problems are also quite active. The study not only provides the evidence that the Analytic network process proved much better in MCDM, but also aids the decision makers and researchers in applying ANP effectively. ANP has thus been successfully applied to a diverse array of problems. A schematic representation of the methodology is given in Fig. 3.3.

3.8. MODEL FOR GREEN SUPPLIER SELECTION

The vendor selection criteria are structured as a hierarchy shown in Fig. 3.4. The hierarchy structure includes goal, criteria, sub-criteria, rating levels, and desired result for the vendor selection problem. The goal for the best vendor selection is put on the highest level of the hierarchy and the criteria on the second level. The hierarchy is easily extended to more detailed levels by breaking down the criteria into sub-criteria. The sub-criterion on the third level and the fourth and last level is desire result.

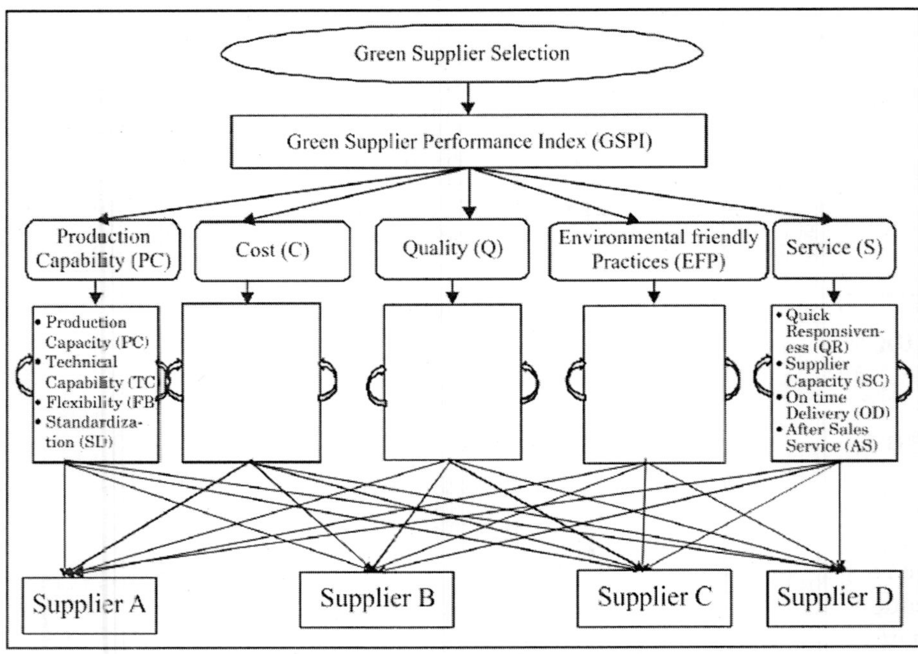

Fig. 3.4: ANP model for green supplier selection.

3.9. DESCRIPTION OF THE MODEL

A thorough analysis of the problem is required along with the identification of the important attributes involved. The selections of attributes have been determined through literature survey and discussions held with experts during industrial visits. The attributes and sub-attributes used in the AHP model for vendor selection in supply chain are given in Table 3.3.

Table 3.3: Attributes and sub-attributes.

Criteria	*Sub-criteria*
Production capability (PDC)	Production capacity (PC)
	Technical capability (TC)
	Flexibility (FB)
	Standardization (SD)
Cost (C)	Price (PV)
	Competitive pricing (CP)
	Transport cost (TC)
	Overhead cost (OC)
Quality (Q)	Defect Rate (DR)
	Commitment to quality (CQ)
	Improved process capability (IP)
	Quality management (QM)
Environmental friendly practices (EFP)	Eco Design (ED)
	RoHS (RH)
	Waste reduction (WR)
	Reuse and recycle (RR)
Service (S)	Quick responsiveness (QR)
	Supplier capacity (SC)
	On time delivery (OD)
	After sales service (AS)

3.10. WORKING STEPS OF CALCULATIONS AND APPLICATIONS FOR ANP IN GREEN SUPPLIER

3.10.1. Model Development and Problem Formulation

In this step, the decision problem is structured into its important components. The relevant criteria and alternatives are chosen based on the review of literature and discussion with few both from industry and academia. The relevant criteria and alternatives are structured in the form of a control hierarchy where the criteria at the top level in the model have the highest strategic value. The Goal in this model is Green Supplier Performance Index (GSPI).

In this ANP model, each of the five dimensions has some enablers, which helps to achieve the required dimension. For example, the dimension Production Capability (PDC) is supported by the enablers • Production Capacity (PC), Technical Capability (TC), Flexibility (FB), Standardization (SD). These enablers also have some interdependency on one another. For example, in the dimension Cost the enablers Price (P) and Competitive pricing (CP) are interdependent. The degree of interdependency may vary from case to case and would be captured in later steps. The strength of the ANP model is that the feedback and the network structure of the ANP makes possible the representation of the decision problem without much concern for what comes first and what comes next in a hierarchy. The objective of this hierarchy is to select the best possible alternative that will best meet the goals of conducting effective reverse logistics in a computer industry. The ANP model so developed is presented in Fig. 3.4. The alternatives that the decision maker wishes to evaluate are shown at the bottom of the model. The opinion of the material manager of the company was sought in the comparisons of the relative importance of the criteria and the formation of pair-wise comparison matrices to be used in the ANP model. In this paper, mainly for brevity, we present and illustrate the results only of the legislation determinant. The results of determinants would be used in the calculation of Green Supplier Performance Index (GSPI).

3.10.2. Pair-wise Comparison of Dimensions

In this step, a pair-wise comparison matrix is prepared for determining the relative importance of each of the dimensions of green supplier on the determinant of green supplier. The matrix Formed is shown below in Table no. 3.4 and 3.5.

From table, the results of the comparison (e-vectors) of the dimensions for the respected determinants are carried as Pja in desirability index matrices.

Table 3.4: Criteria pair-wise comparison matrix.

Criteria	*PDC*	*C*	*Q*	*EFP*	*S*
PDC	1	2	1/7	1/5	1/3
C	1/2	1	1/9	1/7	1/5
Q	7	9	1	2	4
EFP	5	7	1/2	1	2
S	3	5	1/4	1/2	1
Total	16.5000	24.000	2.0040	3.8429	7.5333

Table 3.5: Criteria pair-wise comparison matrix normalized.

Criteria	*PDC*	*C*	*Q*	*EFP*	*S*	*E-vector Wi*
PDC	0.0606	0.0834	0.0713	0.0518	0.0442	0.06114
C	0.0303	0.0417	0.0554	0.0372	0.0264	0.03760
Q	0.4243	0.3750	0.4990	0.5207	0.5312	0.47382
EFP	0.3030	0.2918	0.2495	0.2602	0.2661	0.27353
S	0.1818	0.2083	0.1248	0.1301	0.1321	0.15391

Check of the consistency

Let A1 = Pair-wise comparison matrix

A2 = Weight age matrix

$$A1 = \begin{pmatrix} 1 & 2 & 1/7 & 1/5 & 1/3 \\ 1/2 & 1 & 1/9 & 1/7 & 1/5 \\ 7 & 9 & 1 & 2 & 4 \\ 5 & 7 & 1/2 & 1 & 2 \\ 3 & 5 & 1/4 & 1/2 & 1 \end{pmatrix} \quad A2 = \begin{pmatrix} 0.06114 \\ 0.03760 \\ 0.47382 \\ 0.27353 \\ 0.15391 \end{pmatrix}$$

Then A3 = A1*A2

$$A3 = \begin{pmatrix} 0.3100 \\ 0.1907 \\ 2.4029 \\ 1.3872 \\ 0.7805 \end{pmatrix}$$

$$\text{And} \quad A4 = A3/A2 \quad A4 = \begin{pmatrix} 5.0703 \\ 5.0718 \\ 5.0713 \\ 5.0715 \\ 5.0711 \end{pmatrix}$$

λ max = Average of the elements of A4

λ max = 5.0712

Now Consistency Index (CI) = (λ max-M) / (M-1)

And Consistency Ratio (CR) = CI/RCI corresponding to M.

Where RCI = Random Consistency Index and

M = No. of elements

CI = (5.0712 – 5) / (5.0712 – 1) = 0.0579

CR = 0.0178/1.12 = 0.001590

Here CR is less than 10% (0.1), so the judgment is consistent.

3.10.3. *Pair-wise Comparison Matrices between Component/Enablers Levels*

In this step, the decision maker is asked to respond to a series of pair-wise comparisons where two components would be compared at a time with respect to an upper level control criterion. The pair-wise comparisons of the elements at each level are conducted with respect to their relative influence towards their control criterion. In the case of interdependencies, components within the same level may be viewed as controlling components for each other, or levels may be interdependent on each other. For a determinant, pair-wise comparison is done between the applicable enablers within a given dimension cluster. The number of such pair-wise comparison matrices depends on the number of determinants and the dimensions in the ANP model. In this model, 5 such pair-wise comparison matrices are formed. The e-vectors obtained from these matrices are imported as ADkja. All the pair-wise comparison matrices for the dimensions under each determinant is shown below in Table no. 3.6 to 3.15.

Table 3.6: Sub-criteria production capability pair-wise comparison matrix

Sub-criteria	*PC*	*TC*	*FB*	*SD*
PC	1	4	1	2
TC	1/4	1	1/4	1/3
FB	1	4	1	2
SD	1/2	3	1/2	1

Table 3.7: Sub-criteria production capability pair-wise comparison matrix normalized

Sub-criteria	*PC*	*TC*	*FB*	*SD*	*E-vector Wi*
PC	0.3637	0.3333	0.3637	0.3764	0.35928
TC	0.0908	0.0821	0.0908	0.0625	0.08153
FB	0.3637	0.3333	0.3637	0.3764	0.35928
SD	0.1817	0.2513	0.1817	0.1847	0.19992

Check of the consistency

Let A1 = Pair-wise comparison matrix

A2 = Weight age matrix

$$A1 = \begin{pmatrix} 1 & 4 & 1 & 2 \\ 1/4 & 1 & 1/4 & 1/3 \\ 1 & 4 & 1 & 2 \\ 1/2 & 3 & 1/2 & 1 \end{pmatrix} \quad A2 = \begin{pmatrix} 0.35928 \\ 0.08153 \\ 0.35928 \\ 0.19992 \end{pmatrix}$$

Then A3 = A1*A2 $$A3 = \begin{pmatrix} 1.4445 \\ 0.3278 \\ 1.4445 \\ 0.8038 \end{pmatrix}$$

$$A4 = \begin{pmatrix} 4.0205 \\ 4.0206 \\ 4.0205 \\ 4.0206 \end{pmatrix}$$

And A4 = A3/A2

λmax = Average of the elements of A4

λmax = 4.021675

Now Consistency Index (CI) = (λmax – M) / (M – 1)

And Consistency Ratio (CR) = CI/RCI corresponding to M.

Where RCI = Random Consistency Index and

M = Number of elements

CI = (4.021675-4)/(4 – 1) = 0.0072

CR = 0.0072/0.9 = CR = 0.00772

Here CR is less than 10% (0.1), so the judgment is consistent.

Table 3.8: Sub-criteria cost pair-wise comparison matrix (level 3).

Sub-criteria	*PV*	*CP*	*TC*	*OC*
PV	1	3	4	2
CP	1/3	1	2	1/2
TC	1/4	1/2	1	1/3
OC	1/2	2	3	1

Table 3.9: Sub-criteria cost pair-wise comparison matrix normalized.

Sub-criteria	*PV*	*CP*	*TC*	*OC*	*E-vector Wi*
PV	0.4800	0.4615	0.4000	0.5217	0.46730
CP	0.1600	0.1538	0.2000	0.1304	0.16009
TC	0.1200	0.0769	0.1000	0.0870	0.09543
OC	0.2400	0.3077	0.3000	0.2609	0.27718

Check of the consistency

Let A1 = Pair-wise comparison matrix

A2 = Weight age matrix

$$A1 = \begin{pmatrix} 1 & 3 & 4 & 2 \\ 1/3 & 1 & 2 & 1/2 \\ 1/4 & 1/2 & 1 & 1/3 \\ 1/2 & 2 & 3 & 1 \end{pmatrix} \quad A2 = \begin{pmatrix} 0.46730 \\ 0.16009 \\ 0.16009 \\ 0.27718 \end{pmatrix}$$

Then A3 = A1*A2

$$A3 = \begin{pmatrix} 1.8837 \\ 0.6453 \\ 0.3847 \\ 1.1173 \end{pmatrix}$$

And A4 = A3/A2

$$A4 = \begin{pmatrix} 4.0310 \\ 4.0309 \\ 4.0312 \\ 4.0310 \end{pmatrix}$$

λmax = Average of the elements of A4

λmax = 4.0310

Now Consistency Index (CI) = (λmax – M) / (M – 1)

And Consistency Ratio (CR) = CI/RCI corresponding to M.

Where RCI = Random Consistency Index and

M = Number of elements

CI = (4.0310-4)/(4 – 1) = 0.0104

CR = 0.0104/0.9 = 0.01160

Here CR is less than 10% (0.1), so the judgment is consistent

Table 3.10: Sub-criteria quality pair-wise comparison matrix.

Sub-criteria	*DR*	*CQ*	*IP*	*QM*
DR	1	7	4	6
CQ	1/7	1	1/4	1/2
IP	1/4	4	1	2
QM	1/6	2	1/2	1

Table 3.11: Sub-criteria quality pair-wise comparison matrix (level 3) normalized

Sub-criteria	*DR*	*CQ*	*IP*	*QM*	*E-vector Wi*
DR	0.64125	0.5000	0.6957	0.6316	0.62367
CQ	0.0916	0.0714	0.0435	0.0526	0.06341
IP	0.1603	0.2857	0.1739	0.2105	0.20435
QM	0.1069	0.1429	0.0870	0.1053	0.10856

Check of the consistency

Let A1 = Pair-wise comparison matrix

A2 = Weight age matrix

$$A1 = \begin{pmatrix} 1 & 7 & 4 & 6 \\ 1/7 & 1 & 1/4 & 1/2 \\ 1/4 & 4 & 1 & 2 \\ 1/6 & 2 & 1/2 & 1 \end{pmatrix} \quad A2 = \begin{pmatrix} 0.62367 \\ 0.06341 \\ 0.20435 \\ 0.10856 \end{pmatrix}$$

Then A3 = A1*A2 $A3 = \begin{pmatrix} 2.5363 \\ 0.2579 \\ 0.8310 \\ 0.4415 \end{pmatrix}$

And A4 = A3/A2 $A4 = \begin{pmatrix} 4.0667 \\ 4.0672 \\ 4.0666 \\ 4.0669 \end{pmatrix}$

λmax = Average of the elements of A4

λmax = 4.0668

Now Consistency Index (CI) = (λmax – M) / (M – 1)

And Consistency Ratio (CR) = CI/RCI corresponding to M.

Where RCI = Random Consistency Index and

M = Number of elements

CI= (4.0668-4)/(4 – 1) = 0.0225

CR= 0.0225/0.9 = 0.0250

Here CR is less than 10% (0.1), so the judgment is consistent.

Table 3.12: Sub-criteria environment friendly practices pair-wise comparison matrix

Criteria	*ED*	*RH*	*WR*	*RR*
ED	1	1/2	1/6	1/4
RH	2	1	1/3	1/2
WR	6	3	1	2
RR	4	2	½	1

Table 3.13: Sub-criteria environment friendly practices pair-wise comparison matrix normalized.

Criteria	*ED*	*RH*	*WR*	*RR*	*E-vector Wi*
ED	0.0769	0.0769	0.0833	0.0667	0.07578
RH	0.1537	0.1537	0.1667	0.1333	0.15156
WR	0.4616	0.4616	0.5000	0.5333	0.48992
RR	0.3078	0.3078	0.2500	0.2667	0.28275

Check of the consistency

Let A1 = Pair-wise comparison matrix

A2 = Weight age matrix

$$A1 = \begin{pmatrix} 1 & 1/2 & 1/6 & 1/4 \\ 2 & 1 & 1/3 & 1/2 \\ 6 & 3 & 1 & 2 \\ 4 & 2 & 1/2 & 1 \end{pmatrix} \quad A2 = \begin{pmatrix} 0.07578 \\ 0.15156 \\ 0.48992 \\ 0.28275 \end{pmatrix}$$

Then A3 = A1*A2 $$A3 = \begin{pmatrix} 0.3039 \\ 0.6078 \\ 1.9648 \\ 1.1340 \end{pmatrix}$$

And A4 = A3/A2 $$A4 = \begin{pmatrix} 4.0103 \\ 4.0103 \\ 4.0105 \\ 4.0106 \end{pmatrix}$$

λmax = Average of the elements of A4

λmax = 4.01042

Now Consistency Index (CI) = (λmax -M) / (M – 1)

And Consistency Ratio (CR) = CI/RCI corresponding to M.

Where RCI= Random Consistency Index and

M = Number of elements

CI = (4.01042-4)/(4 – 1) = 0.0035

CR = 0.0035/0.9 = 0.00388

Here CR is less than 10% (0.1), so the judgment is consistent.

Table 3.14: Sub criteria service pair-wise comparison matrix.

Sub-criteria	*QR*	*SC*	*OD*	*AS*
QR	1	1/4	1/3	1/2
SC	4	1	2	3
OD	3	1/2	1	2
AS	2	1/3	1/2	1

Table 3.15: Sub-criteria service pair-wise comparison matrix normalized.

Sub-criteria	*QR*	*SC*	*OD*	*AS*	*E-vector Wi*
QR	0.1000	0.1200	0.0870	0.0769	0.09543
SC	0.4000	0.4800	0.5217	0.4615	0.46730
OD	0.3000	0.2400	0.2609	0.3077	0.27718
AS	0.2000	0.1600	0.1304	0.1538	0.16009

Check of the consistency

Let A1 = Pair-wise comparison matrix

A2 = Weight age matrix

$$A1 = \begin{pmatrix} 1 & 1/4 & 1/3 & 1/2 \\ 4 & 1 & 2 & 3 \\ 3 & 1/2 & 1 & 2 \\ 2 & 1/3 & 1/2 & 1 \end{pmatrix} \quad A2 = \begin{pmatrix} 0.09543 \\ 0.46730 \\ 0.27718 \\ 0.16009 \end{pmatrix}$$

Then A3 = A1*A2 A3 =

$$\begin{pmatrix} 0.3847 \\ 1.8836 \\ 1.1173 \\ 0.6453 \end{pmatrix}$$

And A4 = A3/A2 A4 =

$$\begin{pmatrix} 4.0312 \\ 4.0308 \\ 4.0310 \\ 4.0309 \end{pmatrix}$$

λmax = Average of the elements of A4

λmax = 4.0310

Now Consistency Index (CI) = (λmax – M) / (M – 1)

And Consistency Ratio (CR) = CI/RCI corresponding to M.

Where RCI = Random Consistency Index and

M = Number of elements

CI = (4.0310-4)/(4-1) = 0.0104

CR = 0.0104/0.9 = 0.01160

Here CR is less than 10% (0.1), so the judgment is consistent

3.10.4. Pair-wise Comparison Matrices of Interdependencies

Pair-wise comparisons are done to consider the interdependencies among the enablers. All such comparison is presented below. The e-vectors from these matrices are used in the formation of super matrices. As there are four determinants, 20 such matrices have been formed and shown in

Table no. 3.16 to Table no. 3.36. The e-vectors have been used in sixth column of the super matrices.

Table 3.16: Pair-wise comparison matrix for enablers under production capability & production capacity.

Pair-wise comparison matrix for enablers under production capability & production capacity

Criteria	*TC*	*FB*	*SD*	*E-vector*
TC	1	1/4	1/3	0.12196
FB	4	1	2	0.55842
SD	3	1/2	1	0.31962
		CR = 0.01759		

Table 3.17: Pair-wise comparison matrix for enablers under production capability & technical capability

Pair-wise comparison matrix for enablers under production capability & technical capability

Criteria	*PC*	*FB*	*SD*	*E-vector*
PC	1	2	2	0.49339
FB	1/2	1	2	0.31081
SD	1/2	½	1	0.19580
		CR = 0.005156		

Table 3.18: Pair-wise comparison matrix for enablers under production capability & flexibility.

Pair-wise comparison matrix for enablers under production capability & flexibility

Criteria	*PC*	*TC*	*SD*	*E-vector*
PC	1	4	2	0.55842
TC	1/4	1	1/3	0.12196
SD	1/2	3	1	0.31962
		CR = 0.01759		

Table 3.19: Pair-wise comparison matrix for enablers under production capability & standardization.

Pair-wise comparison matrix for enablers under production capability & standardization

Criteria	*PC*	*TC*	*FB*	*E-vector*
PC	1	4	1	0.45793
TC	1/4	1	1/3	0.12601
FB	1	3	1	0.41606
		CR = 0.000885		

Table 3.20: Pair-wise comparison matrix for enablers under cost & performance value.

Pair-wise comparison matrix for enablers under cost & performance value				
Criteria	***CP***	***TC***	***OC***	***E-vector***
CP	1	2	1/2	0.29696
TC	1/2	1	1/3	0.16342
OC	2	3	1	0.53961
CR = 0.000885				

Table 3.21: Pair-wise comparison matrix for enablers under cost & competitive pricing.

Pair-wise comparison matrix for enablers under cost & competitive pricing				
Criteria	***PV***	***TC***	***OC***	***E-vector***
PV	1	4	2	0.55842
TC	1/4	1	1/3	0.12196
OC	1/2	3	1	0.31962
CR = 0.01759				

Table 3.22: Pair-wise comparison matrix for enablers under cost & transport cost.

Pair-wise comparison matrix for enablers under cost & transport cost				
Criteria	***PV***	***CP***	***OC***	***E-vector***
PV	1	3	2	0.53961
CP	1/3	1	1/2	0.16342
OC	1/2	2	1	0.29696
CR = 0.000885				

Table 3.23: Pair-wise comparison matrix for enablers under cost & overhead cost.

Pair-wise comparison matrix for enablers under cost & overhead cost				
Criteria	***PV***	***CP***	***TC***	***E-vector***
PV	1	3	4	0.62501
CP	1/3	1	2	0.23849
TC	1/4	1/2	1	0.13650
CR = 0.01759				

Table 3.24: Pair-wise comparison matrix for enablers under quality & defect rate.

Pair-wise comparison matrix for enablers under quality & defect rate				
Criteria	***CQ***	***IP***	***QM***	***E-vector***
CQ	1	1/4	1/3	0.12196
IP	4	1	2	0.55842
QM	3	1/2	1	0.31962
CR = 0.01759				

Table 3.25: Pair-wise comparison matrix for enablers under quality & commitment to quality.

Pair-wise comparison matrix for enablers under quality & commitment to quality

Criteria	*DR*	*IP*	*QM*	*E-vector*
DR	1	4	6	0.70097
IP	1/4	1	2	0.19288
QM	1/6	1/2	1	0.10615
CR = 0.000885				

Table 3.25: Pair-wise comparison matrix for enablers under quality & improved process capability.

Pair-wise comparison matrix for enablers under quality & improved process capability

Criteria	*DR*	*CQ*	*QM*	*E-vector*
DR	1	7	6	0.76038
CQ	1/7	1	1/2	0.14407
QM	1/6	2	1	0.09555
CR = 0.0077				

Table 3.26: Pair-wise comparison matrix for enablers under quality & quality management.

Pair-wise comparison matrix for enablers under quality & quality management

Criteria	*DR*	*CQ*	*IP*	*E-vector*
DR	1	7	4	0.69552
CQ	1/7	1	1/4	0.07543
IP	1/4	4	1	0.22905
CR = 0.000885				

Table 3.27: Pair-wise comparison matrix for enablers under environment friendly practices & eco design.

Pair-wise comparison matrix for enablers under environment friendly practices & eco design

Criteria	*RH*	*WR*	*RR*	*E-vector*
RH	1	1/3	1/2	0.16342
WR	3	1	2	0.53961
RR	2	½	1	0.29696
CR = 0.000885				

Table 3.28: Pair-wise comparison matrix for enablers under environment friendly practices & reduction of hazardous substances.

Pair-wise comparison matrix for enablers under environment friendly practices & reduction of hazardous substances

Criteria	*ED*	*WR*	*RR*	*E-vector*
ED	1	1/6	1/4	0.08898
WR	6	1	2	0.58763
RR	4	½	1	0.32339
CR = 0.000885				

Table 3.29: Pair-wise comparison matrix for enablers under environment friendly practices & waste reduction.

Pair-wise comparison matrix for enablers under environment friendly practices & waste reduction

Criteria	*ED*	*RH*	*RR*	*E-vector*
ED	1	1/2	1/5	0.12827
RH	2	1	1/2	0.27635
RR	5	2	1	0.59538
CR = 0.00532				

Table 3.30: Pair-wise comparison matrix for enablers under environment friendly practices & reuse and recycle.

Pair-wise comparison matrix for enablers under environment friendly practices & reuse and recycle

Criteria	*ED*	*RH*	*WR*	*E-vector*
ED	1	1/2	1/7	0.08795
RH	2	1	1/3	0.24264
WR	7	3	1	0.66942
CR = 0.00532				

Table 3.31: Pair-wise comparison matrix for enablers under service & quick responsiveness.

Pair-wise comparison matrix for enablers under service & quick responsiveness

Criteria	*SC*	*OD*	*AS*	*E-vector*
SC	1	2	3	0.53961
OD	1/2	1	2	0.29696
AS	1/3	1/2	1	0.16342
CR = 0.00885				

Table 3.32: Pair-wise comparison matrix for enablers under service & supplier capacity.

Pair-wise comparison matrix for enablers under service & supplier capacity				
Criteria	***QR***	***OD***	***AS***	***E-vector***
QR	1	1/3	1/2	0.16342
OD	3	1	2	0.53961
AS	2	1/2	1	0.29696
CR = 0.01160				

Table 3.33: Pair-wise comparison matrix for enablers under service & on time delivery.

Pair-wise comparison matrix for enablers under service & on time delivery				
Criteria	***QR***	***SC***	***AS***	***E-vector***
QR	1	1/4	1/2	0.13650
SC	4	1	3	0.62501
AS	2	1/3	1	0.23849
CR = 0.01759				

Table 3.34: Pair-wise comparison matrix for enablers under service & after sales service.

Pair-wise comparison matrix for enablers under service & after sales service				
Criteria	***QR***	***SC***	***OD***	***E-vector***
QR	1	1/5	1/3	0.10945
SC	5	1	2	0.58155
OD	3	1/2	1	0.30900
CR = 0.00355				

3.10.5. Evaluation of Alternatives

The final set of pair-wise comparisons is made for the relative impact of each of the alternatives, enablers in influencing the determinants. The number of such pair-wise comparison matrices is dependent on the number of enablers that are included in each of the determinants. In our present case, there are 20 enablers for each of the determinants, which lead to 20 such pair-wise matrices. The entire pair-wise comparison matrix is shown below in Table no. 3.35 to 3.54. The e-vectors from this matrix are used in columns 6–9 of desirability indices matrices. The columns 6–9, correspond to Supplier A, Supplier B and Supplier C and Supplier D respectively.

Table 3.35: Pair-wise comparison matrix between alternatives for production capacity.

Pair-wise comparison matrix between alternatives for production capacity					
PC	***S1***	***S2***	***S3***	***S4***	***E-vector***
S1	1	½	2	½	0.18181
S2	2	1	4	3	0.47784
S3	1/3	¼	1	½	0.10518
S4	2	3	2	1	023517
CR = 0.04417					

Table 3.36: Pair-wise comparison matrix between alternatives for technical capacity.

Pair-wise comparison matrix between alternatives for technical capacity					
TC	***S1***	***S2***	***S3***	***S4***	***E-vector***
S1	1	½	½	1/3	0.12232
S2	2	1	1	½	0.22704
S3	2	1	1	½	0.22704
S4	3	2	2	1	0.42359
CR = 0.0388					

Table 3.37: Pair-wise comparison matrix between alternatives for flexibility.

Pair-wise comparison matrix between alternatives for flexibility					
FB	***S1***	***S2***	***S3***	***S4***	***E-vector***
S1	1	3	5	1	0.39362
S2	1/3	1	2	1/3	0.13747
S3	1/5	1/2	1	1/5	0.07529
S4	1	3	5	1	0.39362
CR = 0.00156					

Table 3.38: Pair-wise comparison matrix between alternatives for standardization.

Pair-wise comparison matrix between alternatives for standardization					
SD	***S1***	***S2***	***S3***	***S4***	***E-vector***
S1	1	5	7	3	0.56501
S2	1/5	1	3	1/3	0.11750
S3	1/7	1/3	1	1/5	0.05529
S4	1/3	3	5	1	0.26220
CR = 0.04381					

Table 3.39: Pair-wise comparison matrix between alternatives for performance value.

Pair-wise comparison matrix between alternatives for performance value

PV	*S1*	*S2*	*S3*	*S4*	*E-vector*
S1	1	1/3	1/7	1/5	0.05529
S2	3	1	1/5	1/3	0.11750
S3	7	5	1	3	0.56501
S4	5	3	1/3	1	0.26220
			CR = 0.04381		

Table 3.40: Pair-wise comparison matrix between alternatives for competitive pricing.

Pair-wise comparison matrix between alternatives for competitive pricing

CP	*S1*	*S2*	*S3*	*S4*	*E-vector*
S1	1	1/3	2	1/5	0.10552
S2	3	1	5	1/3	0.26427
S3	½	1/5	1	1/7	0.06092
S4	5	3	7	1	0.56929
			CR = 0.02547		

Table 3.41: Pair-wise comparison matrix between alternatives for transport cost.

Pair-wise comparison matrix between alternatives for transport cost

TC	*S1*	*S2*	*S3*	*S4*	*E-vector*
S1	1	3	5	1	0.39362
S2	1/3	1	2	1/3	0.13747
S3	1/5	1/2	1	1/5	0.07529
S4	1	3	5	1	0.39362
			CR = 0.00156		

Table 3.42: Pair-wise comparison matrix between alternatives for overhead cost.

Pair-wise comparison matrix between alternatives for overhead cost

OC	*S1*	*S2*	*S3*	*S4*	*E-vector*
S1	1	3	7	5	0.56501
S2	1/3	1	5	3	0.26220
S3	1/7	1/5	1	1/3	0.05529
S4	1/5	1/3	3	1	0.11750
			CR = 0.04381		

Table 3.43: Pair-wise comparison matrix between alternatives for defect rate.

Pair-wise comparison matrix between alternatives for defect rate

DR	*S1*	*S2*	*S3*	*S4*	*E-vector*
S1	1	3	5	1	0.39362
S2	1/3	1	2	1/3	0.13747
S3	1/5	½	1	1/5	0.07529
S4	1	3	5	1	0.39362
CR = 0.04381					

Table 3.44: Pair-wise comparison matrix between alternatives for commitment to quality.

Pair-wise comparison matrix between alternatives for commitment to quality

CQ	*S1*	*S2*	*S3*	*S4*	*E-vector*
S1	1	1/3	1/7	1/7	0.05060
S2	3	1	1/5	1/5	0.10438
S3	7	5	1	1	0.42251
S4	7	5	1	1	0.42251
CR = 0.02752					

Table 3.45: Pair-wise comparison matrix between alternatives for improved process capability.

Pair-wise comparison matrix between alternatives for improved process capability

IP	*S1*	*S2*	*S3*	*S4*	*E-vector*
S1	1	1/3	1/7	1/5	0.05529
S2	3	1	1/5	1/3	0.11750
S3	7	5	1	3	0.56501
S4	5	3	1/3	1	0.26220
CR = 0.04381					

Table 3.46: Pair-wise comparison matrix between alternatives for quality management.

Pair-wise comparison matrix between alternatives for quality management

QM	*S1*	*S2*	*S3*	*S4*	*E-vector*
S1	1	5	7	3	0.56501
S2	1/5	1	3	1/3	0.11750
S3	1/7	1/3	1	1/5	0.05529
S4	1/3	3	5	1	0.26220
CR = 0.04381					

Table 3.47: Pair-wise comparison matrix between alternatives for eco design.

Pair-wise comparison matrix between alternatives for eco design

ED	*S1*	*S2*	*S3*	*S4*	*E-vector*
S1	1	3	7	3	0.52812
S2	1/3	1	5	1	0.21000
S3	1/7	1/5	1	1/5	0.05188
S4	1/3	1	5	1	0.21000
			CR = 0.02752		

Table 3.48: Pair-wise comparison matrix between alternatives for RoHS

Pair-wise comparison matrix between alternatives for RoHS

RH	*S1*	*S2*	*S3*	*S4*	*E-vector*
S1	1	1/3	5	3	0.26220
S2	3	1	7	5	0.56501
S3	1/5	1/7	1	1/3	0.05529
S4	1/3	1/5	3	1	0.11750
			CR = 0.04381		

Table 3.49: Pair-wise comparison matrix between alternatives for waste reduction.

Pair-wise comparison matrix between alternatives for waste reduction

WR	*S1*	*S2*	*S3*	*S4*	*E-vector*
S1	1	3	5	1	0.39362
S2	1/3	1	2	1/3	0.13747
S3	1/5	½	1	1/5	0.07529
S4	1	3	5	1	0.39362
			CR = 0.00156		

Table 3.50: Pair-wise comparison matrix between alternatives for reuse and recycle.

Pair-wise comparison matrix between alternatives for reuse and recycle

RR	*S1*	*S2*	*S3*	*S4*	*E-vector*
S1	1	5	7	3	0.56501
S2	1/5	1	3	1/3	0.11750
S3	1/7	1/3	1	1/5	0.05529
S4	1/3	3	5	1	0.26220
			CR = 0.04381		

Table 3.51: Pair-wise comparison matrix between alternatives for quick responsiveness.

Pair-wise comparison matrix between alternatives for quick responsiveness

QR	*S1*	*S2*	*S3*	*S4*	*E-vector*
S1	1	3	7	3	0.52812
S2	1/3	1	5	1	0.21000
S3	1/7	1/5	1	1/5	0.05188
S4	1/3	1	5	1	0.21000
CR = 0.02752					

Table 3.52: Pair-wise comparison matrix between alternatives for supplier capacity.

Pair-wise comparison matrix between alternatives for supplier capacity

SC	*S1*	*S2*	*S3*	*S4*	*E-vector*
S1	1	1/3	1/7	1/7	0.05060
S2	3	1	1/5	1/5	0.10438
S3	7	5	1	1	0.42251
S4	7	5	1	1	0.42251
CR = 0.02752					

Table 3.53: Pair-wise comparison matrix between alternatives for on time delivery.

Pair-wise comparison matrix between alternatives for on time delivery

OD	*S1*	*S2*	*S3*	*S4*	*E-vector*
S1	1	3	5	1	0.39362
S2	1/3	1	2	1/3	0.13747
S3	1/5	½	1	1/5	0.07529
S4	1	3	5	1	0.39362
CR = 0.00156					

Table 3.54: Pair-wise comparison matrix between alternatives for after sales service.

Pair-wise comparison matrix between alternatives for after sales service

AS	*S1*	*S2*	*S3*	*S4*	*E-vector*
S1	1	½	½	1/3	0.12232
S2	2	1	1	½	0.22704
S3	2	1	1	½	0.22704
S4	3	2	2	1	0.42359
CR = 0.0388					

3.10.6. Super Matrix Formation

The super matrix allows for a resolution of the interdependencies that exist among the elements of a system. It is a partitioned matrix where each sub-matrix is composed of a set of relationships between and within the levels as represented by the decision maker's model. The values of the elements of the super matrix M have been imported from the pair-wise comparison matrices of interdependencies. Super matrices 'M', shown below presents the results of the relative importance measures for the enablers of the Green Supplier Index. The Super-matrix for the green Supplier is shown in Table 3.55.

In the next stage, the super matrix M is made to converge to obtain a long-term stable set of weights. For convergence to occur, super matrix needs to be 'column stochastic', *i.e.*, the sum of each of the columns of the super matrix needs to be one. Raising the super matrix M to the power 2k+1, where k is an arbitrarily large number, allows for the convergence of the interdependent relationships the converged super matrix is shown in Table 3.56.

3.10.7. Selection of the Best Alternative for a Determinant

The selection of the best alternative depends on the outcome of the 'desirability index'. The desirability index, Dia, for the alternative i and the determinant a is defined as.

$$D_{ia} = \sum_{j=1}^{j} \sum_{k=1}^{Kja} P_{ja}\, A^{D}_{kja}\, A^{I}_{kja}\, S_{ikja} \qquad (4)$$

Where

P_{ja} is the relative importance weight of dimension of green supplier j on the determinant of green supplier a,

A^{D}_{kja} is the relative importance weight for reverse logistics attribute enabler k of dimension of green supplier j in the determinant of reverse logistics control hierarchy network a for the dependency (D) relationships between component levels,

A^{I}_{kja} is the stabilized relative importance weight (determined by the super matrix) for green supplier attribute enabler k of dimension of green supplier j in the determinant of green supplier control hierarchy network a for interdependency (I) relationships within the green supplier enablers' component level,

Table 3.55: Super matrix M for green supplier before convergence.

	Super matrix M for green supplier before convergence																			
	PC	*TC*	*FB*	*SD*	*PV*	*CP*	*TC*	*OC*	*DR*	*CQ*	*IP*	*QM*	*ED*	*RH*	*WR*	*RR*	*QR*	*SC*	*OD*	*AS*
PC	0	0.4933	0.5584	0.4579																
TC	0.1219	0	0.1219	0.1260																
FB	0.5584	0.3108	0	0.4160																
SD	0.3196	0.1958	0.3196	0																
PV					0	0.5584	0.5396	0.6250												
CP					0.2969	0	0.1634	0.2384												
TC					0.1634	0.1219	0	0.1365												
OC					0.5396	0.3196	0.2969	0												
DR									0	0.7009	0.7603	0.6552								
CQ									0.1219	0	0.1440	0.0754								
IP									0.5584	0.1928	0	0.2290								
QM									0.3196	0.1061	0.0955	0								
ED										0	0.0889	0.1282	0.0879							
RH										0.1634	0	0.2763	0.2426							
WR													0.5396	0.5876	0	0.6942				
RR													0.2969	0.3233	0.5954	0				
QR																	0	0.1634	0.1365	0.1094
SC																	0.5396	0	0.6250	0.5815
OD																	0.2969	0.5396	0	0.3090
AS																	0.1634	0.2969	0.2384	0

Table 3.56: Super matrix M for green supplier after convergence.

	Super matrix M for green supplier after convergence M35																			
	PC	***TC***	***FB***	***SD***	***PV***	***CP***	***TC***	***OC***	***DR***	***CQ***	***IP***	***QM***	***ED***	***RH***	***WR***	***RR***	***QR***	***SC***	***OD***	***AS***
PC	0.3388	0.3388	0.3388	0.3388																
TC	0.1095	0.1095	0.1095	0.1095																
FB	0.3197	0.3197	0.3197	0.3197																
SD	0.2319	0.2319	0.2319	0.2319																
PV					0.3696	0.3696	0.3696	0.3696												
CP					0.2023	0.2023	0.2023	0.2023												
TC					0.1263	0.1263	0.1263	0.1263												
OC					0.3016	0.3016	0.3016	0.3016												
DR									0.4219	0.4219	0.4219	0.4219								
CQ									0.1073	0.1073	0.1073	0.1073								
IP									0.2963	0.2963	0.2963	0.2963								
QM									0.1745	0.1745	0.1745	0.1745								
ED													0.0953	0.0953	0.0953	0.0953				
RH													0.1996	0.1996	0.1996	0.1996				
WR													0.3838	0.3838	0.3838	0.3838				
RR													0.3214	0.3214	0.3214	0.3214				
QR																	0.1241	0.1241	0.1241	0.1241
SC																	0.3726	0.3726	0.3726	0.3726
OD																	0.3005	0.3005	0.3005	0.3005
AS																	0.2026	0.2026	0.2026	0.2026

Table 3.57: Desirability Indices for green supplier.

						Green supplier performance							
Dimensions	***Enablers***	P_{ja}	A_{kja}^D	A_{kja}^I	$P_{ja}*A_{kja}^D*A_{kja}^I$	S_{1kja}	S_{2kja}	S_{3kja}	S_{4kja}	***Supplier S1***	***Supplier S2***	***Supplier S3***	***Supplier S4***
		1	2	3	4=1*2*3	5	6	7	8	9=4*5	10=4*6	11=4*7	12=4*8
PDC	PC	0.06114	0.35928	0.3388	0.0013531	0.18181	0.47784	0.10518	0.23517	0.0013531	0.0035562	0.000783	0.00175
PDC	TC	0.06114	0.08153	0.1095	0.0000668	0.12232	0.22704	0.22704	0.42359	0.0000668	0.0001239	0.000124	0.000231
PDC	FB	0.06114	0.35928	0.3197	0.0027643	0.39362	0.13747	0.07529	0.39362	0.0027643	0.0009654	0.000529	0.002764
PDC	SD	0.06114	0.19992	0.2319	0.0016015	0.56501	0.1175	0.05529	0.2622	0.0016015	0.0003331	0.000157	0.000743
C	PV	0.03760	0.46730	0.3696	0.0003591	0.05529	0.1175	0.56501	0.2622	0.0003591	0.0007631	0.003669	0.001703
C	CP	0.03760	0.16009	0.2023	0.0001285	0.10552	0.26427	0.06092	0.56929	0.0001285	0.0003218	0.000074	0.000693
C	TC	0.03760	0.09543	0.1263	0.0001784	0.39362	0.13747	0.07529	0.39362	0.0001784	0.0000623	0.000034	0.000178
C	OC	0.03760	0.27718	0.3016	0.0017760	0.56501	0.2622	0.05529	0.1175	0.0017760	0.0008242	0.000174	0.000369
Q	DR	0.47382	0.62367	0.4219	0.0490744	0.39362	0.13747	0.07529	0.39362	0.0490744	0.0171390	0.009387	0.049074
Q	CQ	0.47382	0.06341	0.1073	0.0001631	0.0506	0.10438	0.42251	0.42251	0.0001631	0.0003365	0.001362	0.001362
Q	IP	0.47382	0.20435	0.2963	0.0015862	0.05529	0.1175	0.56501	0.2622	0.0015862	0.0033710	0.016210	0.007522
Q	QM	0.47382	0.10856	0.1745	0.0050715	0.56501	0.1175	0.05529	0.2622	0.0050715	0.0010547	0.000496	0.002353
EFP	ED	0.27353	0.07578	0.0953	0.0010432	0.52812	0.21	0.05188	0.21	0.0010432	0.0004148	0.000102	0.000415
EFP	RH	0.27353	0.15156	0.1996	0.0021696	0.2622	0.56501	0.05529	0.1175	0.0021696	0.0046753	0.000458	0.000972
EFP	WR	0.27353	0.48992	0.3838	0.0202447	0.39362	0.13747	0.07529	0.39362	0.0202447	0.0070704	0.003872	0.020245
EFP	RR	0.27353	0.28275	0.3214	0.0140446	0.56501	0.1175	0.05529	0.2622	0.0140446	0.0029207	0.001374	0.006518
S	QR	0.15391	0.09543	0.1241	0.0009626	0.52812	0.2100	0.05188	0.2100	0.0009626	0.0003828	0.000095	0.000383
S	SC	0.15391	0.46730	0.3726	0.0013560	0.0506	0.10438	0.42251	0.42251	0.0013560	0.0027972	0.011323	0.011323
S	OD	0.15391	0.27718	0.3005	0.0050460	0.39362	0.13747	0.07529	0.39362	0.0050460	0.0017623	0.000965	0.005046
S	AS	0.15391	0.16009	0.2026	0.0006106	0.12232	0.22704	0.22704	0.42359	0.0006106	0.0011334	0.001133	0.002115
								Desirability Indices D_N		0.10294	0.06032	0.05925	0.109674

S_{ikja} is the relative impact of reverse logistics implementation alternative i on green supplier attribute enabler k of dimension of green supplier j of reverse logistics control hierarchy network a,

K_{ja} is the index set of green supplier attribute enablers for dimension of reverse logistics j in for green supplier determinant control hierarchy a, and J is the index set for the dimensions of green supplier (same for all control hierarchies).

Table 3.57 presents the desirability indices for the Green Supplier Performance Index. It is based on the hierarchy using the relative weights obtained from the pair-wise comparison of alternatives, dimensions and weights of enablers from the converged super matrix.

Table 3.58: Green Supplier Overall Performance Index

Suppliers	*Supplier A*	*Supplier B*	*Supplier C*	*Supplier D*
GSPI	0.10294	0.06032	0.05928	0.10967
GSPI normalized	0.30989	0.18159	0.17836	0.330155

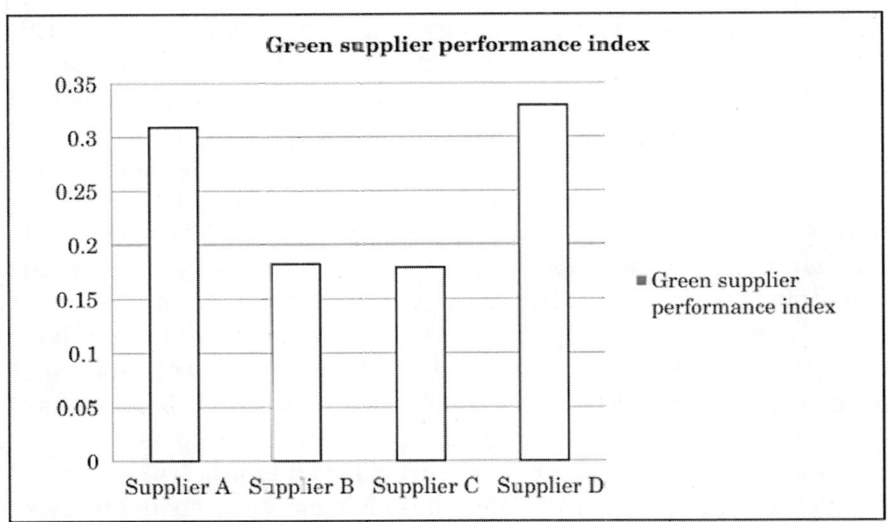

Fig. 3.5: Green supplier overall performance index.

3.11. DISCUSSION AND MANAGERIAL IMPLICATIONS

At the end of this section and in the conclusion, we first discuss the results of the ANP approach. Later, we offer some suggestions for the prospective users of this model. Finally, we discuss the limitations of the ANP approach and directions for further research. The proposed methodology provides for simplification of a complex multi-criteria decision-making problem. It may also be used to quantify many subjective judgments, which are necessary to evaluate different alternative providers. Another advantage of this methodology is that it not only supports group decision-making but also enables us to document the various considerations in the process of decision making. This documentation is useful if the results are to be communicated to various interest groups. For the example undertaken in this study, the results indicate that supplier D is the first choice of the case company and illustrated in Fig. 3.4 and in Table 3.58. This may be attributed to its , Competitive Pricing, low defect rate and waste reduction capiblities. The expertise of Supplier A in the framing of overhead cost and reuse and recycle also supports this result. It is pertinent here to discuss the priority values of the determinants, which influence this decision. From Table 3.4 , it is observed that Quality (Q = 0.47382) is the most important crteria in the selection of a supplier. It is followed by Environmently Friendly servicess (0.27353), Service (0.15391) and Production Capicity (0.06114).

Although the model has been illustrated for four distinct alternative suppliers, it is capable of comparing more than four Suppliers at the cost of complexity. It needs to be emphasized here that despite using a sound algorithm for systematic decision-making, care must be taken in the application of the ANP approach. For example, in its application, the user has to compare the prospective suppliers on a number of pair-wise comparison matrices. In these comparisons, the user must verify the capabilities of the supplier and should not solely rely on the information given by the prospective suppliers. Experts recommend that the user companies should evaluate the supplier by what they have done and not by what they plan to do. Although in this case the input to the comparison matrices is based on the responses to RFI and visits of the material manager to the sites of the supplier, the biasing of the decision-maker towards a particular provider cannot be ruled out. To avoid such situations, group decision-making techniques should be used.

For example, brainstorming and sharing of ideas and insights often lead to a better understanding of the issues than would be possible for a single decision-maker. Scenario building or the Delphi method may also be used for the pair-wise comparisons. In the case of a group decisionmaking process, consensus may be reached by agreeing on the geometric means of individual judgments. In the absence of consensus, voting may also be conducted to arrive at a more acceptable value. Compared to low-level enablers, consensus is more desirable for determinants and dimension as the higher level of the ANP model. This is because of the higher global weights of determinants and dimensions in the ANP model for the calculation of GSPI. Use of software and decision support systems may also reduce the complexities in implementing the group decision-making.

In the light of the results obtained for the case company, it may be noted that these results are valid only for the case company in its own decision environment and should not be generalized to establish the supremacy of one provider over the others. Further, the application of proposed methodology may require significant time and resources from managers and decision-makers. Yet, when seeking to invest in a longterm supplier Sourcing contract that can potentially reach a huge amount, a structured analysis, which is provided by this methodology, may help to reduce the risk of poor decisions.

3.12. CONCLUSIONS

In this chapter, an enterprise's selection of a Green Supplier is presented. In modern era of industrialization and globalization, every organization must ensure that their product must meet the international Environmental standards and quality requirements to remain competent in this rapidly changing industrial environment. To achieve this, it is a necessary requirement that supplier from which the organization is getting the supply of raw material or any other kind of necessary inputs should be selected correctly as per the Green requirements. We argue that the selection of Green supplier (s) should not be broken down into a set of stand-alone selections, the approach traditionally pursued. The selection should be evaluated holistically at the supply chain level, both strategically and tactically, thus requiring an approach that is capable of addressing many interdependent and intangible elements. The thesis also provides for a review of the issues, which influence the selection of a green supplier. The ANP approach, as a part of this methodology, not only leads to a logical result but also enables the decision-makers to visualize the impact of various criteria in the final result. Further, It have been demonstrated that the interdependencies among various criteria can be effectively captured using the ANP technique, which has rarely been applied in the context of supplier selections.

The proposed methodology serves as a guideline to the material managers in supplier related decisions. The ANP approach is capable of taking into consideration both qualitative and quantitative criteria. Similar ANP-based models may also be developed in other contexts as well. But, as the development and evaluation of these models demand significant time and efforts from the decision-makers in the formation of pair-wise comparison matrices, these should be used for long-term strategic decisions only where the investments made in the lengthy and cumbersome process of decisionmaking are recovered in due course of time. Further, though the technique is computationally intensive, the benefits of risk reduction will outweigh the cost and time.

Using this formulation, decision makers can easily test a number of what-if scenarios. For future research, it would be worthwhile to implement the ANP model with a decision maker or a set of decision makers. Such research endeavour could be used to validate the effectiveness of the ANP model. More importantly, managerial implications can be empirically derived regarding the selection of green Suppliers. Such research should include a comprehensive sensitivity analysis to examine the significance of individual attributes to the selection decision. It is also worthwhile to compare the proposed ANP approach with other evaluation approaches. Here ANP approach illustrated in this paper has a few limitations as well. For example, the model result efficiency is dependent on the inputs provided by the material manager of the case company. The possibility of bias of the decision-maker towards any particular provider cannot be ruled out while applying this model. Therefore, group decisions should be preferred in the pair-wise comparison. Moreover, the formation of pair-wise comparison matrices is a time-consuming and tedious task. Inconsistency and human error may also occur in calculating the pair-wise comparison of matrices, which may give wrong results.

4

Modeling the Barriers of Green Supply Chain Practices: An Interpretive Structural Modeling (ISM) Approach

4.1. INTRODUCTION

A green supply chains aims at confining the wastes within the industrial system to conserve energy and prevent the dissipation of dangerous materials into the environment. It recognizes the disproportionate environmental impact of supply chain processes within an organization. It recognizes the disproportionate environmental impact of supply chain processes within an organization. GSCM is the summing up of green purchasing, green manufacturing, green packing, green distribution and marketing. GSCM is to eliminate or minimize waste in the form of energy, emission, hazardous, chemical and solid waste. Concepts and models related to environmental issues have been suggested by different researchers. Some of them have been described. Interpretive structural modeling (ISM) methodology was utilized to understand the mutual influences among the barriers so that those driving barriers, which can aggravate few more barriers and those independent barriers, which are mostly influenced by driving barriers are identified (Shankar *et al.*, 2005) GSCM practices adopted by the electrical and electronic industry in Taiwan were investigated, which was dominated by original equipment manufacturing and original designing and manufacturing. The data were analyzed by using statistical package and the structural equation modeling.

An increased attention for developing environmental management (EM) strategies for the supply chain was suggested. This study analyzed the interaction of criteria that was used to select the green suppliers based upon the environmental performance using ISM and Analytic Hierarchy Process (Saaty, 1990). ISM based model for greening the supply chain in Indian manufacturing industries was explained. Fifteen enablers were identified. A questionnaire-based survey was conducted to rank these enablers. They found contextual relationships among enablers and developed hierarchy-based model for the enablers by using ISM. ISM based model for modeling the barriers of green supply chain Practices in Indian manufacturing industries was put forward. They suggested green businesses practices are not easy to adopt and implement due to the presence of many barriers. A questionnaire-based survey was conducted to analyze and rank these barriers. Fifteen barriers were identified. ISM approach has been used to model and analyze key barriers. Importance of GSCM and factors important to implement GSCM in Indian auto component manufacturing industry was identified and described (Jayant and Azhar, 2014).

Interpretive Structural Modeling (ISM) is a methodology used to identify relationship among specific elements, which define a problem or issue. ISM is an interactive learning process in which a set of dissimilar and directly related elements are structured into a comprehensive systematic model. The model so formed, portrays the structure of a complex issue or problem, a system or a field of study, in a carefully designed pattern implying graphics as well as words. The basic idea of ISM is to use experts' practical experience and knowledge to decompose a complicated system into several sub-systems (elements) and construct a multilevel structural model; it was firstly developed in 1970's.

4.2. OBJECTIVES AND ISSUES

In modern era of industrialization and globalization, every industry must ensure that their product must meet the international standards and quality requirements to remain competent in this rapidly changing industrial environment. To achieve this, it is required that industry must have green product and have green supply chain management. Another condition to achieve this is that the supplier/vendor should be environment conscious.

4.3. PROBLEM DESCRIPTION

According to world statistics, the automobile industry is world's largest single manufacturing sector. The growth in the world's population has also heightened the demand for the vehicles. Increasing trend of demand of automobiles such as cars, bikes and commercial vehicles in India has been noticed in last few years, therefore leading international and domestic automobile manufacturers (like Maruti Suzuki, Hyundai, Tata Motors, General Motors, Honda, Fiat, Bajaj Auto, Hero Honda etc.) are either setting up their new manufacturing plants or increasing their production capacity in their existing plants in India. We have identified various barriers to implement GSCM in Indian industry from the literature reviews and expert opinions. Literature was reviewed to identify barriers to implement GSCM in Indian auto component manufacturing industry. We conducted a brainstorming session, in which two experts from academia and three experts from industry were invited. Three were from industry and two was from academia. Brainstorming session was conducted and twenty barriers relevant to Indian industry were identified. These barriers to implementation of GSCM in Indian auto component manufacturing industry are discussed in Table 4.1.

Table 4.1: Description of green supply chain management barriers.

Barriers	***Description***	***Sources***
1. Cost implication	Cost has been used as the prime performance measure. Usually, high cost is a big pressure in GSCM as compared to conventional SCM. The initial investment requirement by green methodologies such as green design, green manufacturing, green labeling of packing etc. are too high.	Khidir & Zailani, 2009
2. Lack of IT applications	IT systems support collaborative supply chain processes and enhance supply chain performance. An efficient information and technology system is very necessary for supporting the GSCM during various stages of product life cycle. It can be very useful for product development programs encompassing the design for the environment, recovery and reuse.	Shankar *et. al.*, 2005
3. Poor organizational culture in adopting GSCM	It directs towards the participation of top level management in motivating the employee.	Hsu & Hu, 2008
4. Lack of Top management commitment in adopting GSCM	Resistance of the top management of an organization to bring a change to the existing investments, information systems and habits and hence a change to a new supply chain system. Top managements particularly of SMEs, differ from the management behaviors found on large publicly possessed business because they reply to different stakeholders and have different experiences and capabilities.	Zhu & Sarkis, 2007
5. Resistance to advance technology adoption	Technology is a kind of knowledge. An organization with rich experiences in the application and adoption of related technologies will have higher ability in technological innovation. An organization	Gant R. M., 1996

Table 4.1 (*Contd...*)

Table 4.1: (*Contd...*)

Barriers	***Description***	***Sources***
	will have higher innovative capability when knowledge can be shared more easily within the organization.	
6. Lack of government support to adopt GSCM	Legislation and regulation are the instruments very much necessary for the proper governance of business enterprises including the environment in which they operate. Environmental laws and regulations are an important framework, within which the companies must operate.	Porter & Van de Linde, 1995
7. Lack of knowledge about green practice	Lack of Innovative Green Practices means not taking consideration of practices like hazardous solid waste disposal, energy conservation, reusing and recycling materials etc.	Shankar *et. al.*, 2005
8. Lack of Technical expertise	Inability to find an alternative way to design a pollution free product to fulfill environmental requirements. Lack of financial resources for equipment and other Investments (especially long term). Technical support is not updated within industry.	Revell & Ruther Ford, 2003
9. Market competition	The external environment in which a firm conducts its business will also influence the innovative capability as well as intention to adopt innovations. We assume that market competition and uncertainty is most important barrier to achieve GSCM in Indian auto component manufacturing industry.	Yu Lin, C., 2007
10. Less awareness of customer about GSCM	The customer is the essence of any business. Businesses must design and manufacture products and provide services that meet customers' needs and expectations. Environmental consciousness of consumers is one of the most significant driving forces for companies to engage in environmental management.	Sharma *et. al.*, 2012
11. Lack of environmental awareness to the supplier	Industries are unable to maintain the environmental conscious suppliers and suppliers also are concerned to maintain the environmental concepts in their industries.	Gibbon, P., 1997
12. Fear of failure	Involves the fear of failure while adopting green supply chain which could lead to monetary losses for the firm or the fear of failure of the product, hence leading to losing the competitive advantage	Perron G.M., 2005
13.Pollution/Wastage of industries	Industrial pollution and waste encompass the full range of unwanted substances and losses generated by industrial activities.	Self-contributed barrier
14. Non-availability of bank loans to encourage green product	Industries are struggling to get bank loans for environmental initiatives in their industries.	Self-contributed barrier

Table 4.1 (*Contd...*)

Table 4.1: (*Contd...*)

Barriers	***Description***	***Sources***
15. Lack of training courses about implementing GSCM	Industrial professionals need training to adopt GSCM system in their industries and need training to maintain and monitor growth	Sharma *et. al.*, 2012
16. Lack of recycling and reuse efforts of organization	Inability of technology and knowledge to design the reuse and recycle of used products	Self-contributed barrier
17. Lack of sustainability certification (ISO 14001)	It refers to authenticity of quality of products and services as per pre-established norms.	Sharma *et. al.*, 2012
18. Cost of disposal of hazardous products	Habitually disposing of hazardous waste is very difficult and needs high cost	Self-contributed barrier
19. Lack of awareness about reverse logistics adoption	Adoption of reverse logistics practices is critical for organizations from both economic and environmental perspectives	Shankar *et. al.*, 2005
20. Lack of corporate social responsibility	Corporate social responsibility suggests that a firm is willing to go beyond simple compliance and consider the public consequences of organizational actions	Mudgal *et. al.*, 2010 & Sharma S., 2000

4.4. INTERPRETIVE STRUCTURAL MODELING (ISM) TECHNIQUE

Warfield first proposes ISM in 1973. J.N. Warfield has developed a powerful methodology for structuring complex issues. Drawing upon discrete or finite mathematics, Warfield has produced a mathematical language applicable to many complex issues, if they can be analyzed in terms of sets of elements and relations. "Interpretive structural modeling" (ISM) is used here to refer to the systematic application of some elementary notions of graph theory in such a way that theoretical, conceptual, and computational leverage is exploited to efficiently construct a directed graph, or network representation, of the complex pattern of a contextual relationship among a set of elements.

ISM is an interactive learning process. In this technique, a set of different directly and indirectly related elements are structured into a comprehensive systematic model. The model so formed portrays the structure of a complex issue or problem in a carefully designed pattern implying graphics as well as words. Interpretive structural modeling (ISM) is a well-established methodology for identifying relationships among specific items, which define a problem or an issue. For any complex problem under consideration, good number of factors may be related to an issue or problem. However, the direct and indirect relationships between the factors describe the situation far more accurately than the individual factor taken into isolation. Therefore, ISM develops insights into collective understandings of these relationships. ISM is a systematic application of some elementary graph theory in such a way that theoretical, conceptual and computational advantage are exploited to explain the complex pattern of conceptual relations among the variables. The various steps for adopting the ISM methodology are extracted from and their logical flows are shown in Fig. 4.1.

4.5. STEPS FOR ISM TECHNIQUE

Step 1. Variables (criteria) considered for the system under consideration are listed.

Step 2. From the variables identified in step 1, a contextual relationship is established among the variables to identify as to which pairs of variables should be examined.

Step 3. A structural self-interaction matrix (SSIM) is developed for variables, which indicates pair-wise relationships among variables of the system under consideration.

Step 4. Reachability matrix is developed from the SSIM and the matrix is checked for transitivity. The transitivity of the contextual relation is a basic assumption made in ISM. It states that if a variable A is related to B and B is related to C, then A is necessarily related to C.

Step 5. The reachability matrix obtained in step 4 is partitioned into different levels.

Step 6. Based on the relationships given above in the reachability matrix, a directed graph is drawn, and the transitive links are removed.

Step 7. The resultant digraph is converted into an ISM, by replacing variable nodes with statements.

Step 8. The ISM model developed in step 7 is reviewed to check for conceptual inconsistency and necessary modifications are made.

The above steps are shown in Fig. 4.1.

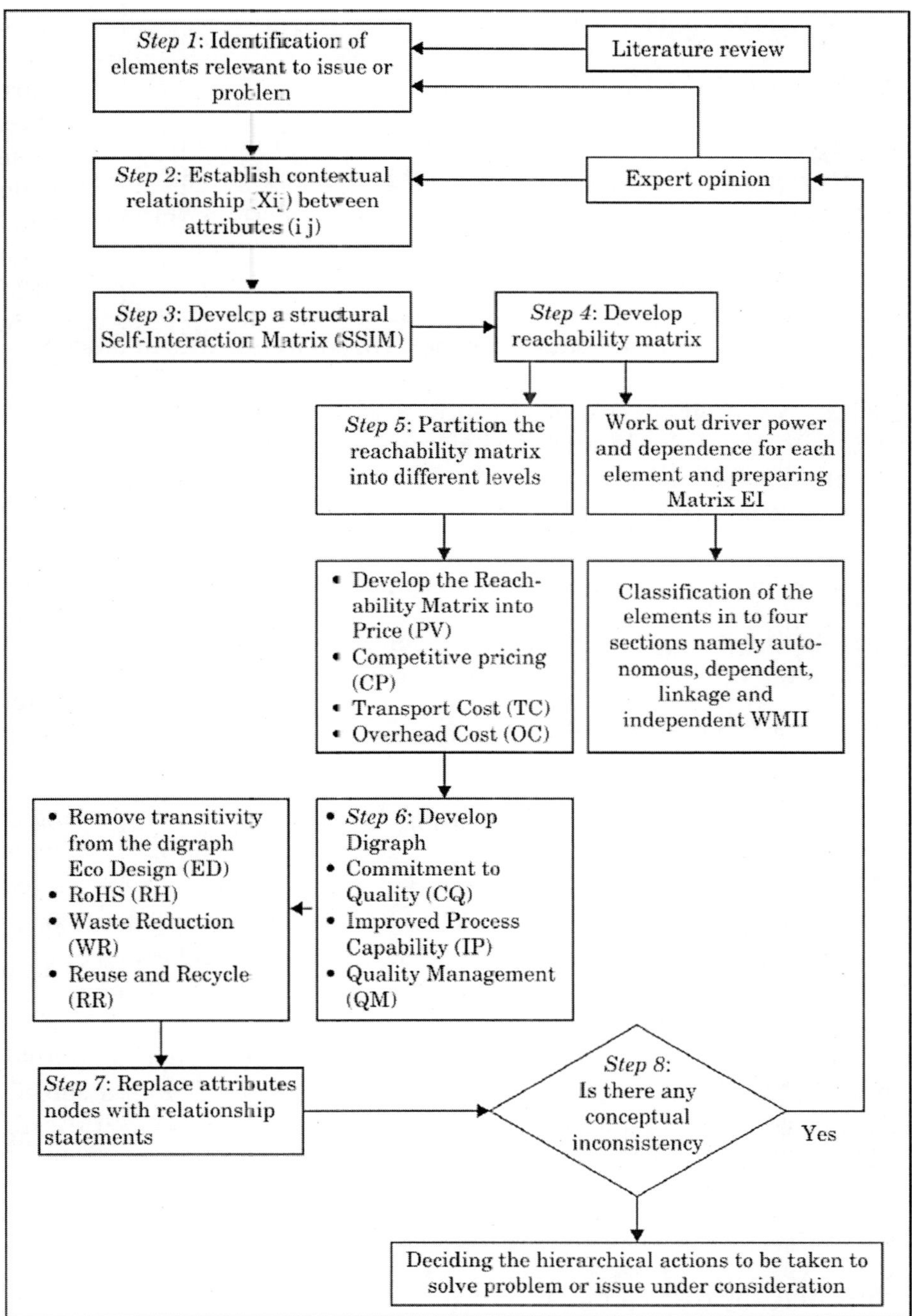

Fig. 4.1: Flow diagram for preparing model by using ISM.

4.6. APPLICATION OF ISM

4.6.1. Data Collection

To analyze the barriers for the adoption of GSCM in industry, twenty barriers were considered. From the literature sixteen barriers were taken and after discussion with industrial experts four barriers were included. The twenty barriers which are considered are main barrier for implementing GSCM in auto component manufacturing industries. These experts from industry were senior managers in manufacturing and procurement departments and experts in environmental in industry. Total experts were five, two from academics and three from the industry. As discussed in above, experts from industry and the academia were consulted during the brainstorming session to identify the nature of contextual relationships among the barriers to implement GSCM in Indian auto component manufacturing industry.

4.6.2. Structural Self-interaction Matrix (SSIM)

In developing SSIM, following four symbols have been used to denote the direction of relationship between two barriers i and j.

V- Barrier i will lead to barrier j;

A- Barrier j will lead to barrier i;

X- Barrier i and j will lead to each other;

O- Barrier i and j are unrelated

Based on the contextual relationships, the SSIM has been developed (Table 4.2). Barrier 1 leads to barrier 7 so symbol 'V' has been given in the cell (1,7); barrier 11 leads to barrier 2 so symbol 'A' has been given in the cell (2, 11); barrier 5 and 8 lead to each other so symbol 'X' has been given in the cell (5,8); barrier 3 and 14 do not lead to each other so symbol 'O' has been given in the cell (3,14) and so on. The number of pair wise comparison question addressed for developing the SSIM are ((N)* (N-1)/2), where N is the number of barriers.

4.6.3. Initial Reachability Matrix

In this step, the reachability matrix is developed from SSIM. The SSIM format is initially converted into an initial reachability matrix format by transforming the information of each cell of SSIM into binary digits (*i.e.,* ones or zeros) in the initial reachability matrix by substituting V,

A, X, O by 1 or 0 applying following rules:

- If (i, j) value in the SSIM is V, (i, j) value in the reachability matrix will be 1 and (j, i) value will be 0; for V (1, 7) in SSIM, '1' has been given in cell (1, 7) and '0' in cell (7, 1) in initial reachability matrix.
- If (i, j) value in the SSIM is A, (i, j) value in the reachability matrix will be 0 and (j, i) value will be 1; for A (2, 11) in SSIM, '0' has been given in cell (2, 11) and '1' in cell (11, 2) in initial reachability matrix.
- If (i, j) value in the SSIM is X, (i, j) value in the reachability matrix will be 1 and (j, i) value will also be 1; for X (5, 8) in SSIM, '1' has been given in cell (5, 8) and '1' in cell (8, 5) also in initial reachability matrix.
- If (i, j) value in the SSIM is O, (i, j) value in the reachability matrix will be 0 and (j, i) value will also be 0; for O (3, 14) in SSIM, '0' has been given in cell (3, 14) and '0' in cell (14, 3) also in initial reachability matrix.

By applying these rules, an initial reachability matrix for the barriers to implement GSCM has been obtained in Table 4.3.

4.6.4. Final Reachability Matrix

The final reachability matrix has been obtained by adding transitivity as explained in Step 4 earlier and shown in Table 4.4. It is a basic assumption made in ISM. It states that, If Attribute 1 is related to 2, and Attribute 2 is related to 3, then criterion 1 is necessarily related to 3.

The reachability and antecedent set for each barrier have been determined from the final reachability matrix. The reachability set for a barrier consists of the barrier itself and the other barriers, which it influences. The antecedent set consists of the barrier itself and other barriers, which may influence it. Reachability and Antecedent set and Intersection sets are found for the all barriers. Barrier having same reachability set and the intersection set is assigned as top-level barrier in the ISM hierarchy or Level 1 is shown in Table 4.6. After finding Level 1, it is then discarded for finding further Levels. The iterative procedure is continued until Level of each barrier is found as shown in Table 4.6, 4.7 and 4.8. We have identified three levels in our study. But fourth level is given to Lack of Government Support Policies because its driving power is very high than other barriers.

Table 4.2: Structured self intersection matrix (SSIM) for barriers to implement GSCM in Indian auto component manufacturing industry.

Barriers	*B20*	*B19*	*B18*	*B17*	*B16*	*B15*	*B14*	*B13*	*B12*	*B11*	*B10*	*B9*	*B8*	*B7*	*B6*	*B5*	*B4*	*B3*	*B2*	*B1*
B1	O	A	A	O	V	O	V	A	O	O	X	A	O	V	A	A	O	V	A	X
B2	O	O	A	V	V	V	O	O	V	A	O	O	X	V	O	V	A	O	X	
B3	V	V	O	O	A	A	O	V	V	V	O	V	X	V	A	A	A	X		
B4	V	V	O	O	V	A	O	X	V	V	O	O	A	V	A	V	X			
B5	O	A	A	O	V	A	A	V	V	X	X	V	X	V	A	X				
B6	V	O	V	V	V	V	O	V	V	A	V	O	V	V	X					
B7	O	O	X	V	X	A	O	V	O	X	X	X	A	X						
B8	V	O	A	O	V	O	O	O	A	A	V	V	X							
B9	O	A	A	O	A	A	O	A	O	A	O	X								
B10	X	O	A	O	V	V	A	O	A	O	X									
B11	V	V	V	V	V	V	V	V	V	X										
B12	O	V	A	A	O	A	V	O	X											
B13	O	O	X	A	A	A	A	X												
B14	O	O	O	V	V	O	X													
B15	O	O	A	O	V	X														
B16	V	O	V	V	X															
B17	O	A	A	X																
B18	V	A	X																	
B19	V	X																		
B20	X																			

Table 4.3: Initial reachability matrix for barriers to implement GSCM in Indian auto component manufacturing industry.

Barriers	*B1*	*B2*	*B3*	*B4*	*B5*	*B6*	*B7*	*B8*	*B9*	*B10*	*B11*	*B12*	*B13*	*B14*	*B15*	*B16*	*B17*	*B18*	*B19*	*B20*
B1	1	0	1	0	0	0	1	0	0	1	0	0	0	1	0	1	0	0	0	0
B2	1	1	0	0	1	0	1	1	0	0	0	1	0	0	1	1	1	0	0	0
B3	0	0	1	0	0	0	1	1	1	0	1	1	1	0	0	0	0	0	1	1
B4	0	1	1	1	1	0	1	0	0	0	1	1	1	0	0	1	0	0	1	1
B5	1	0	1	0	1	0	1	1	1	1	1	1	1	0	0	1	0	0	0	0
B6	1	0	1	1	1	1	1	1	0	1	0	1	1	0	1	1	1	1	0	1
B7	0	0	0	0	0	0	1	0	1	1	1	0	1	0	0	1	1	1	0	0
B8	0	1	1	1	1	0	1	1	1	1	0	0	0	0	0	1	0	0	0	1
B9	1	0	0	0	0	0	1	0	1	0	0	0	0	0	0	0	0	0	0	0
B10	1	0	0	0	1	0	1	0	0	1	0	0	0	0	1	1	0	0	0	1
B11	0	1	0	0	1	1	1	1	1	0	1	1	1	1	1	1	1	1	1	1
B12	0	0	0	0	0	0	0	1	0	1	0	1	0	1	0	0	0	0	1	0
B13	1	0	0	1	0	0	0	0	1	0	0	0	1	0	0	0	0	1	0	0
B14	0	0	0	0	1	0	0	0	0	1	0	0	1	1	0	1	1	0	0	0
B15	0	0	1	1	1	0	1	0	1	1	0	1	1	0	1	1	0	0	0	0
B16	0	0	1	0	0	0	1	0	1	0	0	0	1	0	0	1	1	1	0	1
B17	0	0	0	0	0	0	0	0	0	0	0	1	1	0	0	0	1	0	0	0
B18	1	1	0	0	1	0	1	1	1	1	0	1	1	0	1	0	1	1	0	1
B19	1	0	0	0	1	0	0	0	1	0	0	0	0	0	0	0	1	1	1	1
B20	0	0	0	0	0	0	0	0	0	0	1	0	0	0	0	0	0	0	0	1

Table 4.4: Final reachability matrix for barriers to implement GSCM in Indian auto component manufacturing industry.

Ba-rriers	***1***	***2***	***3***	***4***	***5***	***6***	***7***	***8***	***9***	***10***	***11***	***12***	***13***	***14***	***15***	***16***	***17***	***18***	***19***	***20***
B1	1	0	1	0	1*	0	1	1*	1*	1	1*	1*	1*	1	1*	1	1*	1*	1*	1*
B2	1	1	1*	1*	1	0	1	1	1*	1*	1*	1	1*	1*	1	1	1	1*	1*	1*
B3	1*	1*	1	1*	1*	1*	1	1	1	1*	1	1	1	1*	1*	1*	1*	1*	1	1
B4	1*	1	1	1	1	1*	1	1*	1*	1*	1	1	1	1*	1*	1	1*	1*	1	1
B5	1	1*	1	1*	1	1*	1	1	1	1	1	1	1	1*	1*	1	1*	1*	1*	1*
B6	1	1*	1	1	1	1	1	1	1*	1	1*	1	1	1*	1	1	1	1	1*	1
B7	1*	1*	1*	1*	1*	1*	1	1*	1	1	1	1*	1	1*	1*	1	1	1	1*	1*
B8	1*	1	1	1	1	0	1	1	1	1	1*	1*	1*	0	1*	1	1*	1*	1*	1
B9	1	0	1*	0	0	0	1	0	1	1*	1*	0	1*	1*	0	1*	1*	1*	0	0
B10	1	0	1*	1*	1	0	1	1*	1*	1	1*	1*	1*	1*	1	1	1*	1*	0	1
B11	1*	1	1*	1*	1	1	1	1	1	1*	1	1	1	1	1	1	1	1	1	1
B12	1*	1*	1*	1*	1*	0	1*	1	1*	1	0	1	1*	1	1*	1*	1*	1*	1	1*
B13	1	1*	1*	1	1*	0	1*	1*	1	1*	1*	1*	1	1*	1*	1*	1*	1	1*	1*
B14	1*	0	1*	1*	1	0	1*	1*	1*	1	1*	1*	1	1	1*	1	1	1*	0	1*
B15	1*	1*	1	1	1	0	1	1*	1	1	1*	1	1	0	1	1	1*	1*	1*	1*
B16	1*	1*	1	1*	1*	0	1	1*	1	1*	1*	1*	1	0	1*	1	1	1	1*	1
B17	1*	0	0	1*	0	0	0	1*	1*	1*	0	1	1	1*	0	0	1	1*	1*	0
B18	1	1	1*	0	1	0	1	1	1	1	1*	1	1	1*	1	1*	1	1	1*	1
B19	1	1*	1*	0	1	0	1*	1*	1	1*	1*	1*	1*	1*	1*	1*	1	1	1	1
B20	0	1*	0	0	1*	1*	1*	1*	1*	0	1	1*	1*	1*	1*	1*	1*	1*	1*	1

*Means value after applying transitivity

Final reachability matrix with driving power and the dependence power of each barrier have also been shown in the Table 4.5.

Table 4.5: Final reachability matrix with driving and dependence power.

Ba-rriers	***1***	***2***	***3***	***4***	***5***	***6***	***7***	***8***	***9***	***10***	***11***	***12***	***13***	***14***	***15***	***16***	***17***	***18***	***19***	***20***	***Driving power***
1	1	0	1	0	1*	0	1	1*	1*	1	1*	1*	1*	1	1*	1	1*	1*	1*	1*	17
2	1	1	1*	1*	1	0	1	1	1*	1*	1*	1	1*	1*	1	1	1	1*	1*	1*	19
3	1*	1*	1	1*	1*	1*	1	1	1	1*	1	1	1	1*	1*	1*	1*	1*	1	1	20
4	1*	1	1	1	1	1*	1	1*	1*	1*	1	1	1	1*	1*	1	1*	1*	1	1	20
5	1	1*	1	1*	1	1*	1	1	1	1	1	1	1	1*	1*	1	1*	1*	1*	1*	20
6	1	1*	1	1	1	1	1	1	1*	1	1*	1	1	1*	1	1	1	1	1*	1	20
7	1*	1*	1*	1*	1*	1*	1	1*	1	1	1	1*	1	1*	1*	1	1	1	1*	1*	20
8	1*	1	1	1	1	0	1	1	1	1	1*	1*	1*	0	1*	1	1*	1*	1*	1	18
9	1	0	1*	0	0	0	1	0	1	1*	1*	0	1*	1*	0	1*	1*	1*	0	0	11
10	1	0	1*	1*	1	0	1	1*	1*	1	1*	1*	1*	1*	1	1	1*	1*	0	1	17
11	1*	1	1*	1*	1	1	1	1	1	1*	1	1	1	1	1	1	1	1	1	1	20
12	1*	1*	1*	1*	1*	0	1*	1	1*	1	0	1	1*	1	1*	1*	1*	1*	1	1*	18
13	1	1*	1*	1	1*	0	1*	1*	1	1*	1*	1*	1	1*	1*	1*	1*	1	1*	1*	19
14	1*	0	1*	1*	1	0	1*	1*	1*	1	1*	1*	1	1	1*	1	1	1*	0	1*	17
15	1*	1*	1	1	1	0	1	1*	1	1	1*	1	1	0	1	1	1*	1*	1*	1*	18
16	1*	1*	1	1*	1*	0	1	1*	1	1*	1*	1*	1	0	1*	1	1	1	1*	1	18
17	1*	0	0	1*	0	0	0	1*	1*	1*	0	1	1	1*	0	0	1	1*	1*	0	11
18	1	1	1*	0	1	0	1	1	1	1	1*	1	1	1*	1	1*	1	1	1*	1	18
19	1	1*	1*	0	1	0	1*	1*	1	1*	1*	1*	1*	1*	1*	1*	1	1	1	1	18
20	0	1*	0	0	1*	1*	1*	1*	1*	0	1	1*	1*	1*	1*	1*	1*	1*	1*	1	16
Dependence Power																					
	19	15	18	15	18	7	19	19	20	19	18	19	20	17	18	19	20	20	17	18	355/355

4.6.5. Partitioning of Levels

Table 4.6: First Iteration to find levels of barriers to implement GSCM in Indian auto component manufacturing industry.

Sl. no.	*Reachability Set*	*Antecedent Set*	*Intersection Set*	*Levels*
1	1,3,5,7,8,9,10,11,12,13,14,15,16,17,18,19,20	1,2,3,4,5,6,7,8,9,10,11,12,13,14,15,16,17,18,19	1,3,5,7,8,9,10,11,12,13,14,15,16,17,18,19	
2	1,2,3,4,5,7,8,9,10,11,12,13,14,15,16,17,18,19,20	2,3,4,5,6,7,8,11,12,13,15,16,18,19,20	2,3,4,5,7,8,11,12,13,15,16,18,19,20	
3	1,2,3,4,5,6,7,8,9,10,11,12 13,14,15,16,17,18,19,20	1,2,3,4,5,6,7,8,9,10,11,12,13,14,15,16,18,19	1,2,3,4,5,7,8,9,10,11,12,13,14,15,16,18,19	
4	1,2,3,4,5,6,7,8,9,10,11,12,13,14,15,16,17,18,19,20	2,3,4,5,6,7,8,10,11,12,13,14,15,16,17	2,3,4,5,7,8,10,11,12,13,14,15,16,17	
5	1,2,3,4,5,6,7,8,9,10,11,12,13,14,15,16,17,18,19,20	1,2,3,4,5,6,7,8,10,11,12,13,14,15,16,18,19,20	1,2,3,4,5,7,8,10,11,12,13,14,15,16,18,19,20	
6	1,2,3,4,5,6,7,8,9,10,11,12,13,14,15,16,17,18,19,20	3,4,5,6,7,11,20	3,4,5,6,7,11,20	
7	1,2,3,4,5,6,7,8,9,10,11,12,13,14,15,16,17,18,19,20	1,2,3,4,5,6,7,8,9,10,11,12,13,14,15,16,18,19,20	1,2,3,4,5,6,7,8,9,10,11,12,13,14,15,16,18,19,20	
8	1,2,3,4,5,7,8,9,10,11,12,12,13,15,16,17,18,19,	1,2,3,4,5,7,8,10,11,12,13,14,15,16,17,18,19,20	1,2,3,4,5,7,8,10,11,12,13,14,15,16,17,18,19,20	
9	1,3,7,9,10,11,13,14,16,17,18	1,2,3,4,5,6,7,8,9,10,11,12,13,14,15,16,17,18,19,20	1,3,7,9,10,11,13,14,16,17,18	I
10	1,3,4,5,7,8,9,10,11,12,13,14,15,16,17,18,20	1,2,3,4,5,6,7,8,9,10,11,12,13,14,15,16,17,18,19	1,3,5,7,8,9,10,11,12,13,14,15,16,17,18	
11	1,2,3,4,5,6,7,8,9,10,11,12,13,14,15,16,17,18,19,20	1,2,3,4,5,6,7,8,9,10,11,13,14,15,16,18,19,20	1,2,3,4,5,6,7,8,9,10,11,13,14,15,16,18,19,20	
12	1,2,3,4,5,7,8,9,10,12,13,14,15,16,17,18,19,20	1,2,3,4,5,6,7,8,10,11,12,13,14,15,16,17,18,19,20	1,2,3,4,5,7,8,10,12,13,14,15,16,17,18,19,20	
13	1,2,3,4,5,7,8,9,10,11,12,13,14,15,16,17,18,19,20	1,2,3,4,5,6,7,8,9,10,11,12,13,14,15,16,17,18,19,20	1,2,3,4,5,7,8,9,10,11,12,13,14,15,16,17,18,19,20	I
14	1,3,4,5,7,8,9,10,11,12,13,14,15,16,17,18,20	1,2,3,4,5,6,7,9,10,11,12,13,14,17,18,19,20	1,3,4,5,7,9,10,11,12,13,114,17,18,20	
15	1,2,3,4,5,7,8,9,10,11,12,13,15,16,17,18,19,20	1,2,3,4,5,6,7,8,10,11,12,13,14,15,16,17,18,19,20	1,2,3,4,5,7,8,10,11,12,13,15,16,17,18,19,20	
16	1,2,3,4,5,7,8,9,10,11,12,13,15,16,17,18,19,20	1,2,3,4,5,6,7,8,9,10,11,12,13,14,15,16,18,19,20	1,2,3,4,5,7,8,9,10,11,12,13,15,16,17,18,19,20	
17	1,4,8,9,10,12,13,14,17,18,19	1,2,3,4,5,6,7,8,9,10,11,12,13,14,15,16,17,18,19,20	1,4,8,9,10,12,13,14,17,18,19	I
18	1,2,3,5,7,8,9,10,11,12,13,14,15,16,17,18,19,20	1,2,3,4,5,6,7,8,9,10,11,12,13,14,15,16,17,18,19,20	1,2,3,5,7,8,9,10,11,12,13,14,15,16,17,18,19,20	I
19	1,2,3,5,7,8,9,10,11,12,13,14,15,16,17,18,19,20	1,2,3,4,5,6,7,8,11,12,13,15,16,17,18,19,20	1,2,3,5,7,11,12,13,15,16,17,18,19,20	
20	2,5,6,7,8,9,11,12,13,14,15,16,17,18,19,20	1,2,3,4,5,6,7,8,10,11,12,13,14,15,16,18,19,20	2,5,6,7,8,11,12,13,14,15,16,18,19,20	

Table 4.7: Second iteration to find levels of barriers to implement GSCM in Indian auto component manufacturing industry.

Sl. no.	*Reachability set*	*Antecedent set*	*Intersection set*	*Levels*
1	2,3,4,5,6,7,8,11,12, 15,16,19,20	2,3,4,5,6,7,8,11,12,15, 16,19	2,3,4,5,6,7,8,11,12,15, 16,19	
2	2,3,4,5,7,8,11,12,15, 16,19,20	2,3,4,5,6,7,8,11,12,15, 16,19,20	2,3,4,5,7,8,11,12,15,16, 19,20	II
3	2,3,4,5,6,7,8,11,12,15, 16,19,20	2,3,4,5,6,7,8,11,12,15, 16,19	2,3,4,5,6,7,8,11,12,15, 16,19	
4	2,3,4,5,6,7,8,11,12,15, 16,19,20	2,3,4,5,6,7,8,11,12,15, 16	2,3,4,5,6,7,8,11,12,15,16	
5	2,3,4,5,6,7,8,11,12,15, 16,19,20	2,3,4,5,6,7,8,11,12,15, 16,19,20	2,3,4,5,6,7,8,11,12,15, 16,19,20	II
6	2,3,4,5,6,7,8,11,12,15, 16,19,20	3,4,5,6,7,11,20	2,3,4,5,6,7,11,20	
7	2,3,4,5,6,7,8,11,12,15, 16,19,20	2,3,4,5,6,7,8,11,12,15, 16,19,20	2,3,4,5,6,7,8,11,12,15,16, 19,20	II
8	2,3,4,5,7,8,11,12,15, 16,19,20	2,3,4,5,6,7,8,11,12,15, 16,19,20	2,3,4,5,7,8,11,12,15,16, 19,20	II
10	3,5,7,8,11,12,15,16, 20	2,3,4,5,6,7,8,11,12,15, 16,19	3,5,7,8,11,12,15,16	
11	2,3,4,5,6,7,8,11,12,15, 16,19,20	2,3,4,5,6,7,8,11,15,16, 19,20	2,3,4,5,6,7,8,11,15,16, 19,20	
12	2,3,4,5,7,8,12,15,16, 19,20	2,3,4,5,6,7,8,11,12,15, 16,19,20	2,3,4,5,7,8,12,15,16,19,20	II
14	3,4,5,7,8,11,12,15,16,20	2,3,4,5,6,7,11,12,19,20	3,4,5,7,11,12,20	
15	2,3,4,5,7,8,11,12,15, 16,19,20	2,3,4,5,6,7,8,11,15,16, 19,20	2,3,4,5,7,8,11,15,16,19,20	
16	2,3,4,5,7,8,11,12,15, 16,19,20	2,3,4,5,6,7,8,11,12,15, 16,19,20	2,3,4,5,7,8,11,12,15,16, 19,20	II
19	2,3,5,7,8,11,12,15,16, 19,20	2,3,4,5,6,7,8,11,12,15, 16,19,20	2,3,5,7,8,11,12,15,16,17, 19,20	II
20	2,5,6,7,8,11,12,15,16, 19,20	2,3,4,5,6,7,8,11,12,15, 16,19,20	2,5,6,7,8,11,12,15,16,19, 20	II

Table 4.8: Third iteration to find levels of barriers to implement GSCM in Indian auto component manufacturing industry.

Sl. no	*Reachability set*	*Antecedent set*	*Intersection set*	*Levels*
1	3,4,6,11	3,4,6,11	3,4,6,11	III
3	3,4,6,11	3,4,6,11	3,4,6,11	III
4	3,4,6,11	3,4,6,11	3,4,6,11	III
6	3,4,6,11	3,4,6,11	3,4,6,11	IV
10	3,11	3,4,6,11	3,11	III
11	3,4,6,11	3,4,6,11	3,4,6,11	III
14	3,4,11	3,4,6,11	3,4,11	III
15	3,4,11	3,4,6,11	3,4,11	III

4.6.6. ISM Model Formulation

Once all levels are found, these levels have been summarized in the Table 4.9. From the final reachability matrix in Table 4.5, the structural model is generated and is given in Fig. 4.2. The relationship between the barriers j and i is shown by an arrow pointing from i to j. The resulting graph is called a digraph. Removing the transitivity's as described in the ISM methodology, the digraph is finally converted into the ISM model has been made as shown in Fig. 4.2.

Table 4.9: Various Levels of Barriers to Implement GSCM in Indian auto component manufacturing industry.

Sl. no.	*Levels*	*Important barriers*
1.	1st	Market competition (B9) Pollution/Wastage of industries (B13) Lack of sustainability certification (ISO 14001) (B17) Cost of disposal of hazardous products (B18)
2.	2nd	Lack of IT applications (B2) Resistance to advance technology adoption (B5) Lack of knowledge about green practice (B7) Lack of Technical expertise (B8) Fear of failure (B12) Lack of recycling and reuse efforts of organization (B16) Lack of awareness about reverse logistics adoption (B19) Lack of corporate social responsibility (B20)
3.	3rd	Cost Implication (B1)Poor organizational culture in adopting GSCM (B3) Lack of Top management commitment in adopting GSCM (B4) Less awareness of customer about GSCM (B10) Lack of environmental awareness to the supplier (B11) Non-availability of bank loans to encourage green product (B14) Lack of training courses about implementing GSCM (B15)
4.	4th	Lack of government support to adopt GSCM (B6)

4.6.7. MICMAC Analysis

Variables are classified in to four clusters named as autonomous variables, dependent variables, linkage variables and independent variables. The MICMAC principle is based on multiplication properties of matrices. The purpose of MICMAC analysis is to analyze the drive power and dependence power of barriers. This is done to identify the key barriers that drive the system in various categories. Based on their drive power and dependence power, the barriers, in the present case, have been classified into four categories as follows:

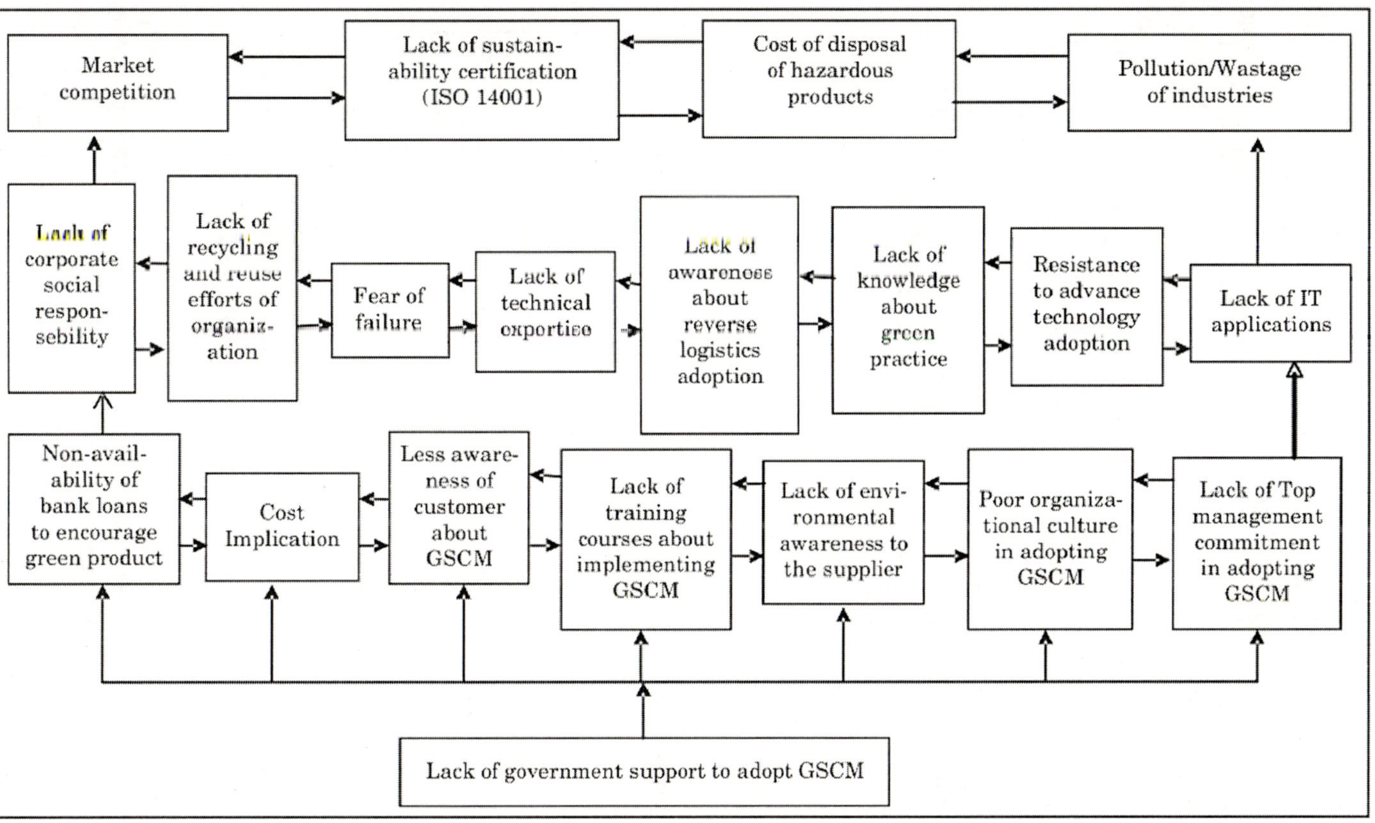

Fig. 4.2: ISM based model for barriers to implement GSCM in Indian auto component manufacturing industry.

1. ***Autonomous enablers*:** These enablers have weak driving power and weak dependence. They are relatively disconnected from the system, with which they have few links, which may be very strong. These enablers are represented in Quadrant – I.
2. ***Dependent enablers*:** This category includes those enablers which have weak drive power but strong dependence power and placed in Quadrant – II.
3. ***Linkage enablers*:** These have strong driving power as well as strong dependence and are placed in Quadrant – III. They are also unstable and so any action on them will influence others and a feedback effect on themselves.
4. ***Independent enablers*:** These have strong driving power but weak dependence power. These are represented in Quadrant – IV.

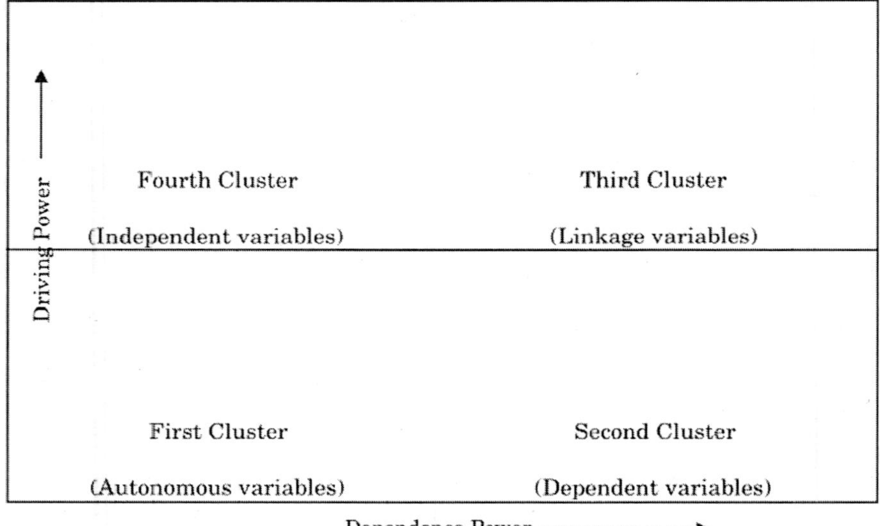

Fig. 4.3: Cluster of MICMAC analysis

1. Autonomous variables (first cluster) have weak driving power and dependence. These variables can be disconnected from the system. In our study, no barrier lies in this range.
2. The second cluster is named dependent variables. They have weak driving power and strong dependence power. In our study, no barrier lies in this range.
3. The third cluster named linkage variables having strong driving power and strong dependence power. In our study, nineteen barriers lies in this region named as Cost Implication, Lack of IT

applications, Poor organizational culture in adopting GSCM, Lack of Top management commitment in adopting GSCM, Resistance to advance technology adoption, Lack of knowledge about green practice, Lack of Technical expertise, Market competition, less awareness of customer about GSCM, Lack of environmental awareness to the supplier, Fear of failure, Pollution/Wastage of industries, Non-availability of bank loans to encourage green product, Lack of training courses about implementing GSCM, Lack of recycling and reuse efforts of organization, Lack of sustainability certification (ISO 14001), Cost of disposal of hazardous products, Lack of awareness about reverse logistics adoption, Lack of corporate social responsibility.

4. The fourth cluster named independent variables has strong driving power and weak dependence power. In our study, one barrier named Lack of government support systems (6) are lying in this range.

The graph between dependence power and driving power for the barriers to implement GSCM in Indian auto component manufacturing industry is given in Fig. 4.3. The aim of this study is to analyze the driving power and the dependency power of barriers.

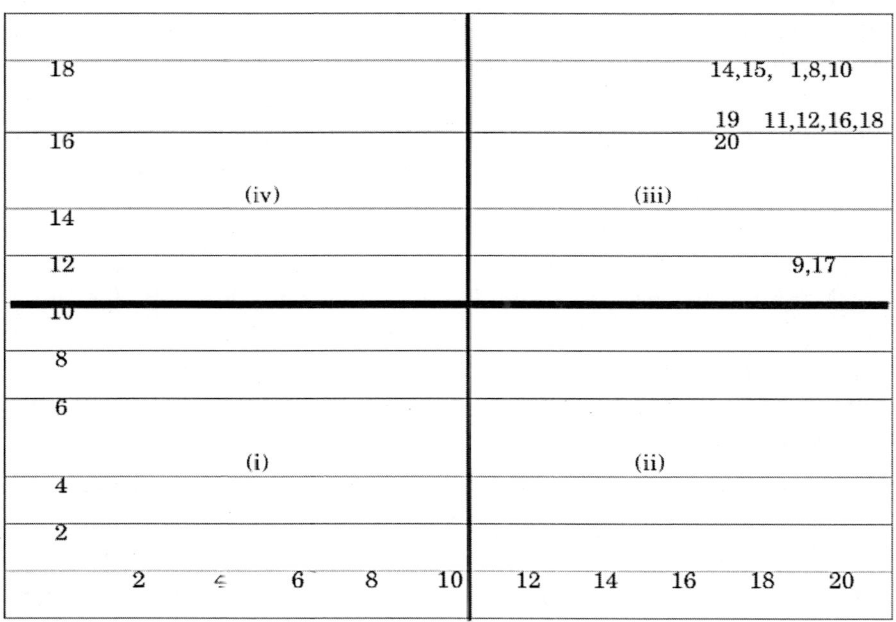

Fig. 4.4: Cluster of barriers to implement GSCM in Indian auto component manufacturing industry.

Without analyzing the barriers, one cannot initiate the implementation of 100% GSCM in Indian industries. In this study, barriers hindering the implementation of GSCM which were obtained from literature and on consultation with industrial experts have been put into an ISM to analyze the interactions between these barriers. This analysis is making GSCM adoption easy by removing the dominant barrier. Higher dependence values for a factor means many barriers to be addressed before its removal and high driving value of a barrier means many barriers that could be removed by its removal.

4.7. RESULTS AND DISCUSSION

Growth of any industries in India is strong but still some gaps like environmental consciousness affect their performance. Environmental improvements are essential for Indian industries to attract customers from Multi-National Companies (MNC). Many of the Indian industries have customers from all over the world, to which they are supplying components, especially auto components as all the involved MNCs (customers) have already implemented green technology they expect the same from their Indian suppliers and so the Indian industries are under great pressure to adopt green technologies in their traditional supply chain management. More over adoption of GSCM is not an easy task and there exists many obstacles during the adoption of green concepts in to the traditional supply chain management. Without analyzing the barriers, one cannot initiate the implementation of 100% GSCM in Indian industries. In this study, barriers hindering the implementation of GSCM which were obtained from literature and on consultation with industrial experts have been put into an ISM to analyze the interactions between these barriers. This analysis is making GSCM adoption easy by removing the dominant barrier. The driver and dependence power diagram obtained from MICMAC analysis gives an insight into the relative importance and interdependencies between these barriers. Some of the major findings of this study have been highlighted here.

- Autonomous variables (first cluster) have weak driving power and dependence. These variables can be disconnected from the system. In our study, no barrier lies in this range.
- The second cluster is named dependent variables. They have weak driving power and strong dependence power. In our study, no barrier lies in this range.

- The third cluster named linkage variables having strong driving power and strong dependence power. In our study, nineteen barriers lies in this region named as Cost Implication (B1), Lack of IT applications (B2), Poor organizational culture in adopting GSCM (B3), Lack of Top management commitment in adopting GSCM (B4), Resistance to advance technology adoption (B5), Lack of knowledge about green practice (B7), Lack of Technical expertise (B8), Market competition (B9), Less awareness of customer about GSCM (B10), Lack of environmental awareness to the supplier (B11), Fear of failure (B12), Pollution/Wastage of industries (B13), Non-availability of bank loans to encourage green product (B14), Lack of training courses about implementing GSCM (B15), Lack of recycling and reuse efforts of organization (B16), Lack of sustainability certification (ISO 14001) (B17), Cost of disposal of hazardous products (B18), Lack of awareness about reverse logistics adoption (B19), Lack of corporate social responsibility (B20).

  ISM formation model Cost Implication (B1) is dominant barrier and it has high driving power (17) and high dependence power (19). Lack of IT applications (B2) barrier has high driving power (19) and dependence power (15). B3 barrier is dominant barrier, because of high driving power (20) and less dependence power (18). Lack of Top management commitment in adopting GSCM (B4) barrier is dominant barrier, because of high driving power (20) and les dependence power (15). Resistance to advance technology adoption (B5) barrier has driving power (20) and dependence power (18). B5 barrier is less dominant compares to B4. Lack of knowledge about green practice (B7) has driving power (20) and dependence power (19) and it less dominant compares to B4. Lack of Technical expertise (B8) barrier is less dominant, because of less driving power and high dependence power (19). Market competition (B9) is less dominant barrier, because of less driving power (11) and high dependence power (20). Less awareness of customer about GSCM (B10) barrier is dominant barrier and its driving power is 17 and dependence power is 19. Lack of environmental awareness to the supplier (B11) barrier is dominant barrier, because of high driving power (20) and less dependence power (18). Fear of failure (B12) barrier is very less dominant compared to B11, because of less driving power (18) and high dependence power (19). Pollution/Wastage of industries (B13) barrier has less driving power (19) and high dependence power (20) and it very less dominant barrier as compares to B11. Non-availability of bank loans to encourage green product (B14) barrier is dominant barrier, because of high driving power (17) and high dependence power (17). Lack of training courses about implementing GSCM (B15) barrier is dominant barrier, because of high driving power (18) and high dependence power (18). Lack of recycling and reuse efforts of organization (B16) barrier is less dominant barrier as compared to B15, because of less driving power (18) and high dependence power (19). Lack of sustainability certification (ISO 14001) (B17) barrier is very less dominant barrier, because of less driving power (11) and high dependence power (20). Cost of disposal of hazardous products (B18) barrier is less dominant barrier, because of less driving power (18) and high dependence power (20). Lack of awareness about reverse logistics adoption (B19) barrier has driving power (18) and dependence power (17). Lack of corporate social responsibility (B20) barrier is less dominant barrier, because of less driving power (16) and high dependence power (18) as shown in Fig. 4.4.

- The fourth cluster named independent variables has strong driving power and weak dependence power. In our study, one barrier named Lack of government support systems (6) are lying in this range. Barrier B6 is very dominant barrier, because of high driving power (20) and very less dependence power (7). It is the only barrier lie in the independent cluster therefore it is most dominant barrier in the implementation of GSCM in the XYZ Ltd. It shows that the government support system plays an important role in the implementation of GSCM in the industry. Therefore, lack of govt. support becomes the most dominant barrier.

4.8. CONCLUSIONS

The environmental consciousness of customers and the increase of environmental image in the market day by day have pushed SMEs indirectly to think about cleaner production by means of GSCM implementation. Auto component manufacturers play an important role in a country's economy and should begin adopting GSCM as their strategy. The barriers that hinder the implementation of GSCM create considerable challenges both for the technical experts and mangers of SME's. Some of the major barriers have been considered in the present study and are analyzed in terms of interaction among them using ISM. Based on the inputs from experts' opinion in one auto component manufacturing industry. A structural self-interaction matrix (SSIM) was formed which was the basis for the Interpretive Structural Modeling (ISM). These barriers are iterated in four (IV) levels.

A structural model was formed using ISM in which Market competition (B9), Pollution/Wastage of industries (B13) and Lack of sustainability certification (ISO 14001) (B17), Cost of disposal of hazardous products (B18) occupied the top most level. These are the barriers that are affected at the lower level and these barriers give less impact as compared to the remaining barriers. It shows that these barriers are comparatively easy to eradicate.

In the second level (2nd) eight barriers are placed namely: Lack of IT applications (B2), Resistance to advance technology adoption (B5), Lack of knowledge about green practice (B7), Lack of Technical expertise (B8), Fear of failure (B12), Lack of recycling and reuse efforts of organization (B16), Lack of awareness about reverse logistics adoption (B19) and Lack of corporate social responsibility (B20). These barriers are more dominant than the 1st level. But barriers are easy to eradicate. In the third level (3rd) seven barriers are placed namely: Cost Implication (B1), Poor organizational culture in adopting GSCM (B3), Lack of Top management commitment in adopting GSCM (B4), Less awareness of customer about GSCM (B10), Lack of environmental awareness to the supplier (B11), Non-availability of bank loans to encourage green product (B14) and Lack of training courses about implementing GSCM (B15). These barriers are very much dominant, and these are not easy to eliminate. These are main obstacle in front of the industry to implement green supply chain. In the fourth level (4th) only one barrier is palace namely: Lack of government support to adopt GSCM (B6). This is most dominant barrier to implement GSCM in the XYZ Ltd. This is main obstacle in front of the industry and it is difficult to eliminate it. This barrier is driving the remaining barriers and not much dependent on other barriers. By diagnosing the dominant barriers for the adoption of GSCM, the fear of implementing GSCM in industries can be eliminated. This also leads to the manufacturing of eco-friendly products in the industries and their ability to sustain in the market is increased. The result of this work shows that Lack of government support to adopt GSCM (B6) barrier is acting the key barrier for the implementation of GSCM. Industries need to give special attention and priority to remove this barrier. The govt. support is must in the implementation of GSCM and govt. should help the industries to make final products and their supply chain green.

4.9. SUMMARY

In the chapter methodology of the Interpretive Structural Modeling (ISM) has been implemented to solve the problem of analysis of barriers. The main work proposes analysis of barriers for implementation of green supply chain management with the help of ISM technology. The main advantage of ISM technique in this work is to get the suitable model for the industry. From the above Fig. 4.4 the behavior of the barriers is identified. There are 4 clusters in the Fig. 4.4 in which only one barrier lies in the 4th cluster. Nineteen barriers lie in the linkage clusters. The other two clusters are empty.

5

Application of Data Envelopment Analysis (DEA) for Performance Evaluation of Suppliers and Manufacturer

5.1. PROBLEM STATEMENT

Consider a product, data envelopment analysis supply chain consisting of manufacturers and suppliers. The suppliers supply a raw material per the customer demand and the manufacturer will be manufacturing the product for the customer demand. There are numbers of suppliers which provide raw material at different cost which affect the manufacturing cost and quality. In the supply chain network, there are several factors which affect the inputs and outputs through which the calculation of these type of problem is very complex and the consideration of each affecting variable should not be possible. As the change in a cost belong to the supplier (raw material cost, transportation cost, labor cost, etc.) will also affect the cost of the manufacturer (machining cost, labor cost, procurement cost, etc.), so we can see there are lot of factors which affect the supply chain network. The DEA is a technique which provide the solution of this type of problem, it includes all the affecting variable and provide the most efficient supply chain network through which we get minimum input and maximum output. Here we introduce a problem of piston manufacturing company which based on DEA technique. Company have n homogeneous supply chain operations. Each supply chain observation is a DMU. It is assumed that each supplier S_j in DMU_j: j = 1,...,n has m inputs x_{ij}: I = 1,...,m and s outputs y_{rj}: r = 1,..., s. These outputs can become the inputs to the manufacturer M_j. The manufacturer M_j has its own inputs Z_{dj}: d = 1,..., D. The final outputs from manufacturer are q_{lj}: l = 1,..., L. This type of problem came into the two-stage supply chain, supplier-manufacturer supply chain as shown in Fig. 5.1

5.2. INTRODUCTION

5.2.1. A Data Envelopment Analysis Approach to Supply Chain Efficiency

Supply chain management is an important competitive strategy used by modern enterprises. Effective design and management of supply chains assists in the production and delivery of a variety of products at low costs, high quality, and short lead times. Recently, data envelopment analysis (DEA) has been extended to examine the efficiency of supply chain operations. The current problem develops a DEA model for measuring the performance of suppliers and manufacturers in supply chain operations. Additive efficiency decomposition for suppliers and manufacturers in supply chain operations is proposed.

Data envelopment analysis (DEA), originated from the work of Charnes 1991. It is a linear programming, non-parametric technique used to measure the relative efficiency of peer decision making units with multiple inputs and outputs. This methodology has been applied in a wide range of applications over the last three decades, in setting that include banks, hospitals, and maintenance. Recently, several studies have looked at production processes that have two-stage network structure, as supply chain operations. Many researchers have applied standard DEA models to measure the performance of supply chain members.

Weber and Desai employed DEA to construct an index of relative supplier performance. Fare and Grosskopf (1991) developed a network DEA approach to model the general multistage process.

Easton *et al.,* Suggested a DEA model to compare the purchasing efficiency of firms in the petroleum industry. Talluri and Baker proposed a multiphase mathematical programming approach for effective supply chain design. Their methodology applies a combination of multi-criteria efficiency models, based on game theory concepts, and linear and integer programming methods. Liang *et al.* and Chen *et al.* developed several DEA-based approaches for characterizing and measuring supply chain efficiency when intermediate measures are incorporated into the performance evaluation. Chen proposed a structured methodology for supplier selection and evaluation in a supply chain, to help enterprises establish a systematic approach for selecting and evaluating potential suppliers in a supply chain. Feng *et al.* defined two types of supply chain production possibility sets, which are proved to be equivalent to each other.

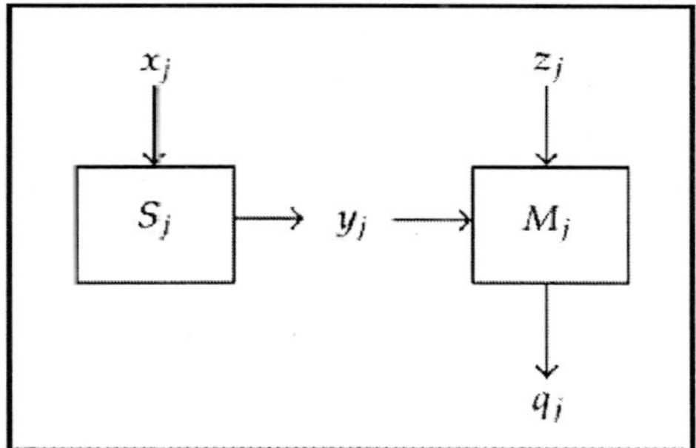

Fig. 5.1: Supplier-manufacturer supply chain.

Some of the above mentioned studies treat the supply chain as a black-box and do not consider the intermediate measures. However, in all studies, the treatment to the intermediate measures is ambiguous. Moreover, some of the above-mentioned studies are not applicable in a more complex supply chain or a network structured case.

If we treat the supply chain operation as a black-box, ignoring the intermediate measure may yield an efficient supply chain with inefficient supplier and/or manufacturer. In the proposed model in this paper, the intermediate measure is considered as a free variable, and it will be reduced to make the whole supply chain as efficient.

In this problem, we develop a DEA model for measuring the performance of suppliers and manufacturers in supply chain operations. Additive efficiency decomposition for suppliers and manufacturers in supply chain operations is proposed and the DEA frontier points for efficient supply chain members are determined.

5.3. DATA ENVELOPMENT ANALYSIS (DEA)

According to William W. Cooper "Data envelopment analysis is a method for evaluating decision making units within an organization, by using imputed shadow prices. These prices are computed using a fractional program that is solved by reducing it to a linear program."

- Data envelopment analysis (DEA) is a nonparametric method in operations research and economics for the estimation of production frontiers. It is used to empirically measure productive efficiency of decision making units (or DMUs).
- Data Envelopment Analysis (DEA) measures the relative efficiencies of organizations with multiple inputs and multiple

outputs. The organizations are called the decision-making units, or DMUs.

- DEA is a mathematical programming approach to provide a relative efficiency assessment (called DEA efficient) for a group of decision making units (DMU) with multiple number of inputs and outputs.
- DEA assigns weights to the inputs and outputs of a DMU that give it the best possible efficiency. It thus arrives at a weighting of the relative importance of the input and output variables that reflects the emphasis that appears to have been placed on them for that DMU.
- It concerned with evaluations of performance and it is especially concerned with evaluating the activities of organizations such as business firms, government agencies, hospitals, educational institutions

DEA has been used for both production and cost data. Utilizing the selected variables, such as unit cost and output, DEA software searches for the points with the lowest unit cost for any given output, connecting those points to form the efficiency frontier. Any company not on the frontier is considered inefficient. A numerical coefficient is given to each firm, defining its relative efficiency. Different variables that could be used to establish the efficiency frontier are: number of employees, service quality, environmental safety, and fuel consumption. An early survey of studies of electricity distribution companies identified more than thirty DEA analyses — indicating widespread application of this technique to that network industry. A number of studies using this technique have been published for water utilities. The main advantage to this method is its ability to accommodate a multiplicity of inputs and outputs. It is also useful because it takes into consideration returns to scale in calculating efficiency, allowing for the concept of increasing or decreasing efficiency based on size and output levels. A drawback of this technique is that model specification and inclusion/exclusion of variables can affect the results."

Under general DEA benchmarking, for example, "if one benchmarks the performance of computers, it is natural to consider different features (screen size and resolution, memory size, process speed, hard disk size, and others). One would then have to classify these features into "inputs" and "outputs" to apply a proper DEA analysis. However, these features may not actually represent inputs and outputs at all, in the standard notion of production. In fact, if one examines the benchmarking literature, other terms, such as "indicators", "outcomes", and "metrics", are used. The issue now becomes one of how to classify these performance measures into inputs and outputs, for use in DEA.

5.3.1. Applications

- It is most useful when a comparison is sought against "best practices" where the analyst doesn't want the frequency of poorly run operations to affect the analysis. DEA has been applied in many situations such as: health care (hospitals, doctors), education (schools, universities), banks, manufacturing, benchmarking, management evaluation, fast food restaurants, and retail stores.
- The analyzed data sets vary in size. Some analysts work on problems with as few as 15 or 20 DMUs while others are tackling problems with over 10,00 DMUs.

5.3.2. Strengths of DEA

DEA can be a powerful tool when used wisely. A few of the characteristics that make it powerful are:

- The main advantage to this method is its ability to accommodate a multiplicity of inputs and outputs.
- It doesn't require an assumption of a functional form relating inputs to outputs.
- DMUs are directly compared against a peer or combination of peers.
- Inputs and outputs can have very different units. For example, X1 could be in units of lives saved and X2 could be in units of lives saved and X2 could be in units of dollars without requiring an a priori trade off between the two.

5.3.3. Drawback

- A drawback of this technique is that model specification and inclusion/exclusion of variables can affect the results.

5.4 MATHEMATICAL MODEL OF DEA

Consider a buyer-seller supply chain as described in Fig.5.1, where *XA* is the input vector of the seller, and *YA* is the seller's output vector. *YA* is also an input vector of the buyer, along with *XB*, with *YB* being the buyer's output vector.

Suppose there are *n* such supply chains or observations on one supply chain. The CCR DEA efficiency of the supply chain is measured as (Charnes *et al.*, 1978; CCR)

Before introducing the method and the models that will be used, we first define the symbols that will be used

X_A = Input of seller

X_B = Input of buyer

Y_A = Output of seller

Y_B = Output of buyer

U = Output weights in fractional program

V = Input weights in fractional program

n = Number of observation

µ = Output weights in linear program

ù = Input weights in linear program

T = Number of decision making unit

E_{AA} = Efficiency of sellers

E_{BB} = Efficiency of buyer

E^*AA = Maximum efficiency of sellers

E^*_{BB} = Maximum efficiency of buyer

e_{AB} = Efficiency of supply chain

$$\text{Max } \frac{U^T Y_{B0}}{V^T(X_{A0}, X_{B0})}$$

$$\text{s.t } \frac{U^T Y_{Bj}}{V^T(X_{Aj}, X_{Bj})} \leq 1 \qquad j = 1, 2\ldots\ldots\ldots, n \tag{1}$$

$$U^T, V^T \geq 0$$

Zhu (2003) shows that DEA model (1) fails to correctly characterize the performance of supply chains, because it only considers the inputs and outputs of the supply chain system and ignores measures *YA* associated with supply chain members. Zhu (2003) also shows that if *YA* are treated as both input and output measures in model (1), all supply chains become efficient.

Zhu (2003) further shows that an efficient performance indicated by model (1) does not necessarily indicate efficient performance in individual supply chain members. Consequently, improvement to the best-practice can be distorted. *i.e.,* the performance improvement of one supply chain member affects the efficiency status of the other, because of the presence of intermediate measures.

Alternatively, we may consider the average efficiency of the buyer and seller as in the following DEA model:

$$\text{Max}\frac{1}{2}\left[\frac{U_A^T Y_{A0}}{V_A^T X_{A0}} + \frac{U_B^T Y_{B0}}{V_B^T (X_{B0}, Y_{A0})}\right]$$

$$\text{s.t.}\frac{U_A^T Y_{Aj}}{V_A^T X_{Aj}} \leq 1 \qquad j = 1,2,\ldots\ldots n \tag{2}$$

$$\frac{U_B^T Y_{Bj}}{V_B^T (X_{Bj}, Y_{Aj})} \leq 1 \qquad j = 1,2,\ldots\ldots n$$

$$U_A^T, V_A^T, U_B^T, V_B^T \geq 0$$

Although model (2) considers Y_A, it does not reflect the relationship between the buyer and the seller. The weights of Y_A as inputs of the buyer may not be equal to the weights of Y_A as outputs of the seller. Model (2) treats the seller and the buyer as two independent units. This does not reflect an ideal supply chain operation.

We next develop several models that can directly evaluate the performance of the supply chain as well as its members while considering the relationship between the buyer and the seller. Our modeling processes are based upon the concept of non-cooperative and cooperative games (see *e.g.,* Simaan and Cruz, 1973; Huang, and Ashley, 1995; Huang, 2000).

Based upon Li, Huang and Ashley (1995), we suppose the seller-buyer interaction is viewed as a two-stage non-cooperative game with the seller as the leader and the buyer as the follower. For example, in the case of non-cooperative advertising between the manufacture (seller) and the retailer (buyer), Toyota automobile company decides that it wants to promote sales of a model and directs and subsidizes its local dealers. The local dealers then react to Toyota's strategy by adjusting the amount they spend on advertising and promotion.

First, we use the CCR model to evaluate the efficiency of the seller, as the leader

$$\text{Max}\frac{U_A^T Y_{A0}}{V_A^T X_{A0}} = \text{E}_{\text{AA}}$$

$$\text{s.t.}\frac{U_A^T Y_{Aj}}{V_A^T X_{Aj}} \le 1 \qquad \text{j} = 1, 2\ldots\ldots\text{n} \tag{3}$$

$$U_A^T, V_A^T \ge 0$$

This model is equivalent to the following standard DEA multiplier model:

$$\text{Max}\,\mu_A^T Y_{A0} = E_{AA}$$

$$\text{s.t.}\,\omega_A^T X_{Aj} - \mu_A^T Y_{Aj} \ge 0 \qquad \text{j=1, 2\ldots\ldots\ n} \tag{4}$$

$$\omega_A^T X_{A0} = 1$$

$$\omega_A^T, \mu_A^T \ge 0$$

Suppose we have an optimal solution of model (4) ω_A^{T*}, μ_A^{T*}, and E_{AA}^* and denote the seller's efficiency as. We then use the following model to evaluate the buyer's efficiency:

$$\text{Max}\,\frac{U_B^T Y_{B0}}{V_B^T X_{B0} + D \times \mu_A^T Y_{A0}} = E_{AB}$$

$$\text{s.t.}\frac{U_B^T Y_{Bj}}{V_B^T X_{Bj} + D \times \mu_A^T Y_{Aj}} \le 1 \qquad \text{j} = 1,2,\ldots\ldots\text{n} \tag{5}$$

$$\mu_A^T Y_{A0} = E_{AA}^*$$

$$\omega_A^T X_{Aj} - \mu_A^T Y_{Aj} \ge 0 \qquad \text{j} = 1,2,\ldots\ldots\text{n}$$

$$\omega_A^T X_{A0} = 1$$

$$\omega_A^T, \mu_A^T, U_B^T, V_B^T, D \geq 0$$

Note that in model (5), we try to determine the buyer's efficiency given that the seller's efficiency remains at E_{AA}^*. Model (5) is equivalent to the following non-linear model:

$$\text{Max } \mu_B^T Y_{B0} = E_{AB}$$

$$\text{s.t. } \omega_B^T X_{Bj} - d\mu_A^T Y_{Aj} - \mu_B^T Y_{Bj} \geq 0 \qquad j = 1,2,\ldots\ldots n \tag{6}$$

$$\omega_B^T X_{B0} + d\mu_A^T Y_{A0} = 1$$

$$\mu_A^T Y_{A0} = E_{AA}^*$$

$$\omega_A^T X_{Aj} + \mu_A^T Y_{Aj} \geq 0 \qquad j = 1,2,\ldots\ldots n$$

$$\omega_A^T X_{A0} = 1$$

$$\omega_A^T, \mu_A^T, \omega_B^T, \mu_B^T, d \geq 0$$

Note that $\omega_B^T X_{B0} + d\mu_A^T Y_{A0} = 1$ and $\mu_A^T Y_{A0} = E_{AA}^*$. Thus, we have $0 \leq d \leq \frac{1}{\mu_A^T Y_{A0}} = \frac{1}{E_{AA}^*}$, *i.e.*, we have the upper and lower bounds on *d*. Therefore, *d* can be treated as a parameter and model (6) can be solved as a linear program. In computation, we set the initial *d* value as the upper bound, namely $d_0 = \frac{1}{E_{AA}^*}$, and solve the resulting linear program.

We then start to decrease *d* according to $d_t = \frac{1}{E_{AA}^*} - \varepsilon \times t$ for each step *t*, where ε is a small positive number. We solve each linear program of model (6) corresponding to d_t and denote the optimal objective value as $E_{BA}^*(d_t)$.

Let $E_{BA}^* = \text{Max } E_{BA}^*(d_t)$. Then we obtain a best heuristic search solution E_{BA}^* to model (6). This E_{BA}^* represents the buyer's efficiency when the seller is given the pre-emptive priority to achieve its best performance. The efficiency of the supply chain can then be defined as

$$e_{AB} = \frac{1}{2}(E_{AA}^{*} + E_{AB}^{*})$$

Similarly, one can develop a procedure for the situation when the buyer is the leader and the seller the follower. For example, in the October 6, 2003 issue of the Business Week, its cover story reports that Walmart dominates its suppliers and not only dictates delivery schedules and inventory levels, but also heavily influences product specifications.

We first evaluate the efficiency of the buyer using the standard CCR ratio model

$$\text{Max } \frac{U_B^T Y_{B0}}{V_B^T X_{B0} + V^T Y_{Aj}} = E_{BB}$$

$$\text{s.t. } \frac{U_B^T Y_{Bj}}{V_B^T X_{Bj} + V^T Y_{Aj}} \leq 1 \qquad j = 1,2,\ldots\ldots n \tag{7}$$

$$U_B^T, V_B^T, V^T \geq 0$$

Model (7) is equivalent to the following standard CCR multiplier model

$$\text{Max } \mu_B^T Y_{B0} = E_{BB}$$

$$\text{s.t. } \omega_B^T X_{Bj} + \mu^T Y_{Aj} - \mu_B^T Y_{Bj} \geq 0 \qquad j = 1,2,\ldots\ldots\ldots,n \tag{8}$$

$$\omega_B^T X_{B0} + \mu^T Y_{A0} = 1$$

$$\omega_B^T, \mu_B^T, \mu^T \geq 0$$

Let $\omega_B^{T*}, \mu_B^{T*}, \mu^{T*}, E_{BB}^{*}$ an optimal solution from model (8) where E_{BB}^{*} represents the buyer's efficiency score.

The efficiency of the supply chain network can then be defined as

$$E_{AB} = \frac{1}{2}(E_{AA}^{*} + E_{BB}^{*})$$

We now illustrate the DEA procedures with ten supply chain operations (DMUs) given in Table 1. The supplier has three inputs, X_1(labor cost), X_2(operating cost) and X_3(shipping cost) and two outputs, Y_1(Shipping of Cast aluminum alloy) and Y_2(Shipping of Titanium). The buyer has another input Z_1(labor cost) in addition to Y_1, Y_2 and two outputs: q_1(sales) and q_2(profit).

Table 2 reports the efficiency scores obtained from the newly developed supply chain efficiency models. It can be seen from models (1) and (2) that the supply chain is rated as efficient while its two members are inefficient (*e.g.,* DMUs 2, 3, 4, 7, 8 and 9). This is because the intermediate measures are ignored in model (1). This also indicates that supply chain efficiency cannot be measure by the conventional DEA approach.

5.5. DEA-FRONTIER SOFTWARE

For evaluating the efficiency of supply chain networks, we used a DEA-frontier software. DEA-frontier software works along with the Microsoft Excel, before we run the DEA-frontier software some arrangement and setting had to be done in Microsoft Excel.

- In the present work MS-Excel 2016 has been used and Fig. 5.2 shows the setting in Microsoft Excel 2016.

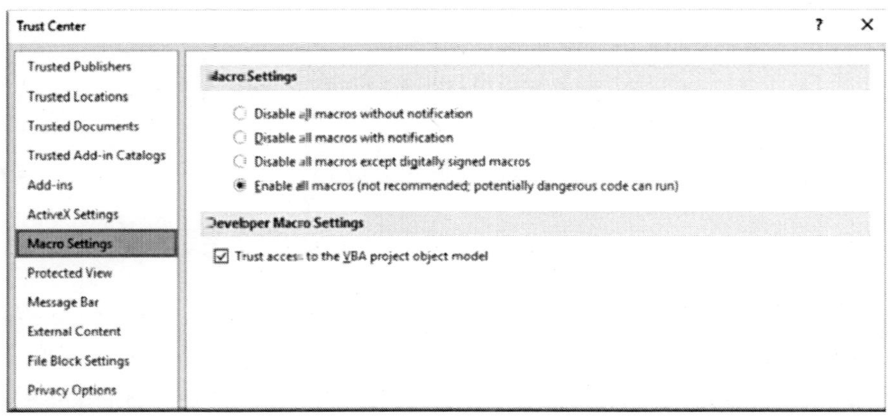

Fig. 5.2: Screenshot of macro setting.

A) Macro Settings

You need to do this in the Trust Center. You may check (1) enable all macros and (2) Trust access to the VBA project object model.

You may also need to add the subdirectory where the DEA-frontier software is located as "Trusted Locations".

B) Install/Load Excel Solver in Excel 2016

1) Click the Microsoft Office Button in Excel 2007 or File in 2010, and then click Excel Options.
2) Click Add-Ins, and then in the Manage box, select Excel Add-ins. (as shown in Fig. 5.3).

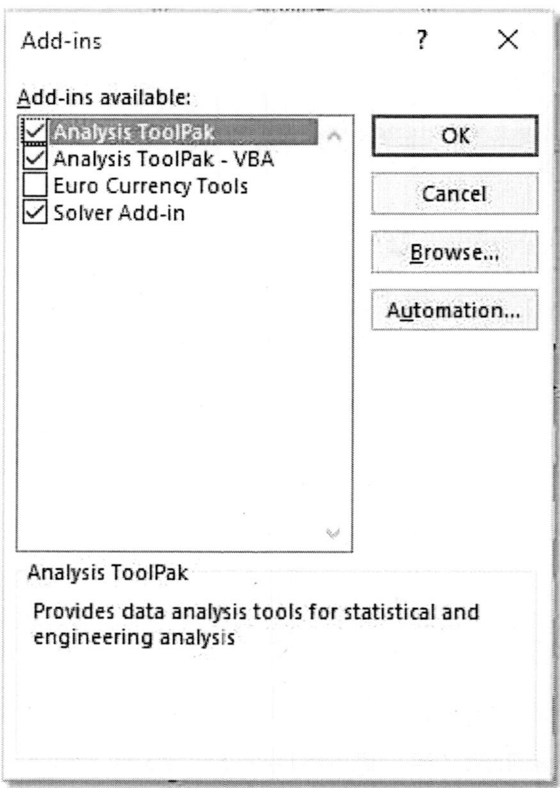

Fig. 5.3: Screenshot of Add-ins tab.

3) Click Go.
4) In the Add-Ins available box, select the Solver Add-in check box, and then click OK.
5) If you get prompted that the Solver Add-in is not currently installed on your computer, click Yes to install it.

After you load the Solver Add-in, the Solver command is available in the Analysis group on the Data tab.

C) Run the DEA-frontier Software

(1) Click on the Add-Ins tab as shown in Fig. 5.4.

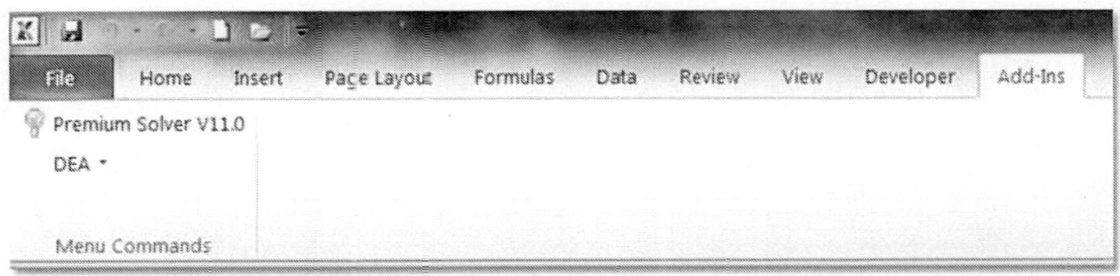

Fig. 5.4: Screenshot of DEA menu

(2) Navigate the DEA menu option.

(3) Click on Envelopment model (as show in Fig. 5.5)

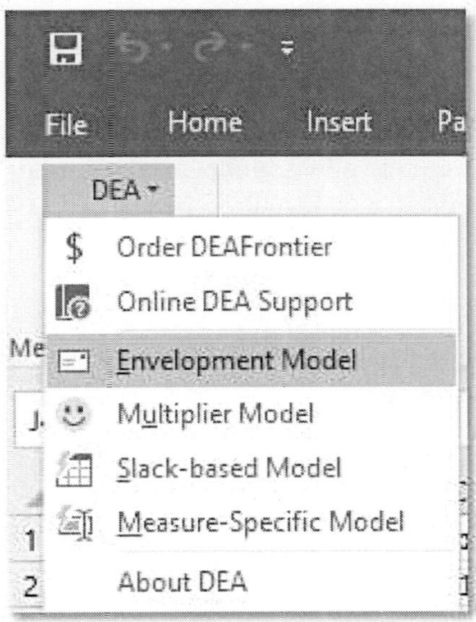

Fig. 5.5: Screenshot of envelopment model.

5.6. EFFICIENCY ANALYSIS OF COMPANY ABC LTD SUPPLY CHAIN MANAGEMENT PROBLEM

X_1 = Labor cost (Lakh Rs)

X_2 = Operating cost (Lakh Rs)

X_3 = Shipping cost (Lakh Rs)

Y_1 = Number of product *A* shipped (Cast aluminum alloy) (Rs/Kg)

Y_2 = Number of product *B* shipped (Titanium) (Rs/Kg)

Z_1 = Labor cost (Lakh Rs)

Q_1 = Sales (Lakh Rs)

Q_2 = Profit (Lakh Rs)

All data related to the case company is given in Table no. 5.1 & 5.2.

Table 5.1: Statistical data of supplier and manufacturer.

DMUs	X_1	X_2	X_3	Y_1	Y_2	Z_1	q_1	q_2
DMU-1	9	50	1	200	1000	8	1000	250
DMU-2	10	18	10	100	1500	10	700	200
DMU-3	9	30	3	80	2000	8	960	300
DMU-4	8	25	1	200	2000	10	800	200
DMU-5	10	40	5	150	2000	15	850	150
DMU-6	7	35	2	350	1000	5	900	350
DMU-7	7	30	3	100	2500	10	1000	300
DMU-8	12	40	4	200	2500	8	1200	100
DMU-9	9	25	2	100	1000	15	1100	150
DMU-10	10	50	1	200	1500	10	800	200

Supplier

Input

X_1 = labor cost (Lakh Rs)

X_2 = operating cost (Lakh Rs)

X_3 = shipping cost (Lakh Rs)

Output

Y_1 = number of product *A* shipped (Cast aluminum alloy) (Rs/Kg)

Y_2 = number of product *B* shipped (Titanium) (Rs/Kg)

Steps for finding the efficiency of suppliers

1) Insert a statistical data of supplier which shown in Table 5.2 in Microsoft Excel.
2) Keep a column empty between input and output.
3) Open DEA-frontier software.

Table 5.2: Statistical data of supplier.

DMUs	*Input*			*Output*	
	X_1	X_2	X_3	Y_1	Y_2
Supplier-1	9	50	1	200	1000
Supplier-2	10	18	10	100	1500
Supplier-3	9	30	3	80	2000
Supplier-4	8	25	1	200	2000
Supplier-5	10	40	5	150	2000
Supplier-6	7	35	2	350	1000
Supplier-7	7	30	3	100	2500
Supplier-8	12	40	4	200	2500
Supplier-9	9	25	2	100	1000
Supplier-10	10	50	1	200	1500

4) Chose DEA-frontier free Solver Platform.
5) Click on the Add-Ins tab.
6) Navigate the DEA menu option.
7) Click on envelopment model.

On clicking it takes a couple of minute and shows an efficiency table, which is shown in Fig. 5.6.

1	Inputs		Outputs
2	X1		Y1
3	X2		Y2
4	X3		
5			
6			Input-Oriented
7			CRS
8	DMU No.	DMU Name	Efficiency
9	1	Supplier-1	1.00000
10	2	Supplier-2	1.00000
11	3	Supplier-3	0.80000
12	4	Supplier-4	1.00000
13	5	Supplier-5	0.67595
14	6	Supplier-6	1.00000
15	7	Supplier-7	1.00000
16	8	Supplier-8	0.77042
17	9	Supplier-9	0.50000
18	10	Supplier-10	1.00000

Fig. 5.6: Screenshot of the efficiency of supplier

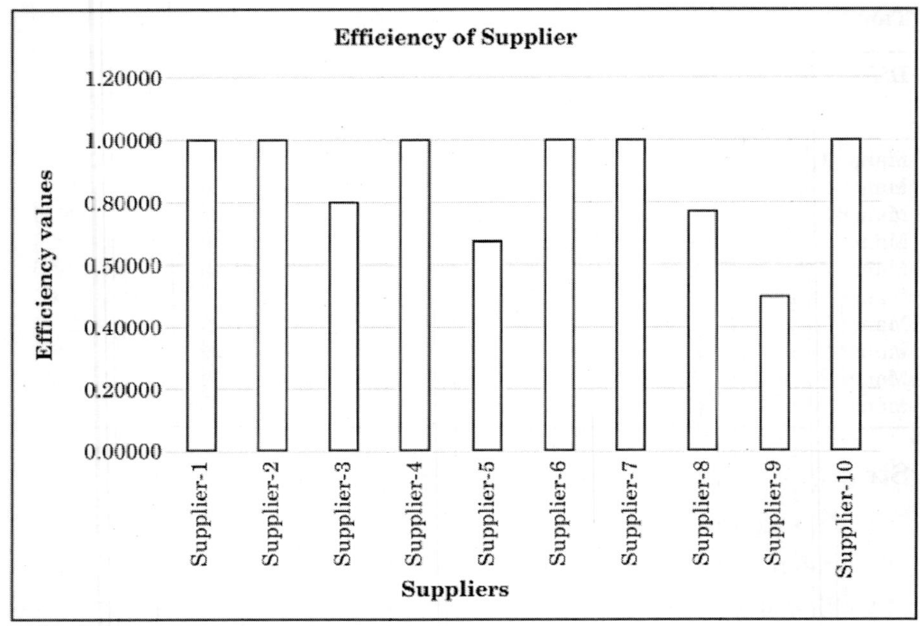

Fig. 5.7: Efficiency of supplier

Fig. 5.6 shows an efficiencies values of suppliers. With the help of the data a two-dimensional graph has been plotted as shown in Fig. 5.7 The graph clearly shows all the suppliers efficiency. Suppliers (1, 2, 4, 6, 7, and 10) has achieved a maximum level of efficiency and supplier-9 has lowest value of efficiency (0.5).

Manufacturer

Input

Y_1 = number of product *A* shipped (Cast aluminum alloy) (Rs/Kg)

Y_2 = number of product *B* shipped (Titanium) (Rs/Kg)

Z_1 = labor cost (Lakh Rs)

Output

Q_1 = sales (Lakh Rs)

Q_2 = profit (Lakh Rs)

Table 5.3: Statistical data of manufacturer.

DMUs	*Input*			*Output*	
	Y_1	Y_2	Z_1	q_1	q_2
Manufacturer-1	200	1000	8	1000	250
Manufacturer-2	100	1500	10	700	200
Manufacturer-3	80	2000	8	960	300
Manufacturer-4	200	2000	10	800	200
Manufacturer-5	150	2000	15	850	150
Manufacturer-6	350	1000	5	900	350
Manufacturer-7	100	2500	10	1000	300
Manufacturer-8	200	2500	8	1200	100
Manufacturer-9	100	1000	15	1100	150
Manufacturer-10	200	1500	10	800	200

Steps for finding the efficiency of manufacturer

1) Insert a statistical data of manufacturer which is shown in Table 5.3 in Microsoft Excel.
2) Keep a column empty between input and output.
3) Open DEA-frontier software.
4) Chose DEA-frontier free Solver Platform.
5) Click on the Add-Ins tab.
6) Navigate the DEA menu option.
7) Click on envelopment model.

On clicking it takes a couple of minute and shows an efficiency table, which is shown in Fig. 5.8.

Fig. 5.8 shows an efficiencies values of manufacturer. With the help of the data we draw a graph which is shown in Fig. 5.9. The graph clearly shows all the manufacturer efficiency. Manufacturer (1, 3, 6, 8, and 9) has achieved a maximum level of efficiency and supplier-5 has lowest value of efficiency (0.59759).

5.7. RESULTS AND DISCUSSION

It should be noted here that the higher the efficiency is the more efficient the supply chain. From Table 5.4 and Fig. 5.10, the following conclusions can be drawn. Firstly, suppliers (1, 2, 4, 6, 7 and 10) and manufacturers (1, 3, 6, 8 and 9) performed well. Their efficiencies are all equal to 1, *i.e.,* they are all efficient. Secondly, supplier-9 and manufactures-5 performs the worst, its average efficiencies are 0.50000 and 0.59759

1	**Inputs**		**Outputs**
2	Y1		q1
3	Y2		q2
4	Z1		
5			
6			Input-Oriented
7			CRS
8	DMU No.	DMU Name	Efficiency
9	1	Manufacturer-1	1.00000
10	2	Manufacturer-2	0.80543
11	3	Manufacturer-3	1.00000
12	4	Manufacturer-4	0.62786
13	5	Manufacturer-5	0.59759
14	6	Manufacturer-6	1.00000
15	7	Manufacturer-7	0.83333
16	8	Manufacturer-8	1.00000
17	9	Manufacturer-9	1.00000
18	10	Manufacturer-10	0.66813

Fig. 5.8: Screenshot of the efficiency of manufacturer

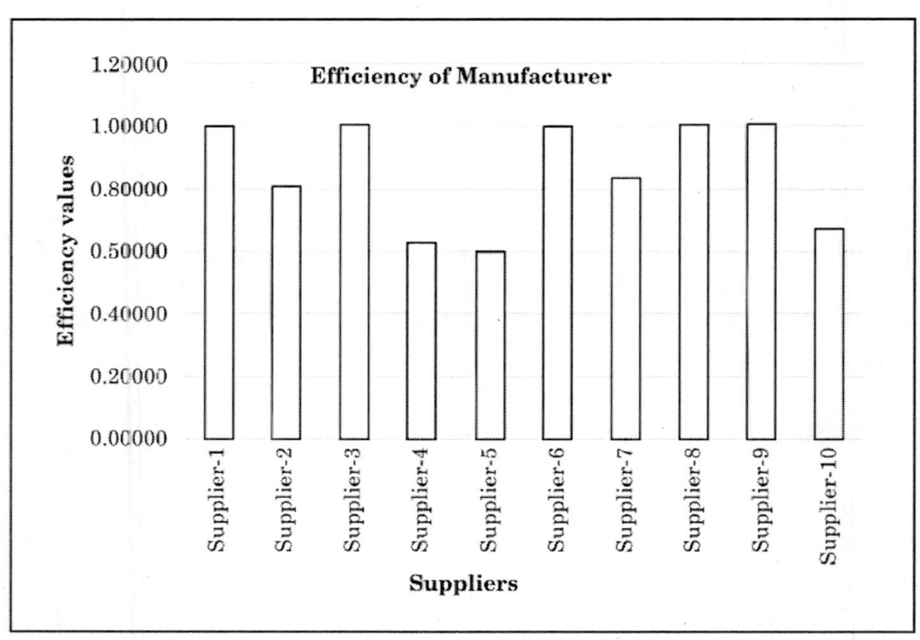

Fig. 5.9: Efficiency of manufacturer

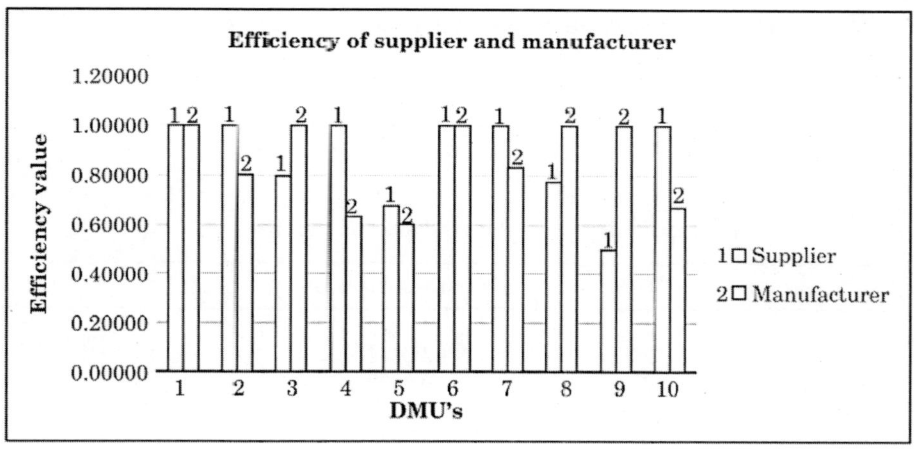

Fig. 5.10: Suppliers and manufacturer efficiency.

which is the smallest among all the DMUs. Thirdly, most of the suppliers and manufactures perform well. Except supplier-9 average efficiencies that have exceeded 0.5. Except manufactures-5 average efficiencies that have exceeded 0.6. Taken DMU-5 as an example its efficiency are 0.67595 and 0.59759 respectively and its average efficiency is 0.63677. For more take DMU-2 as an example its efficiency are 1.0000 and 0.80543 respectively and its average efficiency is 0.902715 and if we see DMU-6 its efficiency is 1.0000 and 1.0000 respectively and its average efficiency is 1.0000 similarly, DMU-1 also having an efficiency are 1.0000 and 1.0000 respectively and its average efficiency is 1.0000. Finally, we find only two DMUs that have an equal efficiency DMU-1 and DMU-6. We must select one of the DMU for the most efficient supply chain network.

Table 5.4: Supplier Efficiency and Manufacturer Efficiency

DMUs	*Supplier*	*Manufacturer*
1	1.00000	1.00000
2	1.00000	0.80543
3	0.80000	1.00000
4	1.00000	0.62786
5	0.67595	0.59759
6	1.00000	1.00000
7	1.00000	0.83333
8	0.77042	1.00000
9	0.50000	1.00000
10	1.00000	0.66813

Estimating the average of the efficiency of supply chain network by using equation number (9)

$$E_{AB} = \frac{1}{2}(E_{AA} + E_{BB}) \tag{9}$$

Table 5.5: Average efficiency of supplier and manufacturer.

DMU's	*Average efficiency*
1	1.0000
2	0.902715
3	0.9000
4	0.81393
5	0.636774
6	1.0000
7	0.91666
8	0.885212
9	0.7500
10	0.834066

The average data of efficiency of a supply chain network are shown in Table 5.5. It should be noted here that the higher the efficiency is the more efficient the supply chain. From Table 5.5 and Fig. 5.11, the following conclusions can be drawn. Firstly, decision making unit number 1 and number 6 performed well. Their average efficiencies are all equal to 1, *i.e.,* they are most efficient among all decision-making unit. Secondly, decision making unit number 5 performs the worst, its average

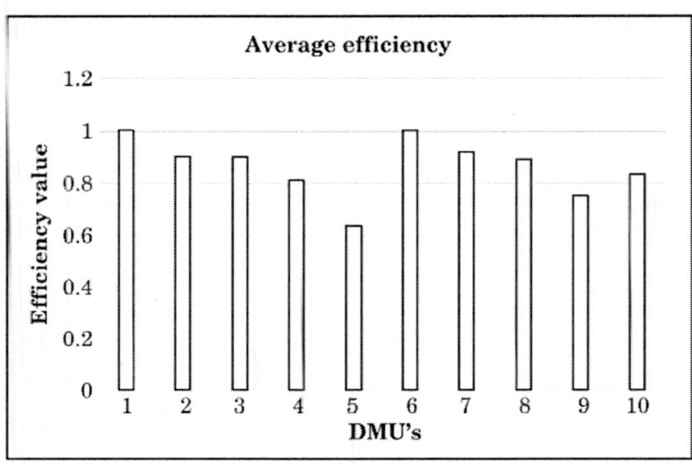

Fig. 5.11: Average efficiency of supplier and manufacturer.

efficiency is 0.636774 which is the smallest among all the DMUs. Thirdly, there are six decision making unit among ten which perform well and having an average efficiency above then 0.8000. Except decision making unit number nine it have a 0.7500 average efficiency value. Taken DMU-5 as an example its average efficiency is 0.636774. For more take DMU-2 as an example its average efficiency is 0.902715 and if we see DMU-6 its average efficiency is 1.0000 similarly, DMU-1 also having an average efficiency is 1.0000. Finally, we find only two DMUs that have an equal efficiency DMU-1 and DMU-6. We must select one of the DMU for the most efficient supply chain network.

5.8. COMPARISON OF THE SUPPLIER EFFICIENCY AND MANUFACTURER EFFICIENCY

Comparison of the Supplier Efficiency and Manufacturer Efficiency are shown in Fig. 5.12. The figure clearly shows the most effective and efficient decision-making unit which have an efficiency value equal to 1. We see in the figure that only decision-making unit number 1 and decision-making unit number 6 having a common point and also attain maximum efficiency. Supplier-1 having efficiency level 1 and Manufacturer-1 also having efficiency level 1, it's means DMU-1 is a most effective. Supplier-2 having efficiency level 1 and Manufacturer-2 having efficiency level 0.80543 which means manufacturer is not efficient. Supplier-3 having efficiency level 0.8 and Manufacturer-3 having efficiency level 1, they are also inefficient. Supplier-4 having efficiency level 1 and Manufacturer-4 having efficiency level 0.62786 which means they did not coincide. Supplier-5 having efficiency level 0.67595 and Manufacturer-5 having efficiency level 0.59759, they both were performed worst. Supplier-6 having efficiency level 1 and Manufacturer-6 also having efficiency level 1, it's means DMU-6 also a most effective. Supplier-7 having efficiency level 1 and Manufacturer-7 having efficiency level 0.8333 which means they did not coincide. Supplier-8 having efficiency level 0.77042 and Manufacturer-8 having efficiency level 1 which means supplier is less efficient. Supplier-9 having efficiency level 0.5 and Manufacturer-9 having efficiency level 1 which means they were not perfect. Supplier-10 having efficiency level 1 and Manufacturer-10 having efficiency level 0.66813 which means they were also not perfect for developing a network. Here we find only two decision making unit which have an efficiency level equal to 1, the DMU's are 1 and 6 so we chose one of them for developing a supply chain network which develop a most efficient and effective network.

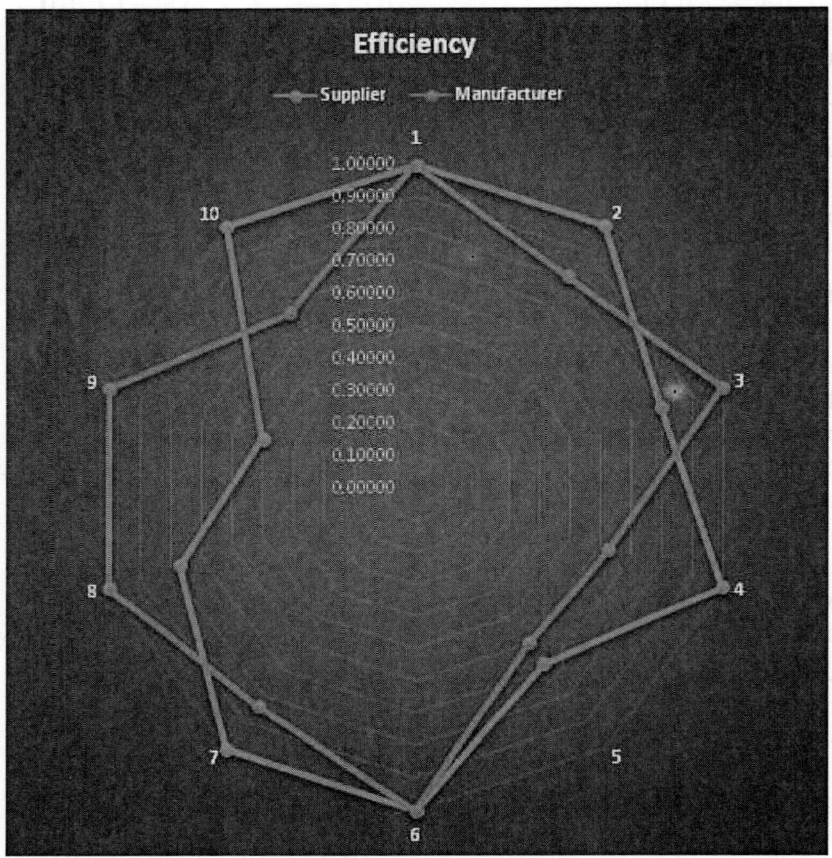

Fig. 5.12 Comparison of the efficiency

5.9. MANAGERIAL IMPLICATIONS

In the present research, the DMU is being identified and analyzed for constructing the most efficient supply chain network. The model developed efforts to suggest that DEA is one of the most effective tool and used for evaluating the efficiency of each network. Then a DMU was selected which have highest efficiency value. In this way, the supply chain management has come to know that which one of the supply chain networks will selected for the most effective supply chain network. Strategic importance of DEA is also analyzed.

The objective of this research was to develop a most effective and efficient supply chain network, the manufacturer gets maximum profit at minimum cost that would help to maximize the profit and provide a best quality product to the customer. These DMU's assume importance because today it is a most effective factor on which efficiency of supply chain network is depend.

The model developed in this research may assist manufacturers and policy makers to comprehend various DMU's and their interdependence towards effective implementation of supply chain management in the supply chain of manufacturing industry.

5.10. CONCLUSIONS

Through careful examinations of the results obtained from DEA-frontier software, among the ten DMUs, we have chosen only one which make supply chain network best. The work of a supply chain management is to construct a network (which start from raw material and end to the customer) which minimize the input cost and maximize the quality and profit. In this research, there are total 10 DMUs which are suitable for the manufacturer but the best choice for any supply chain will directly increase the profit of the company, so it has to choose only one DMU which construct a best supply chain network. As in the present work only suitable for choosing the best supply chain network and find out the all DMUs efficiency. So, it may be a challenge for the future research scholars to explore some new and different types of areas on which DEA technique will apply and that will be beneficial for the future data envelopment analysis system.

6

Application of AHP-GRA Based Hybrid Approach for Green Global Manufacturing Strategy Selection

6.1. INTRODUCTION

Green Manufacturing is a philosophy rather than a standard or a process. It is a method for manufacturing that minimizes waste and pollution through product and process design. The main goal of Green Manufacturing is sustainability. Every manufacturing sector should conserve the resources for future generation. They should also know where, their responsibility ends and what is the acceptable level of toxic emission to the environment. Green manufacturing creates a reputation to public, saves useless cost and promotes research and design. The process of Green Manufacturing involves investing in production process improvements rather than control technology, substitute renewable sources for finite ones, employee recycling and the companies must decide whether to make or buy the product (Maruthi *et al.*, 2015). Environmentally conscious manufacturing (ECM) deals with green principles that are concerned with developing methods for manufacturing products from conceptual design to ûnal delivery to consumers, and ultimately to the EOL disposal, that satisfy environmental standards and requirements. Environmental awareness and recycling regulations have been putting pressure on many manufacturers and consumers, forcing them to produce and dispose of products in an environmentally responsible manner. These have created a need to develop algorithms, models, heuristics, and software for addressing designing, recycling, and other issues (such as the economic viability, logistics, disassembly, recycling, and remanufacturing) for an ever-increasing number of products produced and discarded (Bufardi *et al.*, 2003).

In recent years, environmental awareness and recycling regulations have been putting pressure on many manufacturers and consumers to produce and dispose of products in an environmentally responsible manner. Almost every function within organizations has been inûuence by external and internal pressures to become environmentally sound. Issues such as green consumerism and green product development have impacted marketing. One of the functions that has been profoundly inûuenced by environmental pressures is the organizational operations and manufacturing function. The objective of an ECM program selection procedure is to identify the ECM program selection attributes and obtain the most appropriate combination of the attributes in conjunction with the real requirements of the industrial application. Many precision-based methods for ECM program selection have been developed.

6.2. PROBLEM DESCRIPTION

Industrialization has become a burgeoning virus in today's competitive world. The companies are striving hard to sustain themselves. They tend to give better products and services and improve their manufacturing operations in today's unprecedented global competition. Because of this reason the manufacturing sector consumes lot of energy and other resources and emits large amounts of greenhouse gases which increase environmental problems like climate change - global warming, global dimming etc and environmental degradation. And it also found a large amount of energy is also wasted in many forms. One of the possible ways to strike out these problems is Green Manufacturing. Green Manufacturing can be applied in all manufacturing sectors that minimize waste & pollution, Enables economic progress and conserve resources. This problem is formulated based on literature survey. Literature helped us in defining 37 criteria's. Out of 37 criteria's 12 criteria were selected depending upon the requirement of the theme of the model. Then, these criteria were giving relative importance with the help of a group of experts. The definition of criteria's and their references are discussed in the next section.

6.2.1. Definition of Criterions

Machine utilization

Machine utilization can be defined as the ratio of the output being produced to maximum output that can be produced.

Recycling

Recycling is the process of converting waste material into reusable material to decrease in consumption of fresh material, decreasing waste, decreasing energy waste and environmental pollution.

There are some IS0 standards related to recycling such as ISO 15270: 2008 for plastics waste and ISO 14001: 2004 for environmental management control of recycling practice.

Improved technology

Improved technology here indicates clean means of technology or the technologies which can decrease the harmful emissions. Sometimes use of improved technologies is ignored due to economic constraints, but it should be implemented for sustainability and cleaner means of production can be implemented like wind power plants, hydro power plants and solar power plants.

Optimum resource utilization

Optimum resource utilization as the name signifies is simply the best possible use of available resources like man, machine, money, land etc.

Green training

As every organization spends on the training of their employees, there should be such strategy so that the training gives equal focus to enhance environmental consciousness of the employees. Denisa & Zdenka (2015) stated that enterprises are not interested in educating their employees in this field, which could, in our opinion, enable them to make the processes these employees are responsible for more effective.

Green maintenance

Here green maintenance refers to the maximum utilization of resources and minimum wastage. Things we are using to maintain machinery or equipment should be friendly to the environment and proper disposal of oil, cotton, grease etc. There are several types of maintenance.

- Preventive maintenance, where equipment is maintained before break-down occurs.

- Operational maintenance, where equipment is maintained in using.
- Corrective maintenance, where equipment is maintained after breaking down. This maintenance is often most expensive because worn equipment can damage other parts and cause multiple damages.
- Adaptive maintenance, where equipment is maintained by letting it adapt to new environment.

Green productivity

"Green Productivity (GP) is a strategy for simultaneously enhancing productivity and environmental performance for overall socio-economic development that leads to sustained improvement in the quality of human life. It is the combined application of appropriate productivity and environmental management tools, techniques and technologies that reduce the environmental impact of an organization's activities, products and services while enhancing profitability and competitive advantage". (*Source*: Asian Productivity Organization).

Labor cost

Labor cost is the sum of all wages, which are paid to the employees it consists of two parts direct cost and indirect cost. Direct cost involves the wages of those workers, which are taking part in physical manufacturing means operating machines, while indirect labor cost involves the wages of labor not operating machines while they are acting as support labor.

Green personnel

Hire only those personnel that care about the environment and have an environment friendly mind set. Here green personnel indicate the attitude of employees within the organization. We should select such a manufacturing strategy which emphasis on green personnel recruitment.

Quality control

ISO 9000 defines quality control as *"A part of quality management focused on fulfilling quality requirements".*

Quality is defined as "fitness for use" and to control this quality various tools are used as Statistical quality control (like acceptance sampling, control charts), Total Quality control, Total quality management and Six sigma.

Production control

It is a "set of actions and decision taken during production to regulate output and obtain reasonable assurance that the specification will met. The American Production and Inventory Control Society, nowadays APICS, defined production control in 1959 as:

> *Production control is the task of predicting, planning and scheduling work, considering manpower, materials availability and other capacity restrictions, and cost to achieve proper quality and quantity at the time it is needed and then following up the schedule to see that the plan is carried out, using whatever systems have proven satisfactory for the purpose.*

Green supplier

Green supplier is a supplier who follows all rules and regulations given by governmental bodies. Since, supplier is the first level of any manufacturing system, so it can perform a very important role, making the manufacturing sustainable and green throughout.

6.3. APPLICATION OF ANALYTICAL HIERARCHY PROCESS (AHP)

In this problem we are using AHP process for weight calculation of all the attributes. We will use all the steps as discussed below:

Step 1. Hierarchy model of the problem is constructed as shown in Fig. 6.1.

Step 2. Pair-wise comparison matrix is formed as machine utilization (M.U.) is equally important to machine utilization and machine utilization is 0.200 times important to Recycle (REC) complete comparison is shown in Table 6.1.

Step 3. Column sum of each column is represented in last row of Table 1, obtained by using eq. (1),

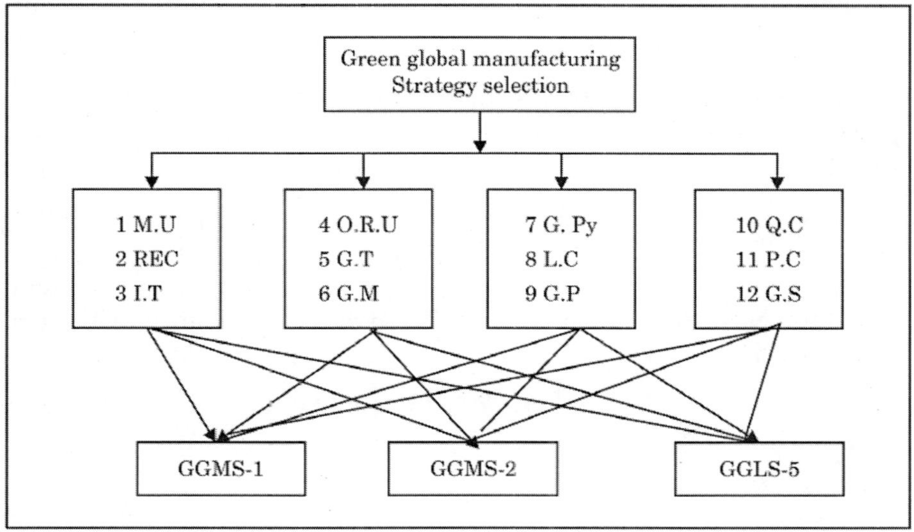

Fig. 6.1: Hierarchical model of green global manufacturing strategy selection.

Table 6.1: Pair-wise comparison matrix for GGMSS.

Criteria	*M.U*	*REC*	*I.T*	*O.R.U*	*G.T*	*G.M*	*G.P*	*L.C*	*G.P*	*Q.C*	*P.C*	*G.S*
M.U	1.000	0.200	0.333	0.250	3.000	1.000	0.333	3.000	2.000	1.000	1.000	2.000
REC	5.000	1.000	3.000	2.000	7.000	5.000	3.000	7.000	6.000	5.000	5.000	6.000
I.T	3.003	0.333	1.000	0.500	5.000	3.000	1.000	5.000	4.000	3.000	3.000	4.000
O.R.U	4.000	0.500	2.000	1.000	6.000	4.000	2.000	6.000	5.000	4.000	4.000	5.000
G.T	0.333	0.143	0.200	0.167	1.000	0.333	0.200	1.000	0.500	0.333	0.333	0.500
G.M	1.000	0.200	0.333	0.250	3.003	1.000	0.333	3.000	2.000	1.000	1.000	2.000
G.P	3.003	0.333	1.000	0.500	5.000	3.003	1.000	5.000	4.000	3.000	3.000	4.000
L.C	0.333	0.143	0.200	0.167	1.000	0.333	0.200	1.000	0.500	0.333	0.333	0.500
G.P	0.500	0.167	0.250	0.200	2.000	0.500	0.250	2.000	1.000	0.500	0.500	1.000
Q.C	1.000	0.200	0.333	0.250	3.003	1.000	0.333	3.003	2.000	1.000	1.000	2.000
P.C	1.000	0.200	0.333	0.250	3.003	1.000	0.333	3.003	2.000	1.000	1.000	2.000
G.S	0.500	0.167	0.250	0.200	2.000	0.500	0.250	2.000	1.000	0.500	0.500	1.000
Col. Sum	20.673	3.586	9.233	5.733	41.009	20.669	9.233	41.006	30.000	20.666	20.666	30.000

$$\text{Column sum} = \sum_{i=1}^{m} C_{ij}, \tag{1}$$

C_{ij}, is the element of Pair-wise Comparison Matrix. Here, for first column of Table 6.1

Column Sum = 1+5+3.003+4+0.333+1+3.003+0.333+0.5+1+1+0.5

= 20.673

Step 4. Normalized matrix is formed by dividing each element of the pair-wise comparison matrix with its corresponding column sum, using eq. (2)

$$N_{ij} = \frac{C_{ij}}{\sum_{i=1}^{m} C_{ij}}, \qquad (2)$$

Where N_{ij} is the element of normalized matrix, as shown in Table 6.2 for example

$$N_{11} = 1/20.673 = 0.024$$

Table 6.2: Normalized matrix

Crit-eria	*M.U*	*REC*	*I.T*	*O.R.U*	*G.T*	*G.M*	*G.P*	*L.C*	*G.P*	*Q.C*	*P.C*	*G.S*
M.U	0.048	0.056	0.036	0.044	0.073	0.048	0.036	0.073	0.067	0.048	0.048	0.067
REC	0.242	0.279	0.325	0.349	0.171	0.242	0.325	0.171	0.200	0.242	0.242	0.200
I.T	0.145	0.093	0.108	0.087	0.122	0.145	0.108	0.122	0.133	0.145	0.145	0.133
O.R.U	0.193	0.139	0.217	0.174	0.146	0.194	0.217	0.146	0.167	0.194	0.194	0.167
G.T	0.016	0.040	0.022	0.029	0.024	0.016	0.022	0.024	0.017	0.016	0.016	0.017
G.M	0.048	0.056	0.036	0.044	0.073	0.048	0.036	0.073	0.067	0.048	0.048	0.067
G.P	0.145	0.093	0.108	0.087	0.122	0.145	0.108	0.122	0.133	0.145	0.145	0.133
L.C	0.016	0.040	0.022	0.029	0.024	0.016	0.022	0.024	0.017	0.016	0.016	0.017
G.P	0.024	0.046	0.027	0.035	0.049	0.024	0.027	0.049	0.033	0.024	0.024	0.033
Q.C	0.048	0.056	0.036	0.044	0.073	0.048	0.036	0.073	0.067	0.048	0.048	0.067
P.C	0.048	0.056	0.036	0.044	0.073	0.048	0.036	0.073	0.067	0.048	0.048	0.067
G.S	0.024	0.046	0.027	0.035	0.049	0.024	0.027	0.049	0.033	0.024	0.024	0.033

Step 5. Weights are calculated taking the average of rows, using the eq. (3),

$$W_{ij} = \frac{\sum_{j=1}^{m} N_{ij}}{m}, \qquad (3)$$

Here for first row it is,

First Row Sum = 0.048 + 0.056 + 0.036 + 0.044 + 0.073 + 0.048 + 0.036 + 0.073 + 0.067 + 0.048 + 0.048 + 0.067 = 0.645

And average will be 0.645/12 = 0.054.

Corresponding values are as given in Table 6.2.

Step 6. Consistency analysis should be done after calculating weights. For this purpose, pair-wise comparison matrix is multiplied by the weight matrix to get weighted vector, using the eq. (4),

$$\begin{bmatrix} C_{11} & \cdots & C_{1m} \\ \vdots & \ddots & \vdots \\ C_{mi} & \cdots & C_{mm} \end{bmatrix} * \begin{bmatrix} W_{11} \\ W_{i1} \\ W_{m1} \end{bmatrix} = \begin{bmatrix} v_{11} \\ v_{i1} \\ v_{m1} \end{bmatrix} \quad (4)$$

As for first row of pair-wise comparison matrix and weighted vector the calculation is

(1.000*0.054) + (0.200*0.249) + (0.333*0.124) + (0.250*0.179) + (3.000*0.022) + (1.000*0.054) + (0.333*0.124) + (3.000*0.022) + (2.000*0.033) + (1.000*0.054) + (1.000*0.054) + (2.000*0.033) = 0.654.

Corresponding values are given in Table 6.3.

Table 6.3: Weight of criterion

Criteria	*M.U*	*REC*	*I.T*	*O.R.U*	*G.T*	*G.M*	*G.P*	*L.C*	*G.P*	*Q.C*	*P.C*	*G.S*
Row sum	0.645	2.987	1.488	2.147	0.259	0.645	1.488	0.259	0.396	0.645	0.645	0.396
Wts.	0.054	0.249	0.124	0.179	0.022	0.054	0.124	0.022	0.033	0.054	0.054	0.033

Step 7. After that consistency vector is calculated using the eq. (5),

$$C_{Vi1} = \frac{1}{W_{i1}}[v_{i1}], \quad \text{for i=1,2,3......m.,} \quad (5)$$

As first consistent vector is 0.654/0.054 = 12.166, values are given in Table 6.4.

Table 6.4: Weighted sum vector (WSM) and Consistency vector (C.V)

Criteria	*M.U*	*REC*	*I.T*	*O.R.U*	*G.T*	*G.M*	*G.P*	*L.C*	*G.P*	*Q.C*	*P.C*	*G.S*
W.S.V	0.654	3.124	1.545	2.248	0.263	0.654	1.564	0.263	0.399	0.654	0.654	0.399
C.V	12.166	12.552	12.463	12.566	12.183	12.166	12.463	12.183	12.079	12.166	12.166	12.079

Step 8. Value of λ_{max} is calculated using eq. (6).

$$\lambda_{max} = \frac{\sum_{i=1}^{m} C_{v_{i1}}}{m} \quad (6)$$

And value is obtained as

$$\lambda_{max} = \frac{\begin{matrix}12.166 + 12.552 + 12.463 + 12.566 + 12.183 + 12.166 + \\ 12.463 + 12.183 + 12.079 + 12.166 + 12.166 + 12.079\end{matrix}}{12} = 12.269$$

Step 9. Consistency Index is calculated using eq. (7)

$$CI = \frac{\lambda_{max} - m}{m - 1} \quad (7)$$

And here $$CI = \frac{12.269 - 12}{11} = 0.024$$

Step 10. Consistency ratio is obtained by $C.R = \frac{C.I}{R.I} = \frac{0.119}{1.48} = 0.017$, which is less than 0.1 hence the results are consistent. The value of R.I is taken from Satty's table for random index. Here we are showing the priorities of attributes.

From the Fig. 6.2, it is clear that recycling is given the highest priority followed by Optimum Resource Utilization (O.R.U), Improved Technology (I.T), Green Productivity (G. Py), Quality Control (Q.C), Production Control (P.C), Machine Utilization (M.U), Green Personnel (G.P), Green Supplier (G.S), Labor Cost (L.C), Green Technology (G.T). Now, these weights will be used for strategy selection.

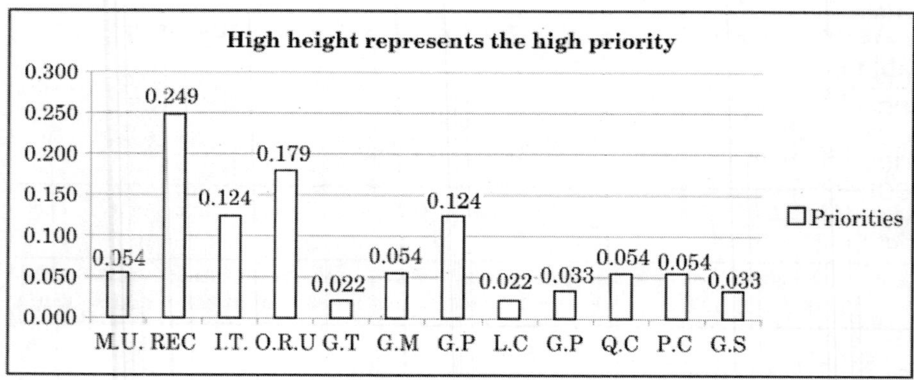

Fig. 6.2: Weights of different attributes.

6.4. APPLICATION OF GREY RELATIONAL ANALYSIS (GRA)

Now application of Grey relation analysis will be made to decide the priority of Green Global Manufacturing Strategies. For this purpose, we will forward our analysis step by step. Decision matrix is shown in Table 6.5 as we can see from the table it is not necessary to normalize the decision matrix because all the entries are between 1 to 9, still for the clarity of the analysis we are normalizing the decision matrix. Steps are as follows:

Step 1. Normalize the decision matrix using eq. (10–12), here all the attributes are benefit attributes. So, we will use eq. (10) only. For *e.g.*, N_{11}= (8–6)/ (8–6) = 1. Normalized values are given in Table 6.6.

Table 6.5: Decision matrix for green global manufacturing strategy selection.

	Attributes / Criterions											
Alter-native	*M.U*	*REC*	*I.T*	*O.R.U*	*G.T*	*G.M*	*G.Py*	*L.C*	*G.P*	*Q.C*	*P.C*	*G.S*
GGMS1	8.000	8.000	7.000	9.000	5.000	7.000	9.000	6.000	7.000	8.000	7.000	5.000
GGMS2	8.000	7.000	6.000	9.000	7.000	8.000	5.000	8.000	5.000	9.000	9.000	6.000
GGMS3	6.000	5.000	8.000	7.000	6.000	6.000	6.000	9.000	7.000	7.000	8.000	7.000
GGMS4	7.000	7.000	7.000	8.000	9.000	8.000	5.000	7.000	6.000	7.000	7.000	6.000
GGMS5	6.000	8.000	7.000	5.000	7.000	6.000	8.000	7.000	6.000	8.000	8.000	8.000
Weights	0.054	0.249	0.124	0.179	0.022	0.054	0.124	0.022	0.033	0.054	0.054	0.033
Max-Min	2.000	3.000	2.000	4.000	4.000	2.000	4.000	3.000	2.000	2.000	2.000	3.000

Table 6.6: Grey relational generating.

	Attributes / Criterions											
Alter-native	*M.U*	*REC*	*I.T*	*O.R.U*	*G.T*	*G.M*	*G.Py*	*L.C*	*G.P*	*Q.C*	*P.C*	*G.S*
GGMS1	1.000	1.000	0.500	1.000	0.000	0.500	1.000	0.000	1.000	0.500	0.000	0.000
GGMS2	1.000	0.667	0.000	1.000	0.500	1.000	0.000	0.667	0.000	1.000	1.000	0.333
GGMS3	0.000	0.000	1.000	0.500	0.250	0.000	0.250	1.000	1.000	0.000	0.500	0.667
GGMS4	0.500	0.667	0.500	0.750	1.000	1.000	0.000	0.333	0.500	0.000	0.000	0.333
GGLS5	0.000	1.000	0.500	0.000	0.500	0.000	0.750	0.333	0.500	0.500	0.500	1.000
reference	1.000	1.000	1.000	1.000	1.000	1.000	1.000	1.000	1.000	1.000	1.000	1.000
	0.000	0.000	0.000	0.000	0.000	0.000	0.000	0.000	0.000	0.000	0.000	0.000
	1.000	1.000	1.000	1.000	1.000	1.000	1.000	1.000	1.000	1.000	1.000	1.000
	0.500	0.500	0.500	0.500	0.500	0.500	0.500	0.500	0.500	0.500	0.500	0.500

Step 2. Reference values are taken as 1 shown in Table 6.7.

Step 3. Grey relational coefficient is calculated using eq. (13), for *e.g.*, C_{11}= (0–0.5*1)/(0+0.5*1), and the value are shown in Table 6.10. ∇_{ij}, the difference between reference and the actual value is shown in Table 6.8. as$\nabla_{11} = 1 - 1 = 0$. andρ is taken as 0.5.

Table 6.7: Table for ∇_{ij}

	Attributes / Criterions											
Alter-native	***M.U***	***REC***	***I.T***	***O.R.U***	***G.T***	***G.M***	***G.Py***	***L.C***	***G.P***	***Q.C***	***P.C***	***G.S***
GGMS1	0.000	0.000	0.500	0.000	1.000	0.500	0.000	1.000	0.000	0.500	1.000	1.000
GGMS2	0.000	0.333	1.000	0.000	0.500	0.000	1.000	0.333	1.000	0.000	0.000	0.667
GGMS3	1.000	1.000	0.000	0.500	0.750	1.000	0.750	0.000	0.000	1.000	0.500	0.333
GGMS4	0.500	0.333	0.500	0.250	0.000	0.000	1.000	0.667	0.500	1.000	1.000	0.667
GGMS5	1.000	0.000	0.500	1.000	0.500	1.000	0.250	0.667	0.500	0.500	0.500	0.000

Table 6.8: Grey relational coefficients.

	Attributes / Criterions											
Alter-native	***M.U***	***REC***	***I.T***	***O.R.U***	***G.T***	***G.M***	***G.Py***	***L.C***	***G.P***	***Q.C***	***P.C***	***G.S***
GGMS1	1.000	1.000	0.500	1.000	0.333	0.500	1.000	0.333	1.000	0.500	0.333	0.333
GGMS2	1.000	0.600	0.333	1.000	0.500	1.000	0.333	0.600	0.333	1.000	1.000	0.429
GGMS3	0.333	0.333	1.000	0.500	0.400	0.333	0.400	1.000	1.000	0.333	0.500	0.600
GGMS4	0.500	0.600	0.500	0.667	1.000	1.000	0.333	0.429	0.500	0.333	0.333	0.429
GGMS5	0.333	1.000	0.500	0.333	0.500	0.333	0.667	0.429	0.500	0.500	0.500	1.000

Table 6.9: Grey relational grade and priorities

Alternatives	***Grey relational grade***	***Priorities***
GGMS1	0.798	1
GGMS2	0.675	2
GGMS3	0.510	5
GGMS4	0.550	4
GGMS5	0.612	3

Step 4. Grey relational grade is calculated using eq. 14, weight values are given in Table 6.9. it is the multiplication of grey relation coefficient matrix and weight vector. Grade calculation for first alternative is as

(1.000*0.054) + (1.000*0.249) + (0.500*0.124) + (1.000*0.179) + (0.033*0.022) + (0.500*0.054) + (1.000*0.124) + (0.333*0.022) + (1.000*0.033) + (0.500*0.054) + (0.333*0.054) + (0.333*0.033) = 0.798

Higher the Grade higher will be the priority to the alternative. As shown in Table 6.9. Here, after the application of GRA, we got the priorities order GGMS1>GGMS2>GGMS5>GGMS4>GGMS3. Now, to validate the results we will apply two another methods VIKOR and TOPSIS to the same problem and we will discuss the results.

6.4.1. Sensitivity Analysis

We can change the weight of each criterion; we are here representing sensitivity analysis giving equal weight to all the criteria's. That means 1/12 = 0.083 to each criterion. Result shows that rank reversal occur. Results are shown in Fig. 6.3 and sensitivity analysis in Fig. 6.4.

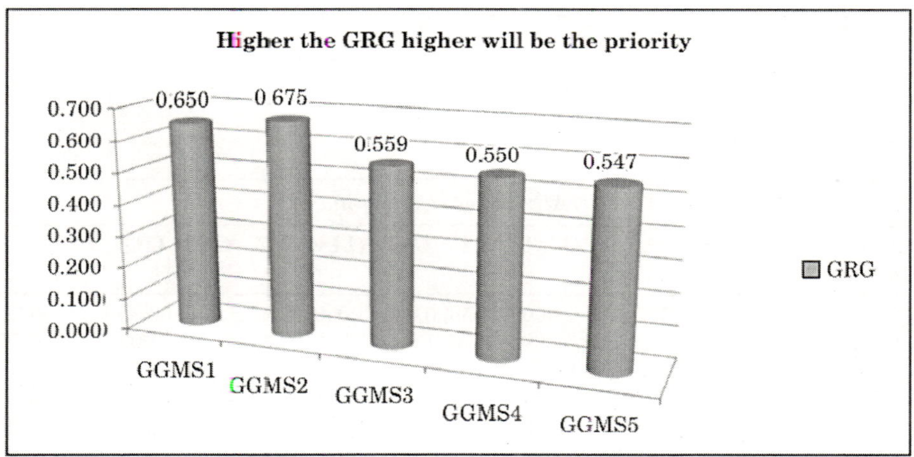

Fig. 6.3: Grey relational grades

6.5. APPLICATION OF VIKOR

Step 1. The analysis of the problem starts with the normalization. This method uses linear normalization. As for first element N_{11}= 8/8=1, and the normalized matrix is shown in Table 6.10.

Step 2. In this step best f_j^* and worst f_j^- values are selected from the criteria's, this shows which alternative is having the best value for criterion. As for first criterion (M.U), GGLS1 and GGLS2 are having the beast normalized value that is 1, and GGMS3 and GGMS5 are having worst values for criterion (M.U) as shown in Table 6.10.

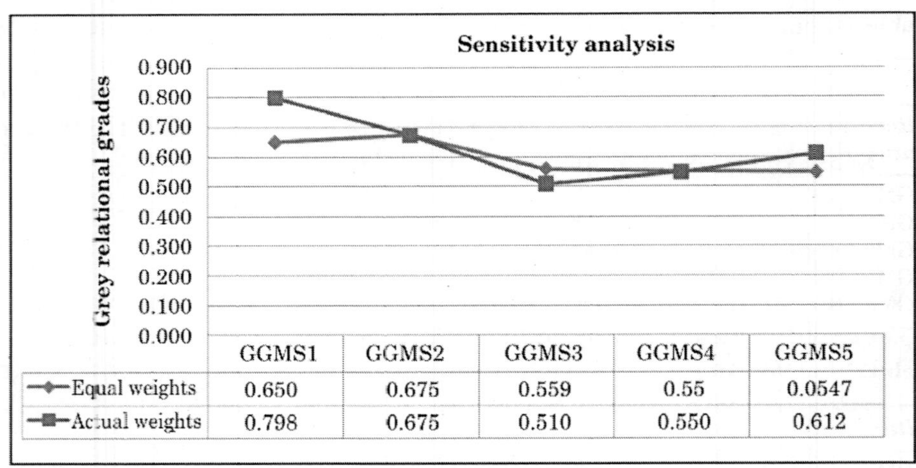

Fig. 6.4: Sensitivity analysis

Table 6.10: Normalized matrix

	Attributes / Criterions											
Alter-native	***M.U***	***REC***	***I.T***	***O.R.U***	***G.T***	***G.M***	***G.Py***	***L.C***	***G.P***	***Q.C***	***P.C***	***G.S***
GGMS1	1.000	1.000	0.875	1.000	0.556	0.875	1.000	0.667	1.000	0.889	0.778	0.625
GGMS2	1.000	0.875	0.750	1.000	0.778	1.000	0.556	0.889	0.714	1.000	1.000	0.750
GGMS3	0.750	0.625	1.000	0.778	0.667	0.750	0.667	1.000	1.000	0.778	0.889	0.875
GGMS4	0.875	0.875	0.875	0.889	1.000	1.000	0.556	0.778	0.857	0.778	0.778	0.750
GGMS5	0.750	1.000	0.875	0.556	0.778	0.750	0.889	0.778	0.857	0.889	0.889	1.000
	1.000	1.000	1.000	1.000	1.000	1.000	1.000	1.000	1.000	1.000	1.000	1.000
	0.750	0.625	0.750	0.556	0.556	0.750	0.556	0.667	0.714	0.778	0.778	0.625
	0.250	0.375	0.250	0.444	0.444	0.250	0.444	0.333	0.286	0.222	0.222	0.375
Weights	0.054	0.249	0.124	0.179	0.022	0.054	0.124	0.022	0.033	0.054	0.054	0.033

Step 3. In this step we calculate value of $\frac{w_j(f_j^* - f_{ij})}{f_j^* - f_j^-}$ is calculated as first element is 0.054(1–1)/ (1–0.750) = 0.

The complete matrix is shown in Table 6.11.

Step 4. Values of S_i and R_i are calculated using the eq. (22), complete values are given in Table 6.12. For *e.g.,* S_1 = 0.000 + 0.000 + 0.062 + 0.000 + 0.022 + 0.027 + 0.000 + 0.200 + 0.000 + 0.027 + 0.054 + 0.033 = 0.247. and R_1 = 0.062.

Table 6.11: Value of $W_j(f_j^* - f_{ij})/(f_j^* - f_j^-)$

	Attributes / Criterions											
Alter-native	***M.U***	***REC***	***I.T***	***O.R.U***	***G.T***	***G.M***	***G.Py***	***L.C***	***G.P***	***Q.C***	***P.C***	***G.S***
GGMS1	0.000	0.000	0.062	0.000	0.022	0.027	0.000	0.022	0.000	0.027	0.054	0.033
GGMS2	0.000	0.083	0.124	0.000	0.011	0.000	0.124	0.007	0.033	0.000	0.000	0.022
GGMS3	0.054	0.249	0.000	0.090	0.017	0.054	0.093	0.000	0.000	0.054	0.027	0.011
GGMS4	0.027	0.083	0.062	0.045	0.000	0.000	0.124	0.015	0.017	0.054	0.054	0.022
GGMS5	0.054	0.000	0.062	0.179	0.011	0.054	0.031	0.015	0.017	0.027	0.027	0.000

Table 6.12: Value of S_i, R_i, Q_i and priorities

Alternatives	***S_i***	***R_i***	***Q_i***	***Priorities***
GGMS1	0.247	0.062	0.000	1
GGMS2	0.404	0.124	0.362	2
GGMS3	0.648	0.249	1.000	5
GGMS4	0.502	0.124	0.484	3
GGMS5	0.476	0.179	0.599	4

Step 5. Value of Q_1 is calculated using the eq. (23). Values are given in Table 6.12.

Here, Q_1 = 0.5(0.247–0.247)/(0.648–0.247) + (1–0.5) (0.062–0.062)/(1.000–0.062) = 0.000.

Lower the value of Q_i higher will be the priority.

Step 6. Optimality test is to be done. Here, $Q_2 - Q_1 > (1/12-1)$ this implies 0.0362 – 0.000 > 0.090.

Here, Q_2 and Q_1, represents value of Q for first two priorities.

Here the values of and are supporting the value of Q_i. Priority values strongly support the results given by GRA. Now we will analyze the same problem with TOPSIS. Fig. 6.5 present analysis of results obtained by VIKOR.

Now we will analyze the same problem with TOPSIS.

6.6. APPLICATION OF TOPSIS

Analysis of the problem consists of the following steps:

Step 1. It consists of normalization of the decision matrix. This technique possesses vector normalization using eq. (15). For *e.g.*, $N_{11} = D_{11}$/square root of sum of square of first column, this implies,

$$N_{11} = 8/15.780 = 0.507$$

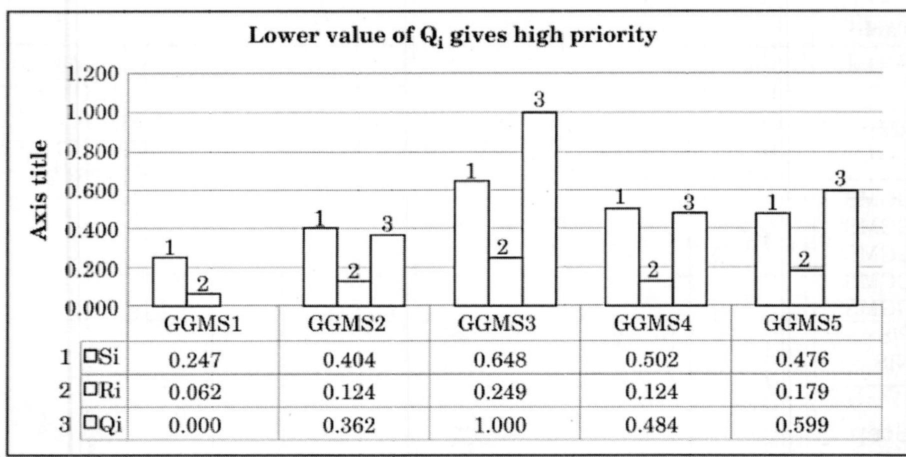

Fig. 6.5: Analysis of results using VIKOR

Normalized matrix is shown in Table 6.13.

Table 6.13: Normalized matrix

	Attributes / Criterions											
Alter-native	***M.U***	***REC***	***I.T***	***O.R.U***	***G.T***	***G.M***	***G.Py***	***L.C***	***G.P***	***Q.C***	***P.C***	***G.S***
GGMS1	0.507	0.549	0.419	0.520	0.446	0.427	0.395	0.340	0.576	0.313	0.425	0.466
GGMS2	0.507	0.481	0.359	0.520	0.390	0.570	0.329	0.454	0.320	0.376	0.478	0.466
GGMS3	0.380	0.343	0.479	0.404	0.446	0.427	0.592	0.510	0.448	0.564	0.478	0.466
GGMS4	0.444	0.343	0.419	0.462	0.502	0.356	0.329	0.510	0.320	0.438	0.425	0.362
GGMS5	0.380	0.481	0.539	0.289	0.446	0.427	0.526	0.397	0.512	0.501	0.425	0.466
Weights	0.054	0.249	0.124	0.179	0.022	0.054	0.124	0.022	0.033	0.054	0.054	0.033

Step 2. Weighted normalized matrix is formed using eq. (16) as for first element it will be like

$$WN_{11} = 0.021 * 0.507 = 0.011$$

These values are shown in Table 6.14.

Step 3. Positive ideal solution and Negative ideal solutions are selected as for first criteria positive ideal solution is the maximum value of the column *i.e.,* 0.011 and Negative ideal solution is the minimum value of the column *i.e.,* 0.008. The idea will remain same for all criterions, because all criterions are benefit criterions. If the criterion is cost criterion, idea will be reversed. These values are shown in Table 6.14.

Table 6.14: Weighted normalized matrix

	Attributes / Criterions											
Alternative	***M.U***	***REC***	***I.T***	***O.R.U***	***G.T***	***G.M***	***G.Py***	***L.C***	***G.P***	***Q.C***	***P.C***	***G.S***
GGMS1	0.027	0.126	0.055	0.093	0.007	0.024	0.073	0.008	0.017	0.025	0.022	0.011
GGMS2	0.027	0.110	0.047	0.093	0.010	0.027	0.041	0.011	0.012	0.028	0.028	0.014
GGMS3	0.021	0.079	0.063	0.072	0.009	0.021	0.049	0.012	0.017	0.022	0.025	0.016
GGMS4	0.024	0.110	0.055	0.083	0.013	0.027	0.041	0.009	0.014	0.022	0.022	0.014
GGMS5	0.021	0.126	0.055	0.052	0.010	0.021	0.065	0.009	0.014	0.025	0.025	0.018
PIS	0.027	0.126	0.063	0.093	0.013	0.027	0.073	0.012	0.017	0.028	0.028	0.018
NIS	0.021	0.079	0.047	0.052	0.007	0.021	0.041	0.008	0.012	0.022	0.022	0.011

Step 4. A matrix for difference between weighted normalized value and positive ideal solution as for first element 0.011 – 0.011 = 0. Table 6.15 shows the matrix values.

Table 6.15: Values of $(PIS - WN_{ij})$

	Attributes / Criterions											
Alternative	***M.U***	***REC***	***I.T***	***O.R.U***	***G.T***	***G.M***	***G.Py***	***L.C***	***G.P***	***Q.C***	***P.C***	***G.S***
GGMS1	0.000	0.000	0.008	0.000	0.006	0.003	0.000	0.004	0.000	0.003	0.006	0.007
GGMS2	0.000	0.016	0.016	0.000	0.003	0.000	0.033	0.001	0.005	0.000	0.000	0.005
GGMS3	0.007	0.047	0.000	0.021	0.004	0.007	0.024	0.000	0.000	0.006	0.003	0.002
GGMS4	0.003	0.016	0.008	0.010	0.000	0.000	0.033	0.003	0.002	0.006	0.006	0.005
GGMS5	0.007	0.000	0.008	0.041	0.003	0.007	0.008	0.003	0.002	0.003	0.003	0.000

Step 5. A matrix for difference between weighted normalized value and negative ideal solution as for first element 0.011 – 0.008 = 0.003. Table 6.16 shows the matrix values.

Step 6. Distance of every alternative from positive ideal solution using eq.17.

As for first alternative (GGLS1) D_1^+ = square root of $[(0.011 - 0.011)^2 + (0.014 - 0.014)^2 + (0.045 - 0.058)^2 + (0.010 - 0.010)^2 + (0.033 - 0.037)^2 + (0.011 - 0.014)^2 + (0.066 - 0.100)^2 + (0.057 - 0.086)^2 + (0.064 - 0.064)^2 + (0.032 - 0.057)^2 + (0.020 - 0.022)^2 + (0.062 - 0.062)^2]$ = 0.053.

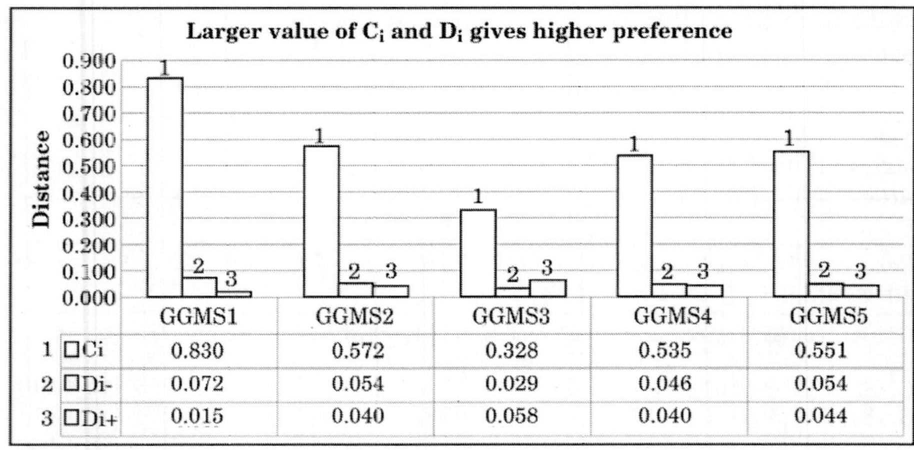

Fig. 6.6: Analyses of results using TOPSIS.

Table 6.16: Values of (WN_{ij} - NIS)

	Attributes / Criterions											
Alter-native	***M.U***	***REC***	***I.T***	***O.R.U***	***G.T***	***G.M***	***G.Py***	***L.C***	***G.P***	***Q.C***	***P.C***	***G.S***
GGMS1	0.007	0.047	0.008	0.041	0.000	0.003	0.033	0.000	0.005	0.003	0.000	0.000
GGMS2	0.007	0.031	0.000	0.041	0.003	0.007	0.000	0.003	0.000	0.006	0.006	0.002
GGMS3	0.000	0.000	0.016	0.021	0.001	0.000	0.008	0.004	0.005	0.000	0.003	0.005
GGMS4	0.003	0.031	0.008	0.031	0.006	0.007	0.000	0.001	0.002	0.000	0.000	0.002
GGMS5	0.000	0.047	0.008	0.000	0.003	0.000	0.024	0.001	0.002	0.003	0.003	0.007

Step 7. Distance of every alternative from negative ideal solution using eq.18.

As for first alternative (GGLS1) D_1^- = square root of $[(0.011 - 0.008)^2 + (0.014 - 0.009)^2 + (0.045 - 0.038)^2 + (0.010 - 0.005)^2 + (0.033 - 0.029)^2 + (0.011 - 0.009)^2 + (0.066 - 0.055)^2 + (0.057 - 0.057)^2 + (0.064 - 0.035)^2 + (0.032 - 0.032)^2 + (0.020 - 0.020)^2 + (0.062 - 0.048)^2]$ = 0.035.

Table 6.17 gives the all values.

Step 8. Calculation of the relative closeness for every alternative using eq. (19), as for first alternative the value is:

$$C_1 = 0.035/(0.035 + 0.053) = 0.400.$$

Larger value of relative closeness shows better alternative. Results are shown in Table 6.17.

Table 6.17: Value of D_i^+, D_i^-, C_i and priorities.

Alternatives	D_i^+	D_i^-	C_i	*Priorities*
GGMS1	0.015	0.072	0.830	1
GGMS2	0.040	0.054	0.572	2
GGMS3	0.058	0.029	0.328	5
GGMS4	0.040	0.046	0.535	4
GGMS5	0.044	0.054	0.551	3

6.7. DISCUSSION

Comparative ranking of alternatives is shown in Fig. 6.7 GRA gives ranking as GGMS1>GGLS2>GGLS5>GGLS4>GGLS3, TOPSIS gives the same results as of GRA,

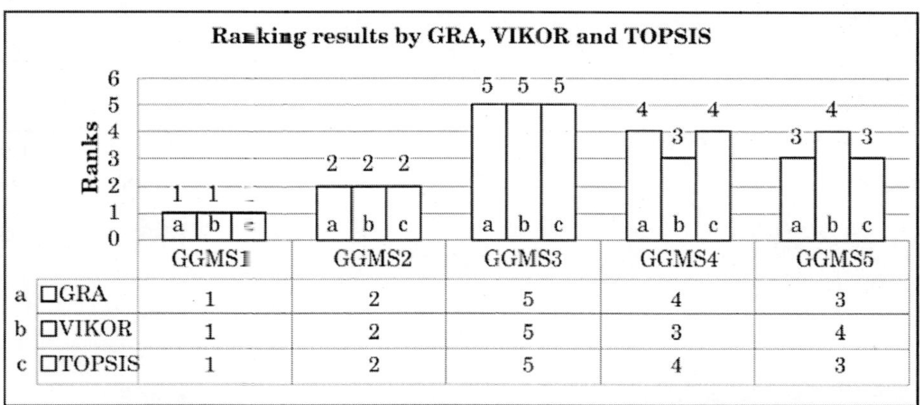

Fig. 6.7: Comparisons of results by GRA, VIKOR and TOPSIS.

And VIKOR gives the ranking as GGMS1>GGMS2>GGMS4>GGMS5>GGMS3, in the decreasing order of their preference. It is clear from the results that alternative GGMS1 is the best alternative under given conditions.

If we look at the problem we find that GGLS1 is going to be the best alternative because Machine utilization and optimum resource utilization is having 9.3% and 10.5% contribution to the strategy and these are also having high weights according to our weight calculation. Sensitivity analysis shows that results are highly sensitive to the weights.

6.8. SUMMARY

In this chapter a model for Green Global Manufacturing Strategy Selection has been developed and used AHP technique for weight calculation of all the 12 attributes. MCDM based analysis using Grey Relational analysis methodology and then applied TOPSIS and VIKOR for the validation of results. A comparison analysis of results obtained by GRA, VIKOR and TOPSIS has been done in the chapter.

7

Evaluation of Remanufacturing Strategies for (EOL)/Used Cell Phones: ANP and Balanced Scorecard Approach

7.1. INTRODUCTION

Remanufacturing of cell phones must meet the challenges of continuously falling prices for new cell phone models, short life cycles, disassembly of unfriendly designs, costly labor and machining costs in developing countries. One of the important problems faced by the top management is strategic decision-making process regarding used/EOL cell phones reverse logistics strategies for end-of-life (EOL) mobile phones management. The most favorable reprocessing/remanufacturing alternative arriving to collection centers has always been a key strategic consideration of any product recovery system. The nature of these types of decisions usually considered to be multidimensional, interdisciplinary, complex, and unstructured due to lack of certainty in environment and information regarding time, quantity and quality of returns, etc. Analytic Network Process (ANP) decision methodology provides an alternative framework to handle these reverse logistics system (RLS) complexities and to determine the decision strategies for best alternative selection for reprocessing. Designing a decision-making model for the same requires quantitative and qualitative evaluation based on criteria such as cost/time, legislative factors, environmental impact, quality, market, etc. Performance must be evaluated based on these criteria to determine a suitable reprocessing option depending on the expert opinion in this domain. In this section, we propose a multiple criteria decision-making (MCDM) model based on ANP and balanced Score Card (BSC) approach. The proposed model can help in designing effective and efficient flexible return policy depending on the various criteria.

In the proposed model, uncertainties regarding quantity and conditions of mobile phones, reliability of capacities, processing times, and demand are considered. The few dimensions of reverse logistics for the EOL mobile phones have been taken from four perspectives derived from balanced scorecard approach, *viz.* finance, social, green business and internal operational perspective. The present approach links the financial and non-financial, tangible and intangible, internal and external factors, thus providing a holistic framework for the selection of an alternative for the reverse logistics operations for EOL cell phones. Many criteria, sub-criteria, determinants, etc. for the selection of reverse manufacturing options are interrelated. The ability of ANP to consider interdependencies among and between levels of decision attributes makes it an attractive multi-criteria decision-making tool. Thus, an ANP-based decision modeling proposed in this chapter provides a more realistic and accurate representation of the problem for conducting remanufacturing logistics operations for EOL cell phones. Finally, a case study has been developed to illustrate the highlight the procedural implementation of the proposed model and data collected for the analysis has been given in Table 7.1.

Table 7.1: Data of the EOL scenario for mobile phones.

Attributes (For non-tangible attributes): on 1-5 scale: 1 for very high, 5 for very low)	*S1*	*S2*	*S3*	*S4*	*S5 (Due to legislation restrictions, this scenario is not possible, hence eliminated)*
Reverse logistics cost (Rs.)	75	85	65	70	---------
Disassembly cost (Rs.)	200	300	300	350	---------
Product value	1	2	3	1	---------
Product cost	4	3	4	2	---------
Manpower involvement	3	2	4	1	---------
Exposure to hazardous materials	5	4	2	3	---------
CO_2 emissions	5	4	1	2	---------
SO_2 emissions	5	4	1	1	---------
Waste reduction (WR)	1	2	3	4	---------
Cost saving (CS)	1	3	4	2	---------
Product recovery option (PRO)	2	3	5	3	---------
Resource consumption (RC)	4	3	2	1	---------
Customer satisfaction (CS)	1	2	5	4	---------
Service quality (SQ)	2	1	4	3	---------

Further, this work also makes efforts to bring ANP-BSC based flexible MCDM and reverse logistics together as a well-suited group decision support tool for alternative selections.

7.2. MOBILE PHONE REMANUFACTURING PROCESS CHAIN

To enable meeting a cost-efficient treatment decision for each phone, a generic remanufacturing process must be selected. Options for phone utilization are reuse, components retrieval and material recycling. In case of reuse, three different quality classes are differentiated, according to the extent of remanufacturing. Upon arrival, phones are separated from accessories and identified. Accessories, *e.g.*, chargers or earphones, are identified and tested to assign them to reuse or recycling. Following identification, phones can be sent directly to recycling or can be tested. Faultless mobile phones are assigned to either reuse (Remanufacturing Class 1) or software update. Updated phones can be assigned to either reuse (Remanufacturing Class 2) or polishing (Remanufacturing Class 3). Defective phones are assigned to alternatively manual or automated disassembly, cleaning and reassembly. Hereafter, phones can undergo additional value-adding processes, *i.e.*, software update or polishing, or be sent directly to reuse. When deciding which phones are to be assigned to which process, certain constraints, *e.g.*, the availability of, and demand for phones and components, the condition of available phones or the cost for and number of remanufacturing resources need to be considered. Due to the high number of existing phone variants at the market — more than 800, including software variants more than 2000 — and the diversity of phone conditions, process complexity is high. The problem of assigning each phone to a cost optimal treatment belongs to the class of combinatorial optimization. Combinatorial optimization deals with models and methods to find the best solution for problems with discrete choices and is well established in disassembly planning (Jayant *et al.*, 2014). In the described problem, the discrete choices are related to the discrete amount of required capacities for remanufacturing processes and the discrete number of mobile phones which are assigned to every remanufacturing process.

7.3. THE DECISION ENVIRONMENT

A graphical representation of the ANP model and decision environment is shown in Fig. 7.1 The overall objective is to carry out reverse logistics operation for EOL mobile phones or evaluation of End-of-Life (EOL) alternatives for used Mobile Phones (RL Operation). The determinants of reverse logistics (Environmentally conscious business practices, legislative factors, organizational performance and operation life cycle/ Logistic focus) are modeled to have dominance over the dimensions of reverse logistics. The reverse logistics operation attribute enablers are those that assist in achieving the controlling dimension of reverse logistics. Thus, these are dependent on the dimension. Also, there are some interdependencies among the enablers, hence the arrow arching back to the enabler's decision level (Fig. 7.1). For example, RLC (reverse logistics cost) and DC (disassembly cost) are interdependent to some degree. To achieve less reverse logistics cost, disassembly cost must be optimized by the companies.

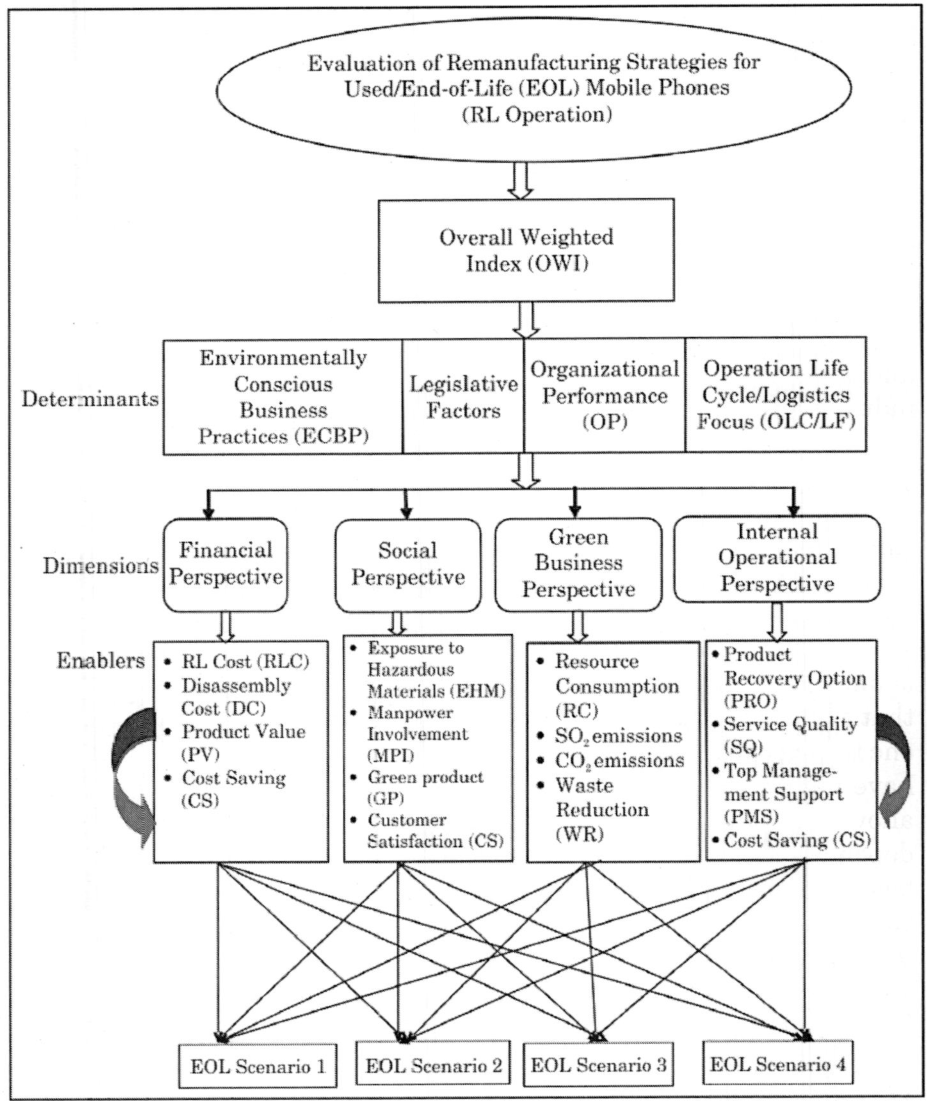

Fig. 7.1: ANP model for reverse logistics operations for used/EOL mobile phones.

***Scenario 1*:** Repair, Refurbishing & Reuse (RRR) of Mobile Phone

***Scenario 2*:** Cannibalization, Remanufacturing & Reuse (CRR) of Mobile Phone

***Scenario 3*:** Incineration with Energy Recovery (INC) for most of the elements & disposal to landfill for a few elements of the mobile phone

***Scenario 4*:** Recycling of complete mobile phone for material recovery (REC)

***Scenario 5*:** Disposal of Whole Product to Landfill (LND)

The various alternatives available to the decision maker in this example include scenario 1: Repair, refurbishing & reuse (RRR) of mobile phone, scenario 2: Cannibalization, remanufacturing & reuse (CRR) of mobile phone, scenario 3: Incineration with energy recovery (INC) for most of the elements & disposal to landfill for a few elements of the mobile phone, scenario 4: Recycling of complete mobile phone for material recovery (REC), scenario 5: Disposal of whole product to landfill (LND).

In this section, we briefly describe the benefits of the ANP process and applying it to mobile phones manufacturing/remanufacturing industries case studies to explain the ANP methodology.

7.3.1. Methodology: The Analytic Network Process

Analytic Network Process (ANP) is a generalized form of the widely used multi-criteria decision-making technique of AHP [93]. ANP offers several advantages over other evaluation techniques such as data envelopment analysis (DEA), expert systems, goal programming etc. It is one of the most comprehensive frameworks for corporate decisions that are available today to the decision-maker. It is a process that allows one to include all the factors and criteria, tangible and intangible, those have bearing on making best decision. The Analytic Network Process allows both interaction and feedback within clusters of elements (inner dependence) and between clusters (outer dependence). Such feedback best captures the complex effects of interplay in corporate world, especially when risk and uncertainty are involved. The ANP, developed by Saaty, provides a way to input judgments and measurements to derive ratio scale priorities for the distribution of influence among the factors and groups of factors in the decision. Because the process is based on deriving ratio scale measurements, it can be used to allocate resources according to their ratio-scale priorities. The well-known decision framework, the Analytic Hierarchy Process (AHP) is a special case of the ANP. Both the AHP and the ANP derive ratio scale priorities for elements and clusters of elements by making paired comparisons of elements on a common property or criterion. Although many decision problems are best studied through the ANP, one may wish to compare the results obtained with it to those obtained using the AHP or any other decision approach with respect to the time it took to obtain the results, the effort involved in making the judgments, and the relevance and accuracy of the results. ANP models have two parts: the first is a control hierarchy or network of objectives and criteria that control the interactions in the system under study; the second are the many sub-networks of influences among the elements and clusters of the problem, one for each control criterion. The ANP has been applied to a large variety of decisions: marketing, medical, political, social, forecasting and prediction and many others. Its accuracy of prediction is impressive in applications that have been made to economic trends, sports and other events for which the outcome later became known. The ANP utilizes the idea of a control hierarchy or a control network to deal with different criteria, eventually leading to the analysis of benefits, opportunities, costs, and risks. By relying on control elements, the ANP parallels what the human brain does in combining different sense data as for example does the thalamus.

The ANP model that is presented in this research has been evaluated for mobile phones manufacturing/remanufacturing industries, these industries were interested in the implementation of the reverse logistics operations for used mobile phones. Due to the limited budget constraints, the industries wanted a systematic way to determine the best possible option for conducting the reverse logistics operations to used mobile phones. The case experience helps us to understand in a better way the advantages and disadvantages of the methodology from a practical point of view. The analysis and the implementation of the ANP model are presented in the following nine steps.

7.3.2. Application of ANP Methodology

Step 1: ***Model development and problem formulation***

In this step, the decision problem is structured into its important components. The relevant criteria and alternatives has been selected through review of literature and discussion with experts from industry and academia. The relevant criteria and alternatives are structured in the form of a control hierarchy where the criteria at the top level in the model have the highest strategic value. The top-level criteria in this model are Environmentally Conscious Business Practices (ECBP), Legislative Factors (LF), Organizational Performance (OP) Operation Life Cycle/Logistics Focus (OLC/LF). These four criteria are termed as the determinants. In the second level of hierarchy, four sub-criteria termed as dimensions of the model is placed which supports all the four determinants at the top level of hierarchy. These are Financial Perspective (FP), Social Perspective (SP), Green Business perspective (GBP) internal Operational perspective (IOP). In this ANP model, each of the four dimensions has some enablers, which helps to achieve that dimension. For example, the dimension FP is supported by the enablers RL Cost (RLC), Disassembly Cost (DC) Product Value (PV) and Cost Saving (CS). These enablers also have some interdependency on one another. For example, in the dimension FP, enablers PV and CS are interdependent as product values and cost saving are interdependent. The degree of interdependency may vary from case to case and would be captured in later steps. The strength of the ANP model is that the feedback and the network structure of the ANP makes possible the representation of the decision problem without much concern for what comes first and what comes next in a hierarchy. The objective of this hierarchy is to select the best possible alternative that will best meet the goals of conducting effective reverse logistics in a mobile phone industry. The ANP model so developed is presented in Fig. 7.1. The alternatives that the decision maker wishes to evaluate are shown at the bottom of the model. The opinion of the logistics manager of the company was sought in the comparisons of the relative importance of the criteria and the formation of pair-wise comparison matrices to be used in the ANP model. In this chapter, mainly for brevity, we present and illustrate the results only of the ECBP determinant. The results of all the four determinants would be used in the calculation of reverse logistics practices overall weighted index (RLP-OWI), which indicates the score assigned to a remanufacturing strategy for used/end-of-life (EOL) mobile phones.

Step 2: ***Pair-wise comparison of four determinants***

In this step, the decision maker is asked to respond to a series of pair-wise comparisons where two components at a time are compared with respect to an upper level 'control' criterion. These comparisons are made to establish the relative importance of determinants in achieving the case company's objectives. In such comparisons, a scale of 1–9 is used to compare two options. In this a score of 1 indicates that the two options under comparison have equal importance, while a score of 9 indicates the overwhelming dominance of the component under consideration (row component) over the comparison component (column component) in a pair-wise comparison matrix. In case, a component has weaker impact than its comparison component, the range of the scores will be from 1 to 1/9, where 1 indicates indifference and 1/9 represents an overwhelming dominance by a column element over the row element. For the reverse comparison between the components already compared, a reciprocal value is automatically assigned within the matrix, so that in a matrix $a_{ij}*a_{ji}=1$. The matrix showing pair-wise comparison of determinants along with the e-vectors of these determinants is shown in Table 7.2. The e-vectors (also referred to as local priority vector) are the weighted priorities of the determinants and shown in the last column of the matrix. In this work, a two-stage algorithm is used for computing e-vector. For the computation of the e-vector, we first add the values in each column of the matrix. Then, dividing each entry in each column by the total of that column, the normalized matrix is obtained which permits the meaningful comparison among elements. Finally, averaging over the rows is performed to obtain the e-vectors. These e-vectors would be used in Table 7.10 for the calculation of reverse logistics overall weighted index (RLP-OWI) for alternatives.

Table 7.2: Pair-wise comparison of determinants.

Determinant	*ECBP*	*LF*	*OP*	*OLC/LF*	*E-vector*
ECBP	1	1/5	1/5	1/4	0.0656
LF	5	1	1	2	0.3638
OP	5	1	1	2	0.3638
OLC/LF	4	1/2	1/2	1	0.2069
			CR = 0.011		

CR = 0.011<0.1 comparison is consistent.

Step 3: ***Pair-wise comparison of dimensions***

In this step, a pair-wise comparison matrix is prepared for determining the relative importance of each of the dimensions of reverse logistics practices (FP, SP, GBP and IOP) on the determinant of reverse logistics. In the model, four such matrices would be formed one for each of the determinant. One such matrix for the pair-wise comparison for dimensions under Environmentally Conscious Business Practices (ECBP) determinant is shown in Table 7.3. From these tables, the results of the comparison (e-vectors) of the dimensions for the respected determinants are carried as Pja in desirability index matrices in Table 7.9.

Table 7.3: Pair-wise comparison for dimensions Under Environmentally Conscious Business Practices (ECBP).

Pair-wise comparison for dimensions under Environmentally Conscious Business Practices (ECBP)					
ECBP	***FP***	***SP***	***GBP***	***IOP***	***E-vector***
FP	1	2	3	1	0.3507
SP	1/2	1	2	1/2	0.1892
GBP	1/3	1/2	1	1/3	0.1093
IOP	1	2	3	1	0.3507
CR = 0.0039					

CR = CR = 0.0039<0.1 comparison is consistent.

Step 4: *Pair-wise comparison matrices between component/enablers levels*

In this step, the decision maker is asked to respond to a series of pair-wise comparisons where two components would be compared at a time with respect to an upper level control criterion. The pair-wise comparisons of the elements at each level are conducted with respect to their relative influence towards their control criterion. In the case of interdependencies, components within the same level may be viewed as controlling components for each other, or levels may be interdependent on each other. For a determinant, pair-wise comparison is done between the applicable enablers within a given dimension cluster. Pair-wise comparison for Financial Perspective under Environmentally Conscious Business Practices (ECBP) determinant is shown in Table 7.4.

Table 7.4: Pair-wise comparison for financial perspective under Environmentally Conscious Business Practices (ECBP) determinant.

Pair-wise comparison for financial perspective under Environmentally Conscious Business Practices (ECBP) determinant					
ECBP/FP	***RLC***	***DC***	***PV***	***CS***	***E-vector***
RLC	1	1/4	1/3	1/2	0.0930
DC	4	1	2	3	0.4478
PV	3	1/2	1	4	0.3224
CS	2	1/3	1/4	1	0.1369
CR = 0.0530					

CR = 0.0530<0.1 comparison is consistent.

In Table 7.4, the relative importance of RLC when compared to DC with respect to PV and CS, in achieving the Environmentally conscious business practices, From Table 7.4 it is also observed that for the case company, the enabler DC has the maximum influence (0.4478) and PV has intermediate influence (0.3224) and CS and has influence (0.1369) on CP for the ECBP. Similarly, RLC has the minimum influence (0.0930) on DC under ECBP. The number of such pair-wise comparison matrices depends on the number of determinants and the dimensions in the ANP model. In this model, 16 such pair-wise comparison matrices are formed. The e-vectors obtained from these matrices are imported as A^{D}_{kja}.

Step 5: *Pair-wise comparison matrices of interdependencies*

Pair-wise comparisons are done to consider the interdependencies among the enablers. Pair-wise comparison matrix for enablers under environmentally conscious business practices, financial perspective and RL cost (RLC) is presented below. From Table 7.5, it is observed that DC (0.5119) has the maximum impact on ECBP/FP cluster with RLC as the control enabler over others. It is also observed that the impact of CS on RLC in ECBP/FP cluster is minimum (0.1279). Therefore, CS is not a problem for the user company and it will have little impact on RL Cost in ECBP/FP cluster. For each determinant, there will be 16 such matrices at this level of relationship will be formed. The e-vectors from these matrices are used in the formation of super matrices. As there are four determinants, 64 such matrices will be formed. The e-vectors have been used in sixth column of the super matrices.

Table 7.5: Pair-wise comparison matrix for enablers under environmentally conscious business practices Financial Perspective and RL Cost (RLC).

Pair-wise comparison matrix for enablers under Environmentally conscious business practices Financial Perspective and RL Cost (RLC)				
ECBP/FP/RLC	*DC*	*PV*	*CS*	*E-vector*
DC	1	2	3	0.5119
PV	1/2	1	4	0.3601
CS	1/3	1/4	1	0.1279

Step 6: *Evaluations of alternatives*

Pair-wise comparison of alternatives under the enablers

The final set of pair-wise comparisons is made for the relative impact of each of the alternatives, enablers in influencing the determinants. The numbers of such pair-wise comparison matrices are dependent on the number of enablers that are included in each of the determinants. In our present case, there are 16 enablers for each of the determinant, which lead to 64 such pair-wise matrices. One such pair-wise comparison matrix between alternatives for ECBP and RLC is shown in Table 7.6. The e-vectors from this matrix are used in columns 6–9 of desirability indices matrices. The columns 6–9, correspond to EOL-S1, EOL-S2 and EOL-S3 & EOL-S4 respectively.

Table 7.6: Pair-wise comparison matrix between alternatives for ECBP and RLC.

Pair-wise comparison matrix between alternatives for ECBP and RLC					
ECBP/RLC	***S1***	***S2***	***S3***	***S4***	***E-vector***
S1	1	1/2	2	1/2	0.2069
S2	2	1	4	3	0.4638
S3	1/3	1/4	1	1/2	0.0956
S4	2	3	2	1	0.2337
			CR = 0.0649		

CR = 0.0649

Step 7: *Super matrix formations*

The super matrix allows for a resolution of the interdependencies that exist among the elements of a system. It is a partitioned matrix where each sub-matrix is composed of a set of relationships between and within the levels as represented by the decision maker's model. In this model, there are four super matrixes for each of the four determinants of reverse logistics hierarchy network, which need to be evaluated. The super matrix 'M', are shown below that presents the results of the relative importance measures for each of the enablers for the ECBP determinant of the reverse logistics practices. In the analysis, the super matrices of the other three determinants have also been generated. The values of the elements of the super matrix M have been imported from the pair-wise comparison matrices of interdependencies (for example, Table 7.7). As there are 16 such pair-wise comparison matrices, one for each of the interdependent enablers in the ECBP, there will be 16 non-zero columns in this super matrix. Each of the non-zero values in the column is the relative importance weight associated with the interdependent pair-wise comparison matrices. In the next stage, the super matrix M is made to converge to obtain a long-term stable set of weights. For convergence to occur, super matrix needs to be 'column stochastic', *i.e.*, the sum of each of the columns of the super matrix needs to be one. Raising the super matrix M to the power 2^{k+1}, where k is an arbitrarily large number, allows for the convergence of the interdependent relationships. One such super matrix before convergence is shown in Table 7.7. One such converged super matrix is shown in Table 7.8.

Step 8: ***Selection of the best alternative for a determinant***

The selection of the best alternative depends on the outcome of the 'desirability index'. The Desirability index, D_{ia}, for the alternative i and the determinant a is defined as:

$$D_{ia} = \sum_{j=1}^{j} \sum_{k=1}^{Kja} P_{ja} A^{D}_{kja} A^{I}_{kja} S_{ikja} \qquad (7.16)$$

Where,

P_{ja} is the relative importance weight of dimension of reverse logistics operation j on the determinant of reverse logistics operation a,

A^{D}_{kja} is the relative importance weight for reverse logistics operation attribute enabler k of dimension of reverse logistics j in the determinant of reverse logistics control hierarchy network a for the dependency (D) relationships between component levels,

A^{I}_{kja} is the stabilized relative importance weight (determined by the super matrix) for reverse logistics operation attribute enabler k of dimension of reverse logistics operation j in the determinant of reverse logistics control hierarchy network a for interdependency (I) relationships within the reverse logistics attribute enablers' component level,

S^{i}_{kja} is the relative impact of reverse logistics implementation alternative i on reverse logistics operation attribute enabler k of dimension of reverse logistics operation j of reverse logistics operation control hierarchy network a,

Kea is the index set of reverse logistics operation attribute enablers for dimension of reverse logistics operation j in for reverse logistics determinant control hierarchy a, and

J is the index set for the dimensions of reverse logistics operation (same for all control hierarchies).

Super matrix formulation

Table 7.7: Super matrix M for Environmentally Conscious Business Practices before convergence.

	Super Matrix M for Environmentally Conscious Business Practices before convergence															
	RLC	***DC***	***PV***	***CS***	***EHM***	***MPI***	***GP***	***CS***	***RC***	***SO_2***	***CO_2***	***WR***	***PRO***	***SQ***	***PMS***	***CS***
RLC	0	0.1560	0.1373	0.1226												
DC	0.5119	0	0.6232	0.5571												
PV	0.3602	0.6196	0	0.3204												
CS	0.1279	0.2243	0.2395	0												
EHM					0	0.3092	0.3339	0.6196								
MPI					0.1822	0	0.0982	0.2243								
GP					0.1149	0.1096	0	0.1560								
CS					0.7028	0.5813	0.5679	0								
RC									0	0.4545	0.4286	0.6479				
SO_2									0.2299	0	0.1428	0.2299				
CO_2									0.1222	0.0910	0	0.1222				
WR									0.6479	0.4545	0.4286	0				
PRO													0	0.1638	0.1222	0.1096
SQ													0.5390	0	0.6479	0.5812
PMS													0.2972	0.5390	0	0.3092
CS													0.1638	0.2972	0.2299	0

Selection of best alternatives

Table 7.9 presents the desirability indices for the Environmentally Conscious Business Practices (Di ECBP). It is based on the hierarchy using the relative weights obtained from the pair-wise comparison of alternatives, dimensions and weights of enablers from the converged super matrix. These weights are used to calculate a score for the determinants of reverse logistics practice overall weighted index (RLP-OWI) for each of the alternative being considered. In Table 7.9, the values of third column are imported from Table 7.3, which are obtained by comparing the relative impact of the dimensions on the Environmentally Conscious Business Practices determinant. For example, in improving the organization performance, the role of financial perspective and internal operation performance is found to be equal and most important (0.3507), which is followed by SP (0.1892) and GBP (0.1093).

The values in the fifth column of Table 7.9 are the stable independent weights of enablers obtained through converged super matrix. The next three columns are from the pair-wise comparison matrices giving the relative impact of each of the alternatives on the enablers. The final three columns represent the weighted values of the alternatives $P_{ja}*A^{D}_{kja}*A^{I}_{kja}*S_{1kja}$ for each of the enablers. For illustration, the value corresponding to reverse logistics practices EOL-S1 for FP is 0.0009392 (0.3507*.0938*0.1380*0.00454*0.2069= 0.0009392).

The summations of these results, for the Environmentally Conscious Business Practices of each of these alternatives, are presented in the final row of Table 7.9. These results indicate that the EOL-S1 with a value of 0.1078929 has maximum influence on the organizational performance. It is followed by EOL-S2 (0.0871902), EOL-S4 (0.071449) and EOL-S3 (0.038042). Similar analysis is carried out for other three determinants. In the next step, an index would be calculated to capture the achievement of overall goal of selecting an alternative of reverse logistics practices.

Step 9: ***Calculation of reverse logistic practices overall weighted index (RLP-OWI)***

The RLP-OWI for an alternative i (RLP_OWIi) is the summation of the products of the desirability indices (Dia) and the relative importance weights of the determinants (Ca) of the reverse logistics overall weighted index. It is represented as:

Table 7.8: Super matrix M25 for Environmentally Conscious Business Practices after convergence.

	Super matrix M25 for Environmentally Conscious Business Practices after convergence															
	RLC	***DC***	***PV***	***CS***	***EHM***	***MPI***	***GP***	***CS***	***RC***	***SO_2***	***CO_2***	***WR***	***PRO***	***SQ***	***PMS***	***CS***
RLC	0.1245	0.1245	0.1245	0.1245												
DC	0.3682	0.3682	0.3682	0.3682												
PV	0.3298	0.3298	0.3298	0.3298												
CS	0.1775	0.1775	0.1775	0.1775												
EHM					0.3316	0.3316	0.3316	0.3316								
MPI					0.1599	0.1599	0.1599	0.1599								
GP					0.1169	0.1169	0.1169	0.1169								
CS					0.3924	0.3924	0.3924	0.3924								
RC									0.3583	0.3583	0.3583	0.3583				
SO_2									0.1796	0.1796	0.1796	0.1796				
CO_2									0.1039	0.1039	0.1039	0.1039				
WR									0.3583	0.3583	0.3583	0.3583				
PRO													0.1206	0.1206	0.1206	0.1206
SQ													0.3771	0.3771	0.3771	0.3771
PMS													0.3013	0.3013	0.3013	0.3013
CS													0.2011	0.2011	0.2011	0.2011

Table 7.9: Desirability indices of Environmentally Conscious Business Practices.

		Environmentally Conscious Business Practices											
ECBP	***Enablers***	P_{ja}	A^D_{kja}	A^1_{kja}	$P_{ja}*A^D_{kja}*A^D_{kja}{}^1$	S_{1kja}	S_{2kja}	S_{3kja}	S_{4kja}	***EOL-S1***	***EOL-S2***	***EOL-S3***	***EOL-S4***
		1	2	3	4 = 1*2*3	5	6	7	8	9 = 4*5	10 = 4*6	11 = 4*7	12 = 4*8
FP	RLC	0.3507	0.0930	0.1245	0.001143	0.2069	0.4638	0.0956	0.2337	0.0002364	0.0005299	0.000109	0.000267
FP	DC	0.3507	0.4478	0.3682	0.023561	0.1225	0.2272	0.2272	0.4231	0.0028863	0.0053532	0.005353	0.009969
FP	PV	0.3507	0.3224	0.3298	0.007884	0.3935	0.1376	0.0754	0.3935	0.0031024	0.0010848	0.000594	0.003102
FP	CS	0.3507	0.1369	0.1775	0.006844	0.5579	0.1219	0.0569	0.2633	0.0038182	0.0008343	0.000389	0.001802
SP	EHM	0.1892	0.2990	0.3316	0.027384	0.0569	0.1219	0.5579	0.2633	0.0015582	0.0033381	0.015278	0.00721
SP	MPI	0.1892	0.1130	0.1599	0.007445	0.1079	0.2671	0.0622	0.5628	0.0008033	0.0019886	0.000463	0.00419
SP	GP	0.1892	0.0842	0.1169	0.001481	0.3935	0.1376	0.0754	0.3935	0.0005827	0.0002038	0.000112	0.000583
SP	CS	0.1892	0.5038	0.3924	0.066305	0.5579	0.2633	0.0569	0.1219	0.0369916	0.0174581	0.003773	0.008083
GBP	RC	0.1093	0.3935	0.3583	0.005513	0.0569	0.1219	0.2633	0.5579	0.0003137	0.0006721	0.001452	0.003076
GBP	SO2	0.1093	0.1376	0.1796	0.002133	0.0518	0.1064	0.4209	0.4209	0.0001105	0.0002269	0.000898	0.000898
GBP	CO2	0.1093	0.0754	0.1039	0.000717	0.0569	0.1219	0.5579	0.2633	0.0000408	0.0000874	0.0004	0.000189
GBP	WS	0.1093	0.3935	0.3583	0.012249	0.5579	0.1219	0.0569	0.2633	0.0068339	0.0014932	0.000697	0.003225
IOP	PRO	0.3507	0.0883	0.1206	0.009008	0.5532	0.1983	0.0502	0.1983	0.0049832	0.0017863	0.000452	0.001786
IOP	SQ	0.3507	0.4824	0.3771	0.081421	0.2633	0.5579	0.0569	0.1219	0.0214382	0.0454249	0.004633	0.009925
IOP	PMS	0.3507	0.2718	0.3013	0.027555	0.3935	0.1376	0.0754	0.3935	0.0108430	0.0037916	0.002078	0.010843
IOP	CS	0.3507	0.1575	0.2011	0.02393	0.5579	0.1219	0.0569	0.2633	0.0133507	0.0029171	0.001362	0.006301
										D_{11} = 0.1078929	D_{12} = 0.0871902	D_{13} = 0.038042	D_{14} = 0.071449

RLP–LOWIi = Σ Dia Ca

For example, the RLP–OWI for EOL-S1 is calculated as:

$RLP\text{–}OWI_{EOL\text{–}S1}$ = [(0.0656*0.1078929) + (0.3638*0.1981808) + (0.3638*0.1030396) + (0.2069*0.1078929)] = 0.138985

The results are shown in Table 7.10.

7.4. ANALYSIS AND RESULTS

It is observed from Table 7.10 that EOL-S1 is the most-suited alternative for the reverse logistics operations for used mobile phones for the mobile manufacturing/remanufacturing industries. Alternatives EOL-S4, EOL-S2 and EOL-S3 in sequence follow alternative EOL-S1 respectively. It is observed from Table 7.10 that legislative factors and operating performance factors plays a major role in the conduct of reverse logistics operations for used mobile phones. It is further observed from the second column of this table that EOL-S1 (0.1078929) performs business in the more environmentally conscious RL operation as compared to EOL-S2 (0.0871902), EOL-S3 (0.038042) and EOL-S4 (0.071449). Different factors influence the results like, uncertainty on configuration, condition and place of origin, thus enabling better planning and control of the reverse logistics networks However, these results should be seen in the light of the mobile phones manufacturing companies and the inputs provided by the decision-makers in the formation of pair-wise comparison matrices.

Table 7.10: Reverse Logistics Practices overall weighted index

			RLP overall weighted index			
Alternatives	***ECBP***	***LF***	***OP***	***OLC/LF***	***RLP-OW I***	***Normalized values for LOW I***
Weights	0.0656	0.3638	0.3638	0.2069		
EOL-S1	0.1078929	0.1981808	0.1030396	0.1078929	0.138985	0.3403
EOL-S2	0.0871902	0.1242377	0.0788480	0.0871902	0.097642	0.2391
EOL-S3	0.038042	0.067707	0.040589	0.038042	0.049765	0.1218
EOL-S4	0.071449	0.187266	0.09468	0.071449	0.122042	0.2988

7.5. DISCUSSION AND MANAGERIAL IMPLICATIONS

The proposed methodology provides for simplification of a complex multi-criteria decision-making problem. Advantage of this methodology is that it not only supports group decision-making but also enables us to document the various considerations in the process of decision making. Proper documentation this work may be useful if the results are to be communicated to various stack holders in the reverse supply chain of mobile phones (Mobile manufacturing & remanufacturing companies, recyclers etc.). For the real case study undertaken in this study, the results indicate that EOL Scenario 1 (Repair, Refurbishing & Reuse (RRR) of Mobile Phone) is the best end-of-life (EOL) remanufacturing strategy for mobile phones collected by the case company as shown in Fig. 7.2. This option may be attributed to its more Environmentally Conscious Business Practices (ECBP), legislative factors, and organizational performance and operation life cycle/logistics focus capabilities. The option of EOL-S4 (recycling of complete mobile phone for material recovery) in the framing of Environmentally Conscious Business Practices (ECBP), legislative factors and operation life cycle/ logistics focus capabilities policy also supports this result. It is pertinent here to discuss the priority values of the dimensions, which influence this decision. From Table 7.17, it is observed that financial perspective (0.3507) and internal operation performance (0.3507) is found to be equal and most important dimensions in the selection of an EOL Scenario for used mobile phones. It is followed by social perspective SP (0.1892) and green business perspective (0.1093).

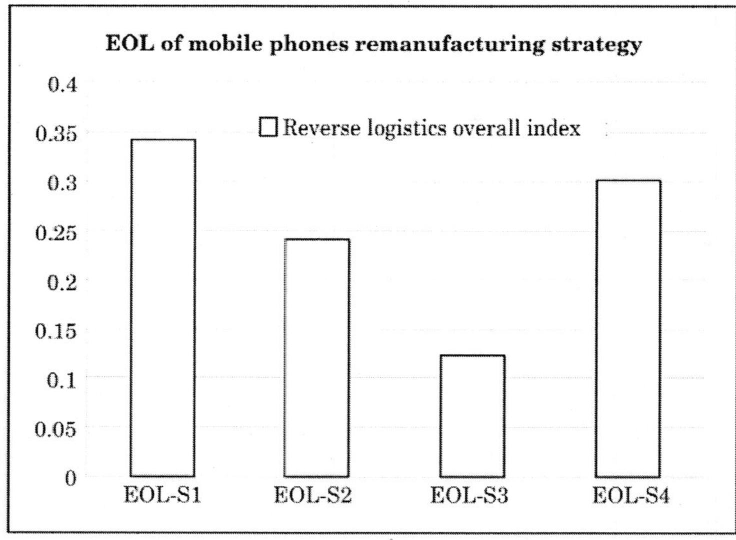

Fig. 7.2: Ranking of End-of-life remanufacturing strategy for Used/EOL mobile phones.

Although the model has been illustrated for four distinct alternative EOL scenarios, it can compare more than four EOL scenarios at the cost of complexity. It needs to be emphasized here that despite using a sound algorithm for systematic decision-making, care must be taken in the application of the ANP approach. In the light of the results obtained for the mobile manufacturing/remanufacturing case companies, it may be noted that these results are valid only for types of companies its own decision environment and should not be generalized to establish the supremacy of one remanufacturing option over the others for the other EOL products. Further, the application of proposed methodology may require significant time and resources from managers and decision-makers.

7.6. SUMMARY

The ANP model presented in this paper structured the problem of conduct of reverse logistics practices for used /EOL mobile phones in a hierarchical form and linked the determinants, dimensions, and enablers of the reverse logistics operations and the alternatives available to the decision maker for mobile phones industries. It can provide support to the top management in the evaluation of the various remanufacturing alternatives available with them as it measures the relative strengths of impacts between elements in the hierarchical model. In this work, we have developed decision support system to evaluate the various alternatives available in the business environment for remanufacturing of used and EOL mobile phones collected by the companies. The proposed methodology serves as a guideline to the various supply chain players who are involved in reverse logistics implementation business for the companies' used and EOL products. The ANP approach is capable enough for taking into consideration both qualitative and quantitative criteria. Similar ANP-based models may also be developed in other contexts as well. But, as the development and evaluation of these models demand significant time and efforts from the decision-makers in the formation of pair-wise comparison matrices, these should be used for long-term strategic decisions only where the investments made in the lengthy and cumbersome process of decision making are recovered in due course of time.

8

Decision Support System Based on MCDM Techniques for Flexible Manufacturing System (FMS) Evaluation

8.1. INTRODUCTION

Globalization, flicking market requirements and modern lifestyle trends has put up tremendous challenge to manufacturing industries. In the current business scenario, the competitiveness of any manufacturing industry is determined by its ability to respond quickly to the rapidly changing market and to produce high quality products at low costs. However, the product cost is no longer the predominant factor affecting the manufacturers' perception. Other competitive factors such as flexibility, quality, efficient delivery and customer satisfaction are drawing the equal attention. Manufacturing industries are striving to achieve these capabilities through automation, robotics and other innovative concepts such as just-in-time (JIT), Production planning and control (PPC), enterprise resource planning (ERP) etc. Flexible manufacturing is a concept that allows manufacturing systems to be built under high customized production requirement. The issues such as reduction of inventories and market-response time to meet customer demands, flexibility to adapt to changes in the market, reducing the cost of products and services to grab more market shares, etc. have made it almost obligatory to many firms to switch over to flexible manufacturing systems (aFMSs) as a viable means to accomplish the above requirements while producing consistently good quality and cost effective products. Flexible Manufacturing Systems (FMSs) present opportunities for manufacturers to improve their technology, competitiveness, and profitability through a highly efficient and focused approach to manufacturing effectiveness (Buzacott and Yao, 1986). The primary reason for implementing FMS lies in its versatility (flexibility).

In general, increased flexibility enables a company to adjust more easily to changes in the market place and in customer requirements, while maintaining high quality standards for its products (Shang & Sueyoshi, 1995). FMS is an automated set of numerically controlled machine tools and material handling systems, capable of performing wide range manufacturing operations with quick tooling and instruction changeovers.

Flexibility is an attribute that allows a mixed model manufacturing system to cope up with a certain level of variations in part or product style, without having any interruption in production due to changeovers between models. Flexibility measures the ability to adapt "to a wide range of possible environment". To be flexible, a manufacturing system must possess the following capabilities:

- Identification of the different production units to perform the correct operation.
- Quick changeover of operating instructions to the computer-controlled production machines.
- Quick changeover of physical setups of fixtures, tools and other working units.

Flexible manufacturing system has come up as a viable mean to achieve these prerequisites. The term flexible manufacturing system, or FMS, refers to a highly automated GT machine cell, consisting of a group of computer numerical control (CNC) machine tools and supporting workstations, interconnected by an automated material handling and storage system, and all controlled by a distributed computer system. The reason, the FMS is called flexible because it is capable for processing a variety of different part styles simultaneously with the quick tooling and instruction changeovers. Also, quantities of productions can be adjusted easily to changing demand patterns.

A flexible manufacturing system (FMS) is designed to combine the efficiency of a mass-production line and the flexibility of a job shop to produce a variety of work pieces on a group of machines (Chan, Kazerooni, and Abhary, 1997).

8.2 RESEARCH METHODOLOGY

8.2.1 AHP Method

One of the most popular analytical techniques for complex decision-making problem is the analytical hierarchy process (AHP). Saaty (1980, 2000) developed AHP, which decomposes a decision-making problem into a system of hierarchies of objectives, attributes (for criteria), and alternatives.

An AHP hierarchy can have as many levels as needed to fully characterize a decision situation. A good number of functional characteristics make AHP a useful methodology. These include the ability to handle decision situations involving subjective judgments, multiple decision makers, and the ability to provide measures of consistency of preference (Triantaphyllou, 2000). Designed to reflect the way people think, AHP continues to be the most highly regarded and widely used decision-making method. AHP can efficiently deal with tangible (*i.e.*, objective) as well as non-tangible (*i.e.*, subjective) attributes, especially where the subjective judgments of different individuals constitute an important part of the decision process. The main procedure of AHP using the radical root method (also called geometric mean method) is as follows:

Step 1: Determine the objective and the evaluation attributes. Develop a hierarchical structure with a goal or objective at the top level, the attributes at the second level and the alternatives at the third level.

Step 2: Determine the relative importance of different attributes with respect to the goal or objective,

- Construct a pair-wise comparison matrix using a scale of relative importance. The judgments are entered using the fundamental scale of the analytic hierarchy process (Saaty 1980, 2000). An attribute compared with itself is always assigned the value 1, so the main diagonal entries of the pair-wise comparison matrix are all 1. The numbers 2, 3, 4, and 5 correspond to the verbal judgments 'moderate importance', 'strong importance', 'very strong importance' and 'absolute importance'. Assuming M attributes, the pair-wise comparison of attribute i with attribute j yields a square matrix R_{m*m} where a_{ij} denotes the comparative importance of attribute i with respect to attribute j. In the matrix, $a_{ij} = 1$ when I = j and $r_{ji} = 1/r_{ij}$.

$$R_{m*m} = \begin{bmatrix} 1 & \cdots & r_{1m} \\ \vdots & \ddots & \vdots \\ r_{m1} & \cdots & 1 \end{bmatrix} \quad (1)$$

Find the relative normalized weight (w_i) of each attribute by (i) calculating the geometric mean of the i-th row, and (ii) normalizing the geometric means of rows in the comparison matrix. This can be represented as

$$GM_i = \left\{ \prod_{j=1}^{m} r_{ij} \right\}^{1/m} \tag{2}$$

$$w_i = \frac{GM_i}{\sum_{i=1}^{m} GM_i} \tag{3}$$

The geometric mean method of AHP is commonly used to determine the relative normalized weights of the attributes, because of its simplicity, easy determination of the maximum Eigen value, and reduction in inconsistency of judgments.

- Calculate matrices A3 and A4 such that A3 = A1 * A2 and A4 = A3/A2, where A2 = $[w_1, w_2, \ldots, w_i]^T$.
- Determine the maximum Eigen value λ_{max} that is the average of matrix A4.
- Calculate the consistency index CI = $(\lambda_{max} - M)/(M - 1)$. The smaller the value of CI, the smaller is the deviation from the consistency.
- Obtain the random index (RI) for the number of attributes used in decision making.
- Calculate the consistency ratio CR = CI/RI. Usually, a CR of 0.1 or less is considered as acceptable, and it reflects an informed judgment attributable to the knowledge of the analyst regarding the problem under study.

8.2.2 VIKOR Method

Multi criteria decision making (MCDM) is one of the most prevalent methods for resolving conflict management issues (Deng & Chan, 2011). MCDM deals with decision and planning problems by consideration of multiple criteria and the importance of each (Haleh & Hamidi, 2011). Among the many MCDM methods, VIKOR is a compromise ranking method to optimize the multi-response process (Opricovic, 1998). It uses a multi criteria ranking index derived by comparing the closeness of each criterion to the ideal alternative. The core concept of VIKOR is the focus on ranking and selecting from a set of alternatives in the presence of conflicting criteria (Opricovic, 2011). In VIKOR, the ranking index is derived by considering both the maximum group utility and minimum individual regret of the opponent (Liou, Tsai, Lin, & Tzeng, 2011).

VIKOR denotes the various n alternatives as $a_1, a_2, . . ., a_n$. For an alternative a_i, the merit of the jth aspect is represented by f_{ij}; that is, f_{ij} is the value of the jth criterion function for the alternative a_i, n being the number of criteria. The VIKOR procedure is divided into the following five steps:

(1) Determine the best f_j^* and worst f_j^- values of all criterion functions. If the jth criterion function represents a merit, then

$$f_j^* = Max_i f_{ij}, f_j^- = Min_i f_{ij} \tag{4}$$

(2) Compute the values S_i and R_i, i = 1, 2, 3,. . .,m, by the relations

$$S_i = \sum_{j=1}^{n} \frac{w_i(f_j^* - f_{ij})}{f_j^* - f_j^-} \tag{5}$$

$$R_i = \max\left[\frac{w_i(f_j^* - f_{ij})}{f_j^* - f_j^-}\right] \tag{6}$$

Where w_i is the weight of the jth criterion which expresses their relative importance of the criteria.

(3) Compute the value Q_i, i = 1, 2, 3,...,m, by the relation

$$Q_i = v\left[\frac{S_i - S^*}{S^- - S^*}\right] + (1\text{-}v)\left[\frac{R_i - R^*}{R^- - R^*}\right] \tag{7}$$

Where $S^* = min_i S_i$, $S^- = max_i S_i$, $R^* = min_i R_i$, $R^- = max_i R_i$, and v is the weight of the strategy of maximum group utility, whereas (1–v) is the weight of the individual regret. Here, when v is larger than 0.5, the index of Q_i follows majority rule.

(4) Rank the alternatives, sorting by the values S, R and Q, in decreasing order.

8.2.3 AHP-PROMETHEE Method

The PROMETHEE method was introduced by Brans *et al.* (1984) and belongs to the category of outranking methods. Like all outranking methods, PROMETHEE proceeds to a pair-wise comparison of alternatives in each single criterion in order to determine partial binary relations denoting the strength of preference of an alternative a1 over alternative a2. In the evaluation table, the alternatives are evaluated on different criteria. The implementation of PROMETHEE requires additional types of information, namely: information on the relative importance or the weights of the criteria considered, and information on the decision maker preference function, which he/she uses when comparing the contribution of the alternatives in terms of each separate criterion.

The PROMETHEE I (partial ranking) and PROMETHEE II (complete ranking) were developed by J.P. Brans and presented for the first time in 1982. A considerable number of successful applications has been treated by the PROMETHEE methodology in various fields such as Banking, Industrial Location, Manpower planning, Water resources, Investments, Medicine, Chemistry, Health care, Tourism, Ethics in OR, Dynamic management. The success of the methodology is basically due to its mathematical properties and to its particular friendliness of use.

PROMETHEE I

The PROMETHEE I give partial ranking and it is obtained from the positive and the negative outranking flows. Both flows do not usually induce the same rankings. PROMETHEE I is their intersection. Where P^I, I^I, R^I respectively stand for preference, indifference and incomparability.

$$\begin{cases} a\,P^I\,b \quad if & \begin{cases} \phi^+(a) > \phi^+(b)\ and\ \phi^-(a) < \phi^-(b), or \\ \phi^+(a) = \phi^+(b)\ and\ \phi^-(a) < \phi^-(b), or \\ \phi^+(a) > \phi^+(b)\ and\ \phi^-(a) = \phi^-(b), or \end{cases} \\ a\,I^I\,b \quad if & \phi^+(a) = \phi^+(b)\ and\ \phi^-(a) = \phi^-(b), or \\ a\,R^I\,b \quad if & \begin{cases} \phi^+(a) > \phi^+(b)\ and\ \phi^-(a) > \phi^-(b), or \\ \phi^+(a) < \phi^+(b)\ and\ \phi^-(a) < \phi^-(b), or \end{cases} \end{cases}$$

When $a\ P^I\ b$, a higher power of a is associated to a lower weakness of a with regard to b. The information of both outranking flows is consistentand may therefore be considered as sure.

When $a\ I^I\ b$, both positive and negative flows are equal.

When $a\ R^I\ b$, a higher power of one alternative is associated to a lower weakness of the other. This often happens when a is good on a set of criteria on which b is weak and reversely b is good on some other criteria on which a is weak. In such a case the information provided by both flows is not consistent. It seems then reasonable to be careful and to consider both alternatives as incomparable. The PROMETHEE I rankingis prudent: it will not decide which action is best in such cases. Itis up to the decision-maker to take his responsibility.

PROMETHEE II

PROMETHEE II consists of the (P^{II}, I^{II}) complete ranking. It is often the case that the decision-maker requests a complete ranking. The netoutranking flow can then be considered. It is the balance between the positive and the negative outranking flows.

The higher the net flow, the better the alternative, so that:

$$\begin{cases} a\,P^{II}\,b & if \quad \phi(a) > \phi(b), \\ a\,I^{II}\,b & if \quad \phi(a) > \phi(b) \end{cases}$$

When PROMETHEE II is considered, all the alternatives are comparable. No incomparability remains, but the resulting information can be more disputable because more information get lost by considering the difference.

AHP-PROMETHEE

The methodology of AHP-PROMETHEE method is described below:

Step 1: Identify the selection criteria for the considered decision making problem and short-list the alternatives on the basis of the identified criteria satisfying the requirements. A quantitative or qualitative value or its range may be assigned to each identified criterion as a limiting value or threshold value for its acceptance for the considered application.

Step 2:

(1) After short-listing the alternatives, prepare a decision table including the measures or values of all criteria for the short-listed alternatives.

(2) The weights of relative importance of the criteria may be assigned using analytic hierarchy process (AHP) method. The steps are explained already in AHP methodology.

Step 3: After calculating the weights of the criteria using AHP method, the next step is to have the information on the decision maker preference function, which he/she uses when comparing the contribution of the alternatives in terms of each separate criterion. The preference function

(P_i) translates the difference between the evaluations obtained by two alternatives ($a1$ and $a2$) in terms of a criterion, into a preference degree ranging from 0 to 1. Let $P_{i,a1,a2}$ be the preference function associated to the criterion c_i.

$$P_{i,a1,a2} = G_i[c_i(a1) - c_i(a2)]$$

$$0 \leq P_{i,a1,a2} \leq 1$$

where G_i, is a non-decreasing function of the observed deviation (d) between two alternatives a1 and a2 over the criterion c_i. In order to facilitate the selection of a specific preference function for a criterion, six basic types were proposed (Brans *et al.*, 1984, Marinoni 2005). These include 'usual function', 'U-shape function', 'V-shape function', 'level function', 'linear function', and 'Gaussian function'. Preference 'usual function' is equal to the simple difference between the values of the criterion c_i for alternatives a1 and a2. For other preference functions, no more than two parameters (threshold q, p or s) have to be fixed (Brans and Mareschall 1994, Wang and Yang 2007). Indifference threshold q is the small value with respect to the scale of measurement. Preference threshold p is the large value with respect to the scale of measurement. Gaussian threshold is only used with the Gaussian preference function. It is usually fixed as an intermediate value between indifference and a preference threshold. The 'usual function' is an easy to use preference function and is generally used with quantitative criteria. 'U-shape function' uses a single indifference threshold and is generally used with qualitative criteria. 'V-shape function' uses a single preference threshold and is often used with quantitative criteria. 'Level function' is similar to U-shape but with an additional preference threshold and it is mostly used with qualitative criteria. 'Linear function' is similar to V-shape but with an additional indifference threshold and is often used with quantitative criteria. 'Gaussian function' is rarely used. In this case, the Usual Criterion is used to demonstrate the calculating processes of PROMETHEE, that is, 1 is used to replace the positive deviation and the rest is replaced with 0.

Step 4: After the decision maker have specified a preference function P_i and weight c_i for each criterion $c_i (i=1, 2, \ldots, M)$ of the problem, the multiple criteria preference index Π_{a1a2}, is then defined as the weighted average of the preference functions P_i:

$$\prod_{a1a2} = \sum_{i=1}^{M} w_i P_{a1a2} \tag{8}$$

Π_{a1a2}, represents the intensity of preference of the decision maker of alternative a1 over alternative a2, when considering simultaneously all the criteria. Its value ranges from 0 to 1. This preference index determines a valued outranking relation on the set of actions. As an example, the schematic calculation of the preference indices for a problem consisting of three alternatives and four criteria is given in Fig. 8.1 (adopted and modified, Marinoni 2005).

$$\prod_{31} = \sum_{i=1}^{4} w_i P_{i,31}$$

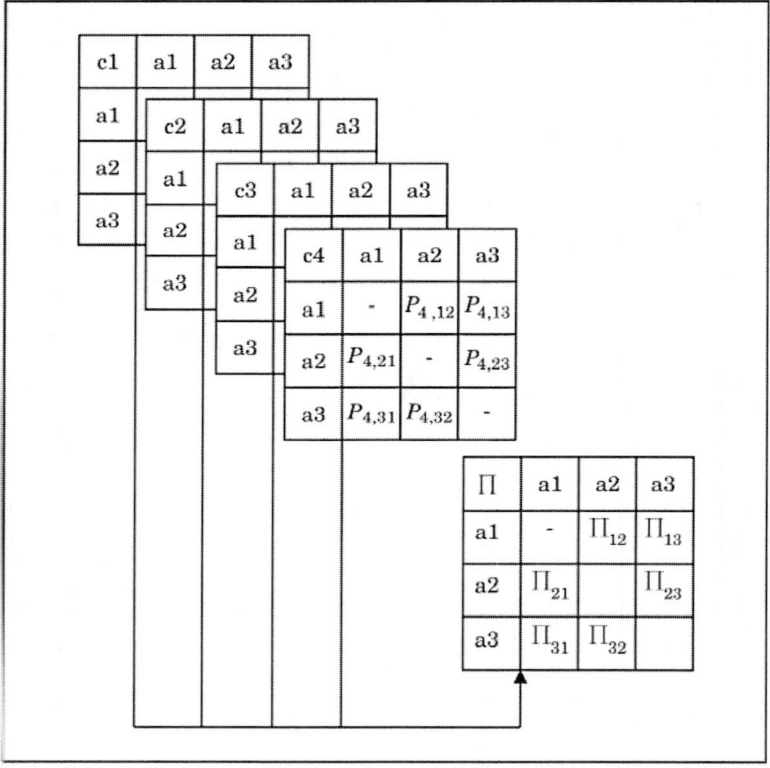

Fig. 8.1: Preference indices for a problem consisting of three alternatives and four criteria

Step 5: For PROMETHEE outranking relations, the leaving flow, entering flow and the net flow for an alternative are defined by the following equations:

$$\varphi^{+}(\mathrm{a}) = \textstyle\sum_{x \in A} \prod_{xa}, \qquad (9)$$

$$\varphi^{-}(a) = \sum_{x \, \varepsilon A} \Pi_{ax}, \tag{10}$$

$$\phi(a) = \varphi^{+}(a) - \varphi^{-}(a) \tag{11}$$

$\varphi^+(a)$ is called the leaving flow, $\varphi^-(a)$ is called the entering flow and $\varphi(a)$ is called the net flow. $\varphi^+(a)$ is the measure of the outranking character of a (i.e. dominance of alternative on overall other alternatives) and $\varphi^-(a)$ gives the outranked character of a (*i.e.* degree to which alternative a is dominated by all other alternatives). The net flow $\varphi(a)$ represents a value function, whereby a higher value reflects a higher attractiveness of alternative a. The net flow values are used to indicate the outranking relationship between the alternatives. Alternative a1 outranks a2 if $\varphi(a1) > \varphi(a2)$ and a1 is said to be indifferent to a2 if $\varphi(a1) = \varphi(a2)$.

The proposed decision making framework using PROMETHEE method provides a complete ranking of the alternatives from the best to the worst one using the net flows.

8.2.4 ELECTRE Method

ELECTRE (in its various forms) was conceived in response to deficiencies of existing decision making solution methods. Different versions of ELECTRE have been developed including ELECTRE I, II, III, IV and TRI. All methods are based on the same fundamental concepts but differ both operationally and according to the type of the decision problem. Specifically, ELECTRE I is designed for selection problems, ELECTRE TRI for assignment problems and ELECTRE II, III and IV for ranking problems.

ELECTRE I

The method does not have a significant practical interest, given the very nature of real world applications, having usually a vast spectrum of quantitative and qualitative elementary consequences, leading to the construction of a contradictory and very heterogeneous set of criteria with both numerical and ordinal scales associated with them. In addition, a certain degree of imprecision, uncertainty or ill-determination is always attached to the knowledge collected from real-world problems. The method is very simple and it should be applied only when all the criteria have been coded in numerical scales with identical ranges.

ELECTRE II

This method was thefirst of ELECTRE methods especially designed to deal with ranking problems.ELECTRE II was also the first method, to use a technique based on the construction of an embedded outranking relations sequence. The construction procedure is much closer to ELECTRE I, in the sense that it is also a true-criteria based procedure. Hence, it is not surprising that the no veto condition remains the same. However, concordance condition is modified in order to take into account the notion of embedded outranking relations. There are two embedded relations: a strong outranking relation followed by a weak outranking relation.

ELECTRE III

ELECTRE III was designed to improve ELECTRE II and thus deal with inaccurate, imprecise, uncertain or ill determination of data. This purpose was actually achieved, and ELECTRE III was applied with success during the last two decades on a broad range of real-life applications. The novelty of this method is the introduction of pseudo criteria instead of true-criteria. In ELECTRE III the outranking relation can be interpreted as a fuzzy relation. The construction of this relation requires the definition of a credibility index, which characterizes the credibility of the assertion "a outranks b".

ELECTRE IV

ELECTRE IV is also a procedure based on the construction of a set of embedded outranking relations. There are five different relations, S^1,..... S^5. The S^{r+1} relation (r = 1, 2, 3, 4) accepts an outranking in a less credible circumstances than the relation S^r. It means (while remaining on a merely ordinal basis) the assignment of a value ρ_r for the credibility index ρ(aSb) to the assertion aSb. The chosen values must be such that $\rho_r > \rho_{r+1}$. Furthermore, the movement from one credibility value ρ_r to another ρ_{r+1} must be perceived as a considerable loss. The ELECTRE IV exploiting procedure is the same as in ELECTRE III.

ELECTRE TRI

ELECTRE TRI is designed to assign a set of actions, objects or items to categories. In ELECTRE TRI categories are ordered; let us assume from the worst (C_1) to the best (C_k). Each category must be characterized by a lower and an upper profile. Let $C = \{C_1, \ldots, C_h, \ldots, C_k\}$ denote the set of categories. The assignment of a given action a to a certain category C_h results from the comparison of a to the profiles defining the lower and upper limits of the categories; b_h being the upper limit of category C_h and the lower limit of category C_{h+1}, for all $h = 1, \ldots, k$.

ELECTRE III description

ELECTRE III has been used in different applications. ELECTRE III outperforms other multi-criteria decision making methods for its ability to deal with inaccurate, imprecise, uncertain data. For criterion j being considered, three associated thresholds are defined which are indifference (q), preference (p), and veto (v). These thresholds produce outranking relations with an allowance for data uncertainly. To use ELECTRE III, the following must be defined by decision makers for all criteria:

1) Criteria indifference (q), preference (p), and veto (v) thresholds; where (v=q=p), and
2) Importance rating (w_j) for each criterion j.

Figure 8.2 Depicts the main steps of ranking using the ELECTRE III model. The complete description of the ELECTRE III model is summarized in the following subsections.

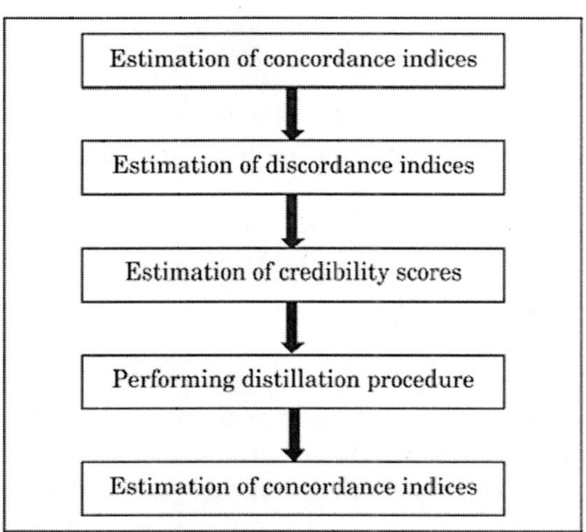

Fig. 8.2: Main steps of ranking using the ELECTRE III model

Concordance index

Concordance index, $C(a,b)$, is estimated for each pair of alternatives a and b based on a general comparison of the performances of alternative a and alternative b for all criteria. Concordance index ranges from 0 to 1; a value of 0 indicates that alternative a is worse than alternative b for all criteria. Concordance index is calculated based on a weighted comparison of the performances over each criterion individually $c_j(a,b)$ as per Eq. (12).

$$C(a,b) = \frac{1}{W}\sum_{j=1}^{n} w_j c_j\,(a,b) \tag{12}$$

Where; $W = \sum_{j=1}^{n} w_j$

The separate comparison indices $c_j(a,b)$ for each criterion are calculated based on one of the following cases:

Case 1: Alternative a is equivalent to or better than b minus the indifference threshold for criteria j as per Eq. (13).

$$c_j(a,b) = 1 \quad \text{if } g_j(a) + g_j(g_j(a)) = g_j(b) \tag{13}$$

Case 2: If the performance of alternative a plus the performance threshold is less than that of alternative b, then alternative a is treated as not better than b for this criteria as per Eq. (14).

$$c_j(a,b) = 0 \quad \text{if } g_j(a) + p_j(g_j(a)) = g_j(b) \tag{14}$$

Case 3: Otherwise the relationship is between these two extremes and is represented as a linear variation as per Eq. (15).

$$c_j(a,b) = \frac{g_j(\mathrm{a}) - g_j(\mathrm{b}) + p_j(g_j(\mathrm{a}))}{p_j\left(g_j(\mathrm{a})\right) - q_j(g_j(\mathrm{a}))} \tag{15}$$

Discordance index

The veto threshold for each criterion is assigned to introduce discordance into the outranking relations. It should be noted that any outranking of b by a indicated by the concordance index can be overruled if there is any criterion for which alternative b outperforms alternative a by at least veto threshold even if all the other criteria favour the outranking of b by a as per Eq. (16).

$$g_j(b) \geq g_j(a) + v_j(g_j(a)) \tag{16}$$

Case 1: Alternative b is not better than alternative a by a margin greater than the veto threshold as per Eq. (17).

$$D_j(a,b) = 0 \quad \text{if } g_j(b) \leq g_j(a) + p_j(g_j(a)) \tag{17}$$

Case 2: Alternative b is better than alternative a by a margin greater than the veto threshold as per Eq. (18).

$$D_j(a,b) = 1 \quad \text{if } g_j(b) \geq g_j(a) + v_j(g_j(a)) \tag{18}$$

Case 3: Otherwise the relationship is linear between the two as per Eq. (19).

$$Dj(a,b) = \frac{g_j(\text{b}) - g_j(\text{a}) - p_j(g_j(\text{a}))}{v_j\left(g_j(\text{a})\right) - p_j(g_j(\text{a}))} \tag{19}$$

Credibility score

The degree of credibility of outranking is calculated based on concordance and discordance indices according to one of the following two cases:

Case 1: The degree of outranking is equal to the concordance index if there is no criterion that is discordant or where no veto threshold is used as per Eq. (20).

$$S(a,b) = C(a,b) \quad \text{if } D_j(a,b) = C_j(a,b), \text{¥}_j \tag{20}$$

Case 2: The degree of outranking is equal to the concordance with a reduction as the level of discordance increases above a threshold value as per Eq. (21).

$$S(a,b) = C(a,b) \prod_{j \varepsilon \Psi(a,b)} \frac{1 - D_j(\text{a, b})}{1 - C(a,b)} \tag{21}$$

Where, $\Psi(a,b)$ is the set of criteria for which $D_j(a,b) > c_j(a,b)$.

Distillation procedure

Alternatives are ranked in two pre-orders which are constructed in different ways. The first pre-order is obtained in a descending manner (Descending Distillation), selecting the best rated alternatives initially, and finishing with the worst. The second pre-order is obtained in an ascending manner (Ascending Distillation), selecting the worst rated alternatives initially, and finishing with the best. The two pre-orders which are set based on a qualification score for each alternative as follows:

Step 1: Set λ_0 equals to the maximum value of $S(a,b)$ in credibility matrix (A) as per Eq. (22).

$$\lambda_0 = max_{a,b\varepsilon A} S(a,b) \tag{22}$$

Step 2: A cut-off level of outranking λ_1 is defined as the largest outranking score which is just less than the maximum outranking score minus the discrimination threshold as per Eq. (13).

$$\lambda_1 = max_{\{S(a,b)<\lambda 0 - s(\lambda 0)\}} S(a,b) \tag{23}$$

Where, $s(\lambda_0)$ is the discrimination threshold at the maximum level of outranking λ_0. At initial cut-off level, a outranks b if $S(a,b)$ is greater than the cut-off level and $S(a,b)$ exceeds $S(b,a)$ by more than the discrimination threshold (see Eq. (23)) satisfying the condition, given in Eq. (24).

$$s(\lambda_0) = 0.3 - 1.5\lambda_0$$

$$aSb \quad \text{if } S(a,b) > \lambda_1 \text{ and } S(a,b) - S(b,a) > s(\lambda) \tag{24}$$

Step 3: Every time a outranks b, a is given a score of +1 (strength) and b is given –1 (weakness). For each alternative, the strengths and weaknesses are added together to give a final qualification score.

Step 4: Within Descending Distillation, the alternative with the highest qualification score is assigned to a rank and removed from the procedure and the process is repeated for all remaining options.

Step 5: Within Ascending Distillation, the alternative with the lowest qualification score is assigned to a rank and removed from the procedure and the process is repeated for all remaining options.

Complete ranking

The results of the two procedures Descending Distillation and Ascending Distillation are combined to form complete ranking that is consistent with the two procedures.

8.3. PROBLEM DESCRIPTION

Karsak and Kuzgunkaya (2002) proposed a fuzzy multiple objective programming approach for the selection of a flexible manufacturing system. The authors had considered eight alternative flexible manufacturing systems and seven criteria. Five criteria were expressed objectively, and two criteria were expressed subjectively. These seven criteria are Reduction of labour cost (RLC, %), Reduction in WIP (RWP, %), Reduction in setup cost (RSC, %), Increase in market response (IMR), Increase in quality (IQ), Capital and maintenance cost (CMC, $1000), Floor space used (FSU, sq. ft.).

Step 1: Data of attributes/criteria of FMS selection problem is given in Table 8.1.

Table 8.1: Data of attributes of FMS

Alternatives				*Criteria*			
FMS	***RLC***	***RWP***	***RSC***	***IMR***	***IQ***	***CMC***	***FSU***
1	30	23	5	Good	Good	1500	5000
2	18	13	15	Good	Good	1300	6000
3	15	12	10	Fair	Fair	950	7000
4	25	20	13	Good	Good	1200	4000
5	14	18	14	Worst	Good	950	3500
6	17	15	9	Good	Fair	1250	5250
7	23	18	20	Fair	Good	1100	3000
8	16	8	14	Worst	Fair	1500	3000

Step 2: Linguistic rating variables are converted into number values based on a 10-point scale. Good as 8, fair as 5 and worst as 2. Objective values of attributes/ criteria are given in Table 8.2.

Table 8.2: Objective data of attributes

Alternatives				*Criteria*			
FMS	***RLC***	***RWP***	***RSC***	***IMR***	***IQ***	***CMC***	***FSU***
1	30	23	5	8	8	1500	5000
2	18	13	15	8	8	1300	6000
3	15	12	10	5	5	950	7000
4	25	20	13	8	8	1200	4000
5	14	18	14	2	8	950	3500
6	17	15	9	8	5	1250	5250
7	23	18	20	5	8	1100	3000
8	16	8	14	2	5	1500	3000

Step 3: Objective values of the FMS selection criteria, which are given in Table 8.2 are normalized. RLC, RWP, RSC, IMR, IQ are beneficial criteria and higher values are desirable. CMC and FSU are non-

beneficial criteria, and lower values are desirable. The normalized values are given in Table 8.3.

Table 8.3: Normalized data of attributes

Alternatives	*Criteria*						
FMS	***RLC***	***RWP***	***RSC***	***IMR***	***IQ***	***CMC***	***FSU***
1	1	1	0.25	1	1	0.63	0.6
2	0.6	0.56	0.75	1	1	0.73	0.5
3	0.5	0.52	0.5	0.62	0.62	1	0.43
4	0.83	0.87	0.65	1	1	0.79	0.75
5	0.47	0.78	0.7	0.25	1	1	0.85
6	0.57	0.65	0.45	1	0.62	0.76	0.57
7	0.77	0.78	1	0.62	1	0.86	1
18	0.53	0.35	0.7	0.25	0.62	0.63	1

Step 4: Relative importance of criteria is decided by decision maker and given Table 8.4.

Table 8.4: Relative importance of criteria of FMS

	RLC	***RWP***	***RSC***	***IMR***	***IQ***	***CMC***	***FSU***
RLC	–	0.5	0.665	0.5	0.335	0.335	0.665
RWP	0.5	–	0.665	0.5	0.335	0.335	0.665
RSC	0.335	0.335	–	0.335	0.255	0.255	0.5
IMR	0.5	0.5	0.665	–	0.335	0.335	0.665
IQ	0.665	0.665	0.745	0.665	–	0.5	0.745
CMC	0.665	0.665	0.745	0.665	0.5	–	0.745
FSU	0.335	0.335	0.5	0.335	0.255	0.255	–

Step 5: The weights of the attributes computed using AHP method (equation 3) is given in Table 8.5.

Table 8.5: Weights of criterions of FMS

	GM_i	W_i
RLC	0.5341	0.14
RWP	0.5341	0.14
RSC	0.3836	0.10
IMR	0.5341	0.14
IQ	0.6990	0.19
CMC	0.6990	0.19
FSU	0.3836	0.10

Step 6: Now VIKOR, improved PROMETHEE and ELECTRE III are applied individually to rank the alternatives and results are compared.

8.3.1. Evaluation of FMS Using VIKOR Method

1) The best f_j^* and worst f_j^- values of all criterion functions are determined using equation (4) and given in Table 8.6.

Table 8.6: Best f_j^* and worst f_j^- values

RLC	RWP	RSC	IMR	IQ	CMC	FSU
1	1	1	1	1	1	1
0.47	0.35	0.25	0.25	0.62	0.63	0.43

2) Value of $\frac{w_i(f_j^* - f_{ij})}{f_j^* - f_j^-}$ is calculated and given in Table 8.7.

Table 8.7: Value of $\frac{w_i(f_j^* - f_{ij})}{f_j^* - f_j^-}$

	RLC	RWP	RSC	IMR	IQ	CMC	FSU
1	0	0	0.10	0	0	0.19	0.070
2	0.10	0.094	0.03	0	0	0.138	0.087
3	0.132	0.103	0.06	0.071	0.19	0	0.10
4	0.045	0.028	0.046	0	0	0.108	0.044
5	0.14	0.047	0.040	0.14	0	0	0.026
6	0.114	0.075	0.073	0	0.19	0.123	0.075
7	0.060	0.047	0	0.071	0	0.072	0
8	0.124	0.14	0.040	0.14	0.19	0.19	0

3) Based on the Table 8.7, Eq. (5), Eq. (6) and Eq. (7) values of S_i, R_i and Q_i are obtained for each alternative, as shown in Table 8.8. Here, the Q_i value of each alternative is calculated using each v value as $v = 0.5$.

$$S^* = 0.250, S^- = 0.824, R^* = 0.072, R^- = 0.19$$

8.3.2. Evaluation of FMS Using Improved PROMETHEE Method

1) Contribution of the alternatives in terms of each separate criterion is compared and preference functions are calculated. 1 for the positive deviation and the rest is replaced with 0 as shown in Fig. 8.9. PROMETHHE Method application for FMS selection is presented in Table no. 8.9 to 8.17.

Table 8.8: Q_i value and ranking for FMS

Alternatives	S_i	R_i	Q_i	***Rank***
1	0.360	0.19	0.596	5
2	0.449	0.138	0.452	4
3	0.656	0.19	0.854	7
4	0.271	0.108	0.170	2
5	0.393	0.14	0.413	3
6	0.656	0.19	0.712	6
7	0.250	0.072	0	1
8	0.824	0.19	1	8

Table 8.9: Preference values related to RLC

	1	***2***	***3***	***4***	***5***	***6***	***7***	***8***
1	–	0	0	0	0	0	0	0
2	1	–	0	1	0	0	1	0
3	1	1	–	1	0	1	1	1
4	1	0	0	–	0	0	0	0
5	1	1	1	1	–	1	1	1
6	1	1	0	1	0	–	1	0
7	1	0	0	1	0	0	–	0
8	1	1	0	1	0	1	1	–

Table 8.10: Preference values related to RWP

	1	***2***	***3***	***4***	***5***	***6***	***7***	***8***
1	–	0	0	0	0	0	0	0
2	1	–	0	1	1	1	1	0
3	1	1	–	1	1	1	1	0
4	1	0	0	–	0	0	0	0
5	1	0	0	1	–	0	0	0
6	1	0	0	1	1	–	1	0
7	1	0	0	1	0	0	–	0
8	1	1	1	1	1	1	1	–

Table 8.11: Preference values related to RSC

	1	**2**	**3**	**4**	**5**	**6**	**7**	**8**
1	–	1	1	1	1	1	1	1
2	0	–	0	0	0	0	1	0
3	0	1	–	1	1	0	1	1
4	0	1	0	–	1	0	1	1
5	0	1	0	0	–	0	1	0
6	0	1	1	1	1	–	1	1
7	0	1	0	0	0	0	–	0
8	0	0	0	0	0	0	1	–

Table 8.12: Preference values related to IMR

	1	*2*	*3*	*4*	*5*	*6*	*7*	*8*
1	–	0	0	0	0	0	0	0
2	0	–	0	0	0	0	0	0
3	1	1	–	1	0	1	0	0
4	0	1	0	–	0	0	0	0
5	1	1	1	1	–	1	1	0
6	0	1	0	0	0	–	0	0
7	1	1	0	1	0	1	–	0
8	1	1	1	1	0	1	1	–

Table 8.13: Preference values related to IQ

	1	*2*	*3*	*4*	*5*	*6*	*7*	*8*
1	–	0	0	0	0	0	0	0
2	0	–	0	0	0	0	0	0
3	1	1	–	1	1	0	1	0
4	0	0	0	–	0	0	0	0
5	0	0	0	0	–	0	0	0
6	1	1	0	1	1	–	1	0
7	0	0	0	0	0	0	–	0
8	1	1	0	1	1	0	1	–

Table 8.14: Preference values related to CMC

	1	*2*	*3*	*4*	*5*	*6*	*7*	*8*
1	–	1	1	1	1	1	1	0
2	0	–	1	1	1	1	1	0
3	0	0	–	0	0	0	0	0
4	0	0	1	–	1	0	1	0
5	0	0	0	0	–	0	0	0
6	0	0	1	1	1	–	1	0
7	0	0	1	0	1	0	–	0
8	0	1	1	1	1	1	1	–

Table 8.15: Preference values related to FSU

	1	*2*	*3*	*4*	*5*	*6*	*7*	*8*
1	–	0	0	1	1	0	1	1
2	1	–	0	1	1	1	1	1
3	1	1	–	1	1	1	1	1
4	0	0	0	–	1	0	1	1
5	0	0	0	0	–	0	1	1
6	1	0	0	1	1	–	1	1
7	0	0	0	0	0	0	–	0
8	0	0	0	0	0	0	0	–

2) The multiple criteria preference index Π_{a1a2}, is then defined as the weighted average of the preference functions P_i is calculated using equation (8)

Table 8.16: Multiple criteria preference index Π_{a1a2}, for FMS

Π	*1*	*2*	*3*	*4*	*5*	*6*	*7*	*8*
1	–	0.29	0.29	0.39	0.39	0.19	0.39	0.20
2	0.38	–	0.19	0.57	0.43	0.43	0.67	0.10
3	0.71	0.81	–	0.81	0.53	0.52	0.67	0.34
4	0.28	0.24	0.19	–	0.39	0	0.39	0.20
5	0.42	0.38	0.28	0.42	–	0.28	0.48	0.24
6	0.57	0.57	0.29	0.86	0.72	–	0.86	0.20
7	0.42	0.14	0.19	0.42	0.19	0.14	–	0
8	0.61	0.90	0.47	0.80	0.52	0.61	0.90	–

3) The leaving flow, entering flow and the net flow for an alternative are calculated by equations (9), (10) and (11).

Table 8.17: Flow values and ranking for FMS

Alternatives	ω^+	ω^-	ω	*Rank*
1	3.39	2.14	1.25	3
2	3.33	2.77	0.56	5
3	1.90	4.39	–2.49	7
4	4.27	1.69	2.58	2
5	3.17	2.50	0.67	4
6	2.17	4.07	–1.90	6
7	4.36	1.50	2.86	1
8	1.28	4.81	–3.53	8

4) Based on the net flow values, flexible manufacturing systems are ranked, and rank is given in Table 8.17.

8.3.3. Evaluation of FMS Using ELECTRE III Method

1) Indifference, preference and veto threshold for problem taken is given in Table 8.18 below.

Table 8.18: Criteria threshold values for FMS

Threshold	*Value*
Indifference (q)	$0.05\,g_j(a)$
Preference (p)	$0.05\,g_j(a)$
Veto (v)	Not used

2) Preference threshold values and indifference threshold values are calculated and listed in Tables 8.19 and 8.20.

Table 8.19: Preference threshold values (p = $0.05g_j(a)$) for FMS.

	RLC	*RWP*	*RSC*	*IMR*	*IQ*	*CMC*	*FSU*
1	0.05	0.05	0.0125	0.05	0.05	0.0315	0.03
2	0.03	0.028	0.0375	0.05	0.05	0.0365	0.025
3	0.025	0.026	0.025	0.031	0.031	0.05	0.0215
4	0.0415	0.0435	0.0325	0.05	0.05	0.0395	0.0375
5	0.0235	0.039	0.035	0.0125	0.05	0.05	0.0425
6	0.0285	0.0325	0.0225	0.05	0.031	0.038	0.0285
7	0.0385	0.039	0.05	0.031	0.05	0.043	0.05
8	0.0265	0.0175	0.035	0.0125	0.031	0.0315	0.05

Table 8.20: Indifference threshold values (q = $0.05g_j(a)$) for FMS.

	RLC	*RWP*	*RSC*	*IMR*	*IQ*	*CMC*	*FSU*
1	0.05	0.05	0.0125	0.05	0.05	0.0315	0.03
2	0.03	0.028	0.0375	0.05	0.05	0.0365	0.025
3	0.025	0.026	0.025	0.031	0.031	0.05	0.0215
4	0.0415	0.0435	0.0325	0.05	0.05	0.0395	0.0375
5	0.0235	0.039	0.035	0.0125	0.05	0.05	0.0425
6	0.0285	0.0325	0.0225	0.05	0.031	0.038	0.0285
7	0.0385	0.039	0.05	0.031	0.05	0.043	0.05
8	0.0265	0.0175	0.035	0.0125	0.031	0.0315	0.05

3) Concordance index, C(a,b), is calculated for each pair of alternatives according to the equation (12). The separate comparison indices c_j(a,b) used in equation (12) for each criterion are calculated according to the equation (13), (14) and (15). Step-wise implementation of ELECTRE technique for FMS selection is illustrated in Table no. 8.21 to Table no. 8.76.

Table 8.21: Separate comparison indices for 1 and 2.

	$g_j(1) + p_j(g_j(1))$	$g_j(2)$	$c_j(1, 2)$	W	$wc_j(1, 2)$
RLC	1.05	0.6	1	0.14	0.14
RWP	1.05	0.56	1	0.14	0.14
RSC	0.2625	0.75	0	0.10	0
IMR	1.05	1	1	0.14	0.14
IQ	1.05	1	1	0.19	0.19
CMC	0.6615	0.73	0	0.19	0
FSU	0.63	0.5	1	0.10	0.10
C(1, 2)			**0.71**		

Table 8.22: Separate comparison indices for 1 and 3.

	$g_j(1) + p_j(g_j(1))$	$g_j(3)$	$c_j(1, 3)$	W	$wc_j(1, 3)$
RLC	1.05	0.5	1	0.14	0.14
RWP	1.05	0.52	1	0.14	0.14
RSC	0.2625	0.5	0	0.10	0
IMR	1.05	0.62	1	0.14	0.14
IQ	1.05	0.62	1	0.19	0.19
CMC	0.6615	1	0	0.19	0
FSU	0.63	0.43	1	0.10	0.10
C(1,3)			0.71		

Table 8.23: Separate comparison indices for 1 and 4.

	$g_j(1) + p_j(g_j(1))$	$g_j(4)$	$c_j(1, 4)$	W	$wc_j(1, 4)$
RLC	1.05	0.83	1	0.14	0.14
RWP	1.05	0.87	1	0.14	0.14
RSC	0.2625	0.65	0	0.10	0
IMR	1.05	1	1	0.14	0.14
IQ	1.05	1	1	0.19	0.19
CMC	0.6615	0.79	0	0.19	0
FSU	0.63	0.75	0	0.10	0
C(1, 4)			**0.61**		

Table 8.24: Separate comparison indices for 1 and 5.

	$g_j(1) + p_j(g_j(1))$	$g_j(5)$	$c_j(1, 5)$	W	$wc_j(1, 5)$
RLC	1.05	0.47	1	0.14	0.14
RWP	1.05	0.78	1	0.14	0.14
RSC	0.2625	0.7	0	0.10	0
IMR	1.05	0.25	1	0.14	0.14
IQ	1.05	1	1	0.19	0.19
CMC	0.6615	1	0	0.19	0
FSU	0.63	0.85	0	0.10	0
C(1, 5)			**0.61**		

Table 8.25: Separate comparison indices for 1 and 6.

	$g_j(1) + p_j(g_j(1))$	$g_j(6)$	$c_j(1, 6)$	W	$wc_j(1, 6)$
RLC	1.05	0.57	1	0.14	0.14
RWP	1.05	0.65	1	0.14	0.14
RSC	0.2625	0.45	0	0.10	0
IMR	1.05	1	1	0.14	0.14
IQ	1.05	0.62	1	0.19	0.19
CMC	0.6615	0.76	0	0.19	0
FSU	0.63	0.57	1	0.10	0.10
C(1, 6)			**0.71**		

Table 8.26: Separate comparison indices for 1 and 7.

	$g_j(1) + p_j(g_j(1))$	$g_j(7)$	$c_j(1, 7)$	W	$wc_j(1, 7)$
RLC	1.05	0.77	1	0.14	0.14
RWP	1.05	0.78	1	0.14	0.14
RSC	0.2625	1	0	0.10	0
IMR	1.05	0.62	1	0.14	0.14
IQ	1.05	1	1	0.19	0.19
CMC	0.6615	0.86	0	0.19	0
FSU	0.63	1	0	0.10	0
C(1, 7)			**0.61**		

Table 8.27: Separate comparison indices for 1 and 8.

	$g_j(1) + p_j(g_j(1))$	$g_j(8)$	$c_j(1, 8)$	W	$wc_j(1, 8)$
RLC	1.05	0.53	1	0.14	0.14
RWP	1.05	0.35	1	0.14	0.14
RSC	0.2625	0.7	0	0.10	0
IMR	1.05	0.25	1	0.14	0.14
IQ	1.05	0.62	1	0.19	0.19
CMC	0.6615	0.63	1	0.19	0.19
FSU	0.63	1	0	0.10	0
C(1, 8)			**0.80**		

Table 8.28: Separate comparison indices for 2 and 1.

	$g_j(2) + p_j(g_j(2))$	$g_j(1)$	$c_j(2, 1)$	W	$wc_j(2, 1)$
RLC	0.63	1	0	0.14	0
RWP	0.588	1	0	0.14	0
RSC	0.7875	0.25	1	0.10	0.10
IMR	1.05	1	1	0.14	0.14
IQ	1.05	1	1	0.19	0.19
CMC	0.7665	0.63	1	0.19	0.19
FSU	0.525	0.6	0	0.10	0
C(2, 1)			**0.62**		

Table 8.29: Separate comparison indices for 2 and 3.

	$g_j(2) + p_j(g_j(2))$	$g_j(3)$	$c_j(2, 3)$	W	$wc_j(2, 3)$
RLC	0.63	0.5	1	0.14	0.14
RWP	0.588	0.52	1	0.14	0.14
RSC	0.7875	0.5	1	0.10	0.10
IMR	1.05	0.62	1	0.14	0.14
IQ	1.05	0.62	1	0.19	0.19
CMC	0.7665	1	0	0.19	0
FSU	0.525	0.43	1	0.10	0.10
C(2, 3)			**0.81**		

Table 8.30: Separate comparison indices for 2 and 4.

	$g_j(2) + p_j(g_j(2))$	$g_j(4)$	$c_j(2, 4)$	W	$wc_j(2, 4)$
RLC	0.63	0.83	0	0.14	0
RWP	0.588	0.87	0	0.14	0
RSC	0.7875	0.65	1	0.10	0.10
IMR	1.05	1	1	0.14	0.14
IQ	1.05	1	1	0.19	0.19
CMC	0.7665	0.79	0	0.19	0
FSU	0.525	0.75	0	0.10	0
C(2, 4)			**0.43**		

Table 8.31: Separate comparison indices for 2 and 5.

	$g_j(2) + p_j(g_j(2))$	$g_j(5)$	$c_j(2, 5)$	W	$wc_j(2, 5)$
RLC	0.63	0.47	1	0.14	0.14
RWP	0.588	0.78	0	0.14	0
RSC	0.7875	0.7	1	0.10	0.10
IMR	1.05	0.25	1	0.14	0.14
IQ	1.05	1	1	0.19	0.19
CMC	0.7665	1	0	0.19	0
FSU	0.525	0.85	1	0.10	0.10
C(2, 5)			**0.67**		

Table 8.32: Separate comparison indices for 2 and 6.

	$g_j(2) + p_j(g_j(2))$	$g_j(6)$	$c_j(2, 6)$	W	$wc_j(2, 6)$
RLC	0.63	0.57	1	0.14	0.14
RWP	0.588	0.65	0	0.14	0
RSC	0.7875	0.45	1	0.10	0.10
IMR	1.05	1	1	0.14	0.14
IQ	1.05	0.62	1	0.19	0.19
CMC	0.7665	0.76	1	0.19	0.19
FSU	0.525	0.57	0	0.10	0

Table 8.33: Separate comparison indices for 2 and 7.

	$g_j(2) + p_j(g_j(2))$	$g_j(7)$	$c_j(2, 7)$	W	$wc_j(2, 7)$
RLC	0.63	0.77	0	0.14	0
RWP	0.588	0.78	0	0.14	0
RSC	0.7875	1	0	0.10	0
IMR	1.05	0.62	1	0.14	0.14
IQ	1.05	1	1	0.19	0.19
CMC	0.7665	0.86	0	0.19	0
FSU	0.525	1	0	0.10	0
C(2, 7)			**0.33**		

Table 8.34: Separate comparison indices for 2 and 8.

	$g_j(2) + p_j(g_j(2))$	$g_j(8)$	$c_j(2, 8)$	W	$wc_j(2, 8)$
RLC	0.63	0.53	1	0.14	0.14
RWP	0.588	0.35	1	0.14	0.14
RSC	0.7875	0.7	1	0.10	0.10
IMR	1.05	0.25	1	0.14	0.14
IQ	1.05	0.62	1	0.19	0.19
CMC	0.7665	0.63	1	0.19	0.19
FSU	0.525	1	0	0.10	0
C(2, 8)			**0.90**		

Table 8.35: Separate comparison indices for 3 and 1.

	$g_j(3) + p_j(g_j(3))$	$g_j(1)$	$c_j(3, 1)$	W	$wc_j(3, 1)$
RLC	0.525	1	0	0.14	0
RWP	0.546	1	0	0.14	0
RSC	0.525	0.25	1	0.10	0.10
IMR	0.651	1	0	0.14	0
IQ	0.651	1	0	0.19	0
CMC	1.05	0.63	1	0.19	0.19
FSU	0.4515	0.6	0	0.10	0
C(3, 1)			**0.29**		

Table 8.36: Separate comparison indices for 3 and 2.

	$g_j(3) + p_j(g_j(3))$	$g_j(2)$	$c_j(3, 2)$	W	$wc_j(3, 2)$
RLC	0.525	0.6	0	0.14	0
RWP	0.546	0.56	0	0.14	0
RSC	0.525	0.75	0	0.10	0
IMR	0.651	1	0	0.14	0
IQ	0.651	1	0	0.19	0
CMC	1.05	0.73	1	0.19	0.19
FSU	0.4515	0.5	0	0.10	0
C(3, 2)			**0.19**		

Table 8.37: Separate comparison indices for 3 and 4.

	$g_j(3) + p_j(g_j(3))$	$g_j(4)$	$c_j(3, 4)$	W	$wc_j(3, 4)$
RLC	0.525	0.83	0	0.14	0
RWP	0.546	0.87	0	0.14	0
RSC	0.525	0.65	0	0.10	0
IMR	0.651	1	0	0.14	0
IQ	0.651	1	0	0.19	0
CMC	1.05	0.79	1	0.19	0.19
FSU	0.4515	0.75	0	0.10	0
C(3, 4)			**0.19**		

Table 8.38: Separate comparison indices for 3 and 5.

	$g_j(3)+p_j(g_j(3))$	$g_j(5)$	$c_j(3, 5)$	W	$wc_j(3, 5)$
RLC	0.525	0.47	1	0.14	0.14
RWP	0.546	0.78	0	0.14	0
RSC	0.525	0.7	0	0.10	0
IMR	0.651	0.25	1	0.14	0.14
IQ	0.651	1	0	0.19	0
CMC	1.05	1	1	0.19	0.19
FSU	0.4515	0.85	0	0.10	0
C(3, 5)			**0.47**		

Table 8.39: Separate comparison indices for 3 and 6.

	$g_j(3)+p_j(g_j(3))$	$g_j(6)$	$c_j(3, 6)$	W	$wc_j(3, 6)$
RLC	0.525	0.57	0	0.14	0
RWP	0.546	0.65	0	0.14	0
RSC	0.525	0.45	1	0.10	0.10
IMR	0.651	1	0	0.14	0
IQ	0.651	0.62	1	0.19	0.19
CMC	1.05	0.76	1	0.19	0.19
FSU	0.4515	0.57	0	0.10	0
C(3, 6)			**0.48**		

Table 8.40: Separate comparison indices for 3 and 7.

	$g_j(3)+p_j(g_j(3))$	$g_j(7)$	$c_j(3, 7)$	W	$wc_j(3, 7)$
RLC	0.525	0.77	0	0.14	0
RWP	0.546	0.78	0	0.14	0
RSC	0.525	1	0	0.10	0
IMR	0.651	0.62	1	0.14	0.14
IQ	0.651	1	0	0.19	0
CMC	1.05	0.86	1	0.19	0.19
FSU	0.4515	1	0	0.10	0
C(3, 7)			**0.33**		

Table 8.41: Separate comparison indices for 3 and 8.

	$g_j(3)+p_j(g_j(3))$	$g_j(8)$	$c_j(3, 8)$	W	$wc_j(3, 8)$
RLC	0.525	0.53	0	0.14	0
RWP	0.546	0.35	1	0.14	0.14
RSC	0.525	0.7	0	0.10	0
IMR	0.651	0.25	1	0.14	0.14
IQ	0.651	0.62	1	0.19	0.19
CMC	1.05	0.63	1	0.19	0.19
FSU	0.4515	1	0	0.10	0
C(3, 8)			**0.66**		

Table 8.42: Separate comparison indices for 4 and 1.

	$g_j(4) + p_j(g_j(4))$	$g_j(1)$	$c_j(4, 1)$	W	$wc_j(4, 1)$
RLC	0.8715	1	0	0.14	0
RWP	0.9135	1	0	0.14	0
RSC	0.6825	0.25	1	0.10	0.10
IMR	1.05	1	1	0.14	0.14
IQ	1.05	1	1	0.19	0.19
CMC	0.8295	0.63	1	0.19	0.19
FSU	0.7875	0.6	1	0.10	0.10
C(4, 1)			**0.72**		

Table 8.43: Separate comparison indices for 4 and 2.

	$g_j(4) + p_j(g_j(4))$	$g_j(2)$	$c_j(4, 2)$	W	$wc_j(4, 2)$
RLC	0.8715	0.6	1	0.14	0.14
RWP	0.9135	0.56	1	0.14	0.14
RSC	0.6825	0.75	0	0.10	0
IMR	1.05	1	1	0.14	0.14
IQ	1.05	1	1	0.19	0.19
CMC	0.8295	0.73	1	0.19	0.19
FSU	0.7875	0.5	1	0.10	0.10
C(4, 2)			**0.90**		

Table 8.44: Separate comparison indices for 4 and 3.

	$g_j(4) + p_j(g_j(4))$	$g_j(3)$	$c_j(4, 3)$	W	$wc_j(4, 3)$
RLC	0.8715	0.5	1	0.14	0.14
RWP	0.9135	0.52	1	0.14	0.14
RSC	0.6825	0.5	1	0.10	0.10
IMR	1.05	0.62	1	0.14	0.14
IQ	1.05	0.62	1	0.19	0.19
CMC	0.8295	1	0	0.19	0
FSU	0.7875	0.43	1	0.10	0.10
C(4, 3)			**0.81**		

Table 8.45: Separate comparison indices for 4 and 5.

	$g_j(4) + p_j(g_j(4))$	$g_j(5)$	$c_j(4, 5)$	W	$wc_j(4, 5)$
RLC	0.8715	0.47	1	0.14	0.14
RWP	0.9135	0.78	1	0.14	0.14
RSC	0.6825	0.7	0	0.10	0
IMR	1.05	0.25	1	0.14	0.14
IQ	1.05	1	1	0.19	0.19
CMC	0.8295	1	0	0.19	0
FSU	0.7875	0.85	0	0.10	0
C(4, 5)			**0.61**		

Table 8.46: Separate comparison indices for 4 and 6.

	$g_j(4)+p_j(g_j(4))$	$g_j(6)$	$c_j(4, 6)$	W	$wc_j(4, 6)$
RLC	0.8715	0.57	1	0.14	0.14
RWP	0.9135	0.65	1	0.14	0.14
RSC	0.6825	0.45	1	0.10	0.10
IMR	1.05	1	1	0.14	0.14
IQ	1.05	0.62	1	0.19	0.19
CMC	0.8295	0.76	1	0.19	0.19
FSU	0.7875	0.57	1	0.10	0.10
C(4, 6)			**1**		

Table 8.47: Separate comparison indices for 4 and 7.

	$g_j(4)+p_j(g_j(4))$	$g_j(7)$	$c_j(4, 7)$	W	$wc_j(4, 7)$
RLC	0.8715	0.77	1	0.14	0.14
RWP	0.9135	0.78	1	0.14	0.14
RSC	0.6825	1	0	0.10	0
IMR	1.05	0.62	1	0.14	0.14
IQ	1.05	1	1	0.19	0.19
CMC	0.8295	0.86	0	0.19	0
FSU	0.7875	1	0	0.10	0
C(4, 7)			**0.61**		

Table 8.48: Separate comparison indices for 4 and 8.

	$g_j(4)+p_j(+g_j(4))$	$g_j(8)$	$c_j(4, 8)$	W	$wc_j(4, 8)$
RLC	0.8715	0.53	1	0.14	0.14
RWP	0.9135	0.35	1	0.14	0.14
RSC	0.6825	0.7	0	0.10	0
IMR	1.05	0.25	1	0.14	0.14
IQ	1.05	0.62	1	0.19	0.19
CMC	0.8295	0.63	1	0.19	0.19
FSU	0.7875	1	0	0.10	0
C(4, 8)			**0.80**		

Table 8.49: Separate comparison indices for 5 and 1.

	$g_j(5)+p_j(+g_j(5))$	$g_j(1)$	$c_j(5, 1)$	W	$wc_j(5, 1)$
RLC	0.4935	1	0	0.14	0
RWP	0.819	1	0	0.14	0
RSC	0.735	0.25	1	0.10	0.10
IMR	0.2625	1	0	0.14	0
IQ	1.05	1	1	0.19	0.19
CMC	1.05	0.63	1	0.19	0.19
FSU	0.8925	0.6	1	0.10	0.10
C(5, 1)			**0.58**		

Table 8.50: Separate comparison indices for 5 and 2.

	$g_j(5) + p_j(+g_j(5))$	$g_j(2)$	$c_j(5, 2)$	W	$wc_j(5, 2)$
RLC	0.4935	0.6	0	0.14	0
RWP	0.819	0.56	1	0.14	0.14
RSC	0.735	0.75	0	0.10	0
IMR	0.2625	1	0	0.14	0
IQ	1.05	1	1	0.19	0.19
CMC	1.05	0.73	1	0.19	0.19
FSU	0.8925	0.5	1	0.10	0.10
C(5,2)			**0.62**		

Table 8.51: Separate comparison indices for 5 and 3.

	$g_j(5) + p_j(+g_j(5))$	$g_j(3)$	$c_j(5, 3)$	W	$wc_j(5, 3)$
RLC	0.4935	0.5	0	0.14	0
RWP	0.819	0.52	1	0.14	0.14
RSC	0.735	0.5	1	0.10	0.10
IMR	0.2625	0.62	0	0.14	0
IQ	1.05	0.62	1	0.19	0.19
CMC	1.05	1	1	0.19	0.19
FSU	0.8925	0.43	1	0.10	0.10
C(5, 3)			**0.72**		

Table 8.52: Separate comparison indices for 5 and 4.

	$g_j(5) + p_j(+g_j(5))$	$g_j(4)$	$c_j(5, 4)$	W	$wc_j(5, 4)$
RLC	0.4935	0.83	0	0.14	0
RWP	0.819	0.87	0	0.14	0
RSC	0.735	0.65	1	0.10	0.10
IMR	0.2625	1	0	0.14	0
IQ	1.05	1	1	0.19	0.19
CMC	1.05	0.79	1	0.19	0.19
FSU	0.8925	0.75	1	0.10	0.10
C(5, 4)			**0.58**		

Table 8.53: Separate comparison indices for 5 and 6.

	$g_j(5) + p_j(+g_j(5))$	$g_j(6)$	$c_j(5, 6)$	W	$wc_j(5, 6)$
RLC	0.4935	0.57	0	0.14	0
RWP	0.819	0.65	1	0.14	0.14
RSC	0.735	0.45	1	0.10	0.10
IMR	0.2625	1	0	0.14	0
IQ	1.05	0.62	1	0.19	0.19
CMC	1.05	0.76	1	0.19	0.19
FSU	0.8925	0.57	1	0.10	0.10
C(5, 6)			**0.72**		

Table 8.54: Separate comparison indices for 5 and 7.

	$g_j(5) + p_j(+g_j(5))$	$g_j(7)$	$c_j(5, 7)$	W	$wc_j(5, 7)$
RLC	0.4935	0.77	0	0.14	0
RWP	0.819	0.78	1	0.14	0.14
RSC	0.735	1	0	0.10	0
IMR	0.2625	0.62	0	0.14	0
IQ	1.05	1	1	0.19	0.19
CMC	1.05	0.86	1	0.19	0.19
FSU	0.8925	1	0	0.10	0
C(5, 7)			**0.52**		

Table 8.55: Separate comparison indices for 5 and 8.

	$g_j(5) + p_j(+g_j(5))$	$g_j(8)$	$c_j(5, 8)$	W	$wc_j(5, 8)$
RLC	0.4935	0.53	0	0.14	0
RWP	0.819	0.35	1	0.14	0.14
RSC	0.735	0.7	1	0.10	0.10
IMR	0.2625	0.25	1	0.14	0.14
IQ	1.05	0.62	1	0.19	0.19
CMC	1.05	0.63	1	0.19	0.19
FSU	0.8925	1	0	0.10	0
C(5, 8)			**0.76**		

Table 8.56: Separate comparison indices for 6 and 1.

	$g_j(6) + p_j(+g_j(6))$	$g_j(1)$	$c_j(6, 1)$	W	$wc_j(6, 1)$
RLC	0.5985	1	0	0.14	0
RWP	0.6825	1	0	0.14	0
RSC	0.4725	0.25	1	0.10	0.10
IMR	1.05	1	1	0.14	0.14
IQ	0.651	1	0	0.19	0
CMC	0.498	0.63	0	0.19	0
FSU	0.5985	0.6	0	0.10	0

Table 8.57: Separate comparison indices for 6 and 2.

	$g_j(6) + p_j(+g_j(6))$	$g_j(2)$	$c_j(6, 2)$	W	$wc_j(6, 2)$
RLC	0.5985	0.6	0	0.14	0
RWP	0.6825	0.56	1	0.14	0.14
RSC	0.4725	0.75	0	0.10	0
IMR	1.05	1	1	0.14	0.14
IQ	0.651	1	0	0.19	0
CMC	0.498	0.73	0	0.19	0
FSU	0.5985	0.5	1	0.10	0.10
C(6, 2)			**0.38**		

Table 8.58: Separate comparison indices for 6 and 3.

	$g_j(6) + p_j(+g_j(6))$	$g_j(3)$	$c_j(6, 3)$	W	$wc_j(6, 3)$
RLC	0.5985	0.5	1	0.14	0.14
RWP	0.6825	0.52	1	0.14	0.14
RSC	0.4725	0.5	0	0.10	0
IMR	1.05	0.62	1	0.14	0.14
IQ	0.651	0.62	1	0.19	0.19
CMC	0.498	1	0	0.19	0
FSU	0.5985	0.43	1	0.10	0.10
C(6, 3)			**0.71**		

Table 8.59: Separate comparison indices for 6 and 4.

	$g_j(6) + p_j(+g_j(6))$	$g_j(4)$	$c_j(6, 4)$	W	$wc_j(6, 4)$
RLC	0.5985	0.83	0	0.14	0
RWP	0.6825	0.87	0	0.14	0
RSC	0.4725	0.65	0	0.10	0
IMR	1.05	1	1	0.14	0.14
IQ	0.651	1	0	0.19	0
CMC	0.498	0.79	0	0.19	0
FSU	0.5985	0.75	0	0.10	0

Table 8.60: Separate comparison indices for 6 and 5.

	$g_j(6) + p_j(+g_j(6))$	$g_j(5)$	$c_j(6, 5)$	W	$wc_j(6, 5)$
RLC	0.5985	0.47	1	0.14	0.14
RWP	0.6825	0.78	0	0.14	0
RSC	0.4725	0.7	0	0.10	0
IMR	1.05	0.25	1	0.14	0.14
IQ	0.651	1	0	0.19	0
CMC	0.498	1	0	0.19	0
FSU	0.5985	0.85	0	0.10	0
C(6,5)			**0.28**		

Table 8.61: Separate comparison indices for 6 and 7.

	$g_j(6) + p_j(+g_j(6))$	$g_j(7)$	$c_j(6, 7)$	W	$wc_j(6, 7)$
RLC	0.5985	0.77	0	0.14	0
RWP	0.6825	0.78	0	0.14	0
RSC	0.4725	1	0	0.10	0
IMR	1.05	0.62	1	0.14	0.14
IQ	0.651	1	0	0.19	0
CMC	0.498	0.86	0	0.19	0
FSU	0.5985	1	0	0.10	0
C(6, 7)			**0.14**		

Table 8.62: Separate comparison indices for 6 and 8.

	$g_j(6) + p_j(+g_j(6))$	$g_j(8)$	$c_j(6, 8)$	W	$wc_j(6, 8)$
RLC	0.5985	0.53	1	0.14	0.14
RWP	0.6825	0.35	1	0.14	0.14
RSC	0.4725	0.7	0	0.10	0
IMR	1.05	0.25	1	0.14	0.14
IQ	0.651	0.62	1	0.19	0.19
CMC	0.498	0.63	0	0.19	0
FSU	0.5985	1	0	0.10	0
C(6, 8)			**0.61**		

Table 8.63: Separate comparison indices for 7 and 1.

	$g_j(7) + p_j(+g_j(7))$	$g_j(1)$	$c_j(7, 1)$	W	$wc_j(7, 1)$
RLC	0.8085	1	0	0.14	0
RWP	0.819	1	0	0.14	0
RSC	1.05	0.25	1	0.10	0.10
IMR	0.651	1	0	0.14	0
IQ	1.05	1	1	0.19	0.19
CMC	0.903	0.63	1	0.19	0.19
FSU	1.05	0.6	1	0.10	0.10
C(7, 1)			**0.58**		

Table 8.64: Separate comparison indices for 7 and 2.

	$g_j(7) + p_j(+g_j(7))$	$g_j(2)$	$c_j(7, 2)$	W	$wc_j(7, 2)$
RLC	0.8085	0.6	1	0.14	0.14
RWP	0.819	0.56	1	0.14	0.14
RSC	1.05	0.75	1	0.10	0.10
IMR	0.651	1	0	0.14	0
IQ	1.05	1	1	0.19	0.19
CMC	0.903	0.73	1	0.19	0.19
FSU	1.05	0.5	1	0.10	0.10
C(7, 2)			**0.86**		

Table 8.65: Separate comparison indices for 7 and 3.

	$g_j(7) + p_j(+g_j(7))$	$g_j(3)$	$c_j(7, 3)$	W	$wc_j(7, 3)$
RLC	0.8085	0.5	1	0.14	0.14
RWP	0.819	0.52	1	0.14	0.14
RSC	1.05	0.5	1	0.10	0.10
IMR	0.651	0.62	1	0.14	0.14
IQ	1.05	0.62	1	0.19	0.19
CMC	0.903	1	0	0.19	0
FSU	1.05	0.43	1	0.10	0.10
C(7, 3)			**0.81**		

Table 8.66: Separate comparison indices for 7 and 4.

	$g_j(7) + p_j(+g_j(7))$	$g_j(4)$	$c_j(7, 4)$	W	$wc_j(7, 4)$
RLC	0.8085	0.83	0	0.14	0
RWP	0.819	0.87	0	0.14	0
RSC	1.05	0.65	1	0.10	0.10
IMR	0.651	1	0	0.14	0
IQ	1.05	1	1	0.19	0.19
CMC	0.903	0.79	1	0.19	0.19
FSU	1.05	0.75	1	0.10	0.10
C(7, 4)			**0.58**		

Table 8.67: Separate comparison indices for 7 and 5.

	$g_j(7) + p_j(+g_j(7))$	$g_j(5)$	$c_j(7, 5)$	W	$wc_j(7, 5)$
RLC	0.8085	0.47	1	0.14	0.14
RWP	0.819	0.78	1	0.14	0.14
RSC	1.05	0.7	1	0.10	0.10
IMR	0.651	0.25	1	0.14	0.14
IQ	1.05	1	1	0.19	0.19
CMC	0.903	1	0	0.19	0
FSU	1.05	0.85	1	0.10	0.10
C(7, 5)			**0.81**		

Table 8.68: Separate comparison indices for 7 and 6.

	$g_j(7) + p_j(+g_j(7))$	$g_j(6)$	$c_j(7, 6)$	W	$wc_j(7, 6)$
RLC	0.8085	0.57	1	0.14	0.14
RWP	0.819	0.65	1	0.14	0.14
RSC	1.05	0.45	1	0.10	0.10
IMR	0.651	1	0	0.14	0
IQ	1.05	0.62	1	0.19	0.19
CMC	0.903	0.76	1	0.19	0.19
FSU	1.05	0.57	1	0.10	0.10
C(7, 6)			**0.86**		

Table 8.69: Separate comparison indices for 7 and 8.

	$g_j(7) + p_j(+g_j(7))$	$g_j(8)$	$c_j(7, 8)$	W	$wc_j(7, 8)$
RLC	0.8085	0.53	1	0.14	0.14
RWP	0.819	0.35	1	0.14	0.14
RSC	1.05	0.7	1	0.10	0.10
IMR	0.651	0.25	1	0.14	0.14
IQ	1.05	0.62	1	0.19	0.19
CMC	0.903	0.63	1	0.19	0.19
FSU	1.05	1	1	0.10	0.10
C(7, 8)			**1**		

Table 8.70: Separate comparison indices for 8 and 1.

	$g_j(8) + p_j(+g_j(8))$	$g_j(1)$	$c_j(8, 1)$	W	$wc_j(8, 1)$
RLC	0.5565	1	0	0.14	0
RWP	0.3675	1	0	0.14	0
RSC	0.735	0.25	1	0.10	0.10
IMR	0.2625	1	0	0.14	0
IQ	0.651	1	0	0.19	0
CMC	0.6615	0.63	1	0.19	0.19
FSU	1.05	0.6	1	0.10	0.10
C(8, 1)			**0.39**		

Table 8.71: Separate comparison indices for 8 and 2.

	$g_j(8) + p_j(+g_j(8))$	$g_j(2)$	$c_j(8, 2)$	W	$wc_j(8, 2)$
RLC	0.5565	0.6	0	0.14	0
RWP	0.3675	0.56	0	0.14	0
RSC	0.735	0.75	0	0.10	0
IMR	0.2625	1	0	0.14	0
IQ	0.651	1	0	0.19	0
CMC	0.6615	0.73	0	0.19	0
FSU	1.05	0.5	1	0.10	0.10
C(8, 2)			**0.10**		

Table 8.72: Separate comparison indices for 8 and 3.

	$g_j(8) + p_j(+g_j(8))$	$g_j(3)$	$c_j(8, 3)$	W	$wc_j(8, 3)$
RLC	0.5565	0.5	1	0.14	0.14
RWP	0.3675	0.52	0	0.14	0
RSC	0.735	0.5	1	0.10	0.10
IMR	0.2625	0.62	0	0.14	0
IQ	0.651	0.62	1	0.19	0.19
CMC	0.6615	1	0	0.19	0
FSU	1.05	0.43	1	0.10	0.10
C(8, 3)			**0.53**		

Table 8.73: Separate comparison indices for 8 and 4.

	$g_j(8) + p_j(+g_j(8))$	$g_j(4)$	$c_j(8, 4)$	W	$wc_j(8, 4)$
RLC	0.5565	0.83	0	0.14	0
RWP	0.3675	0.87	0	0.14	0
RSC	0.s735	0.65	1	0.10	0.10
IMR	0.2625	1	0	0.14	0
IQ	0.651	1	0	0.19	0
CMC	0.6615	0.79	0	0.19	0
FSU	1.05	0.75	1	0.10	0.10
C(8, 4)			**0.20**		

Table 8.74: Separate comparison indices for 8 and 5.

	$g_j(8) + p_j(+g_j(8))$	$g_j(5)$	$c_j(8, 5)$	W	$wc_j(8, 5)$
RLC	0.5565	0.47	1	0.14	0.14
RWP	0.3675	0.78	0	0.14	0
RSC	0.735	0.7	1	0.10	0.10
IMR	0.2625	0.25	1	0.14	0.14
IQ	0.651	1	0	0.19	0
CMC	0.6615	1	0	0.19	0
FSU	1.05	0.85	1	0.10	0.10
C(8, 5)			**0.48**		

Table 8.75: Separate comparison indices for 8 and 6.

	$g_j(8) + p_j(+g_j(8))$	$g_j(6)$	$c_j(8, 6)$	W	$wc_j(8, 6)$
RLC	0.5565	0.57	0	0.14	0
RWP	0.3675	0.65	0	0.14	0
RSC	0.735	0.45	1	0.10	0.10
IMR	0.2625	1	0	0.14	0
IQ	0.651	0.62	1	0.19	0.19
CMC	0.6615	0.76	0	0.19	0
FSU	1.05	0.57	1	0.10	0.10
C(8, 6)			**0.39**		

Table 8.76: Separate comparison indices for 8 and 7.

	$g_j(8) + p_j(+g_j(8))$	$g_j(7)$	$c_j(8, 7)$	W	$wc_j(8, 7)$
RLC	0.5565	0.77	0	0.14	0
RWP	0.3675	0.78	0	0.14	0
RSC	0.735	1	0	0.10	0
IMR	0.2625	0.62	0	0.14	0
IQ	0.651	1	0	0.19	0
CMC	0.6615	0.86	0	0.19	0
FSU	1.05	1	1	0.10	0.10
C(8, 7)			**0.10**		

4) From the values concordance index table is made and given in Table 8.77.

As no veto threshold is used, so degree of credibility of outranking is equal to concordance index. Credibility index is given Table 8.78.

Table 8.77: Concordance index table for FMS.

	1	*2*	*3*	*4*	*5*	*6*	*7*	*8*
1	1	0.71	0.71	0.61	0.61	0.71	0.61	0.80
2	0.62	1	0.81	0.43	0.67	0.76	0.33	0.90
3	0.29	0.19	1	0.19	0.47	0.48	0.33	0.66
4	0.72	0.90	0.81	1	0.61	1	0.61	0.80
5	0.58	0.62	0.72	0.58	1	0.72	0.52	0.76
6	0.24	0.38	0.71	0.14	0.28	1	0.14	0.61
7	0.58	0.86	0.81	0.58	0.81	0.86	1	1
8	0.39	0.10	0.53	0.20	0.48	0.39	0.10	1

Table 8.78: Credibility index table for FMS.

	1	*2*	*3*	*4*	*5*	*6*	*7*	*8*
1	1	0.71	0.71	0.61	0.61	0.71	0.61	0.80
2	0.62	1	0.81	0.43	0.67	0.76	0.33	0.90
3	0.29	0.19	1	0.19	0.47	0.48	0.33	0.66
4	0.72	0.90	0.81	1	0.61	1	0.61	0.80
5	0.58	0.62	0.72	0.58	1	0.72	0.52	0.76
6	0.24	0.38	0.71	0.14	0.28	1	0.14	0.61
7	0.58	0.86	0.81	0.58	0.81	0.86	1	1
8	0.39	0.10	0.53	0.20	0.48	0.39	0.10	1

Value of is λ_0 find out according to equation (22)

$$\lambda_0 = 1$$

And a cut-off level of outranking λ_1 is find out using equation (22) and (23)

$$S\ (\lambda_0) = 0.3 - 0.15\lambda_0 = 0.3 - 0.15(1) = 0.15$$

$$\lambda_0 - s(\lambda_0) = 0.85$$

$$\lambda_1 = 0.82$$

5) Now using the eq. (24) it is find out that which alternative outranks the other alternatives and a table is made for strength, weakness and for the final score or qualification.

   For checking if A outranks B, according to equation (24)

   $S(a,b) > \lambda_1$ and $S(a,b) - S(b,a) > s(\lambda)$

   $S\ (1, 2) = 0.68$ and $\lambda_1 = 0.81$

   It does not satisfy the condition so 1 does not outrank 2.

   Similarly, it is checked for every pair of alternatives and strength, weakness and qualification score are calculated for each alternative. It is given in Table 8.79.

Table 8.79: Qualification score for FMS.

Outranks	*Strength*	*Weakness*	*Qualification score*	
1	None	0	0	0
2	8	1	2	–1
3	None	0	0	0
4	2,6	2	0	2
5	None	0	0	0
6	None	0	2	–2
7	2,6,8	3	0	3
8	None	0	5	–5

Based on the qualification score, FMS are ranked, and ranking is given in Table 8.80 below.

Table 8.80: Ranking of alternatives of FMS.

Alternatives	*Ranking*
1	3
2	6
3	4
4	2
5	5
6	7
7	1
8	8

8.4. DISCUSSION AND PROPOSED SOLUTION

Comparative ranking of alternatives based on VIKOR, improved PROMETHEE and ELECTRE III is given in Table 8.81. According to VIKOR alternatives are ranked as 7 > 4 > 5 > 2 > 1 > 6 > 3 > 8 in the decreasing order of preference, according to improved PROMETHEE alternatives are ranked as 7 > 4 > 1 > 5 > 2 > 6 > 3 > 8 in the decreasing order of preference and according to ELECTRE III alternatives are ranked as 7 > 4 > 1 > 3 > 5 > 2 > 6 > 8 in the decreasing order of preference. It is clear from the ranking results of all three multi criteria decision making approaches VIKOR, improved PROMETHEE and ELECTRE III that flexible manufacturing system numbered as 7 is the best choice for the given industrial application under the given conditions and the second-best choice is flexible manufacturing system numbered as 4 while flexible manufacturing system numbered as 8 is the worst choice.

Flexible manufacturing system numbered as 7 is the ideal according to the criteria Reduction in setup cost (RSC, %), Increase in quality (IQ), Floor space used (FSU, sq. ft.) and closet to ideal according to the criteria Increase in market response (IMR), Capital and maintenance cost (CMC, $1000) and closer to ideal according to criteria Reduction of labor cost (RLC, %) and Reduction in WIP (RWP, %).

Flexible manufacturing system numbered as 4 is the ideal according to the criteria Increase in market response (IMR) and Increase in quality (IQ) and closet to ideal according to the criteria Reduction of labor cost (RLC, %) and Reduction in WIP (RWP, %) and closer to ideal according to criteria Reduction in setup cost (RSC, %), Capital and maintenance cost (CMC, $1000) and Floor space used (FSU, sq. ft.).

As an alternative for a final solution, Flexible manufacturing system numbered as 7 could be considered the best compromise. A comparison of alternatives 4 and 7 is presented in Table 8.82 and it shows that for criteria Reduction of labor cost (RLC, %), Reduction in WIP (RWP, %) alternative 4 is slightly preferred over alternative 7, for Increase in market response (IMR) alternative 4 is highly preferred over alternative 7, for Capital and maintenance cost (CMC, $1000) alternative 7 is slightly preferred over alternative 4, for Reduction in setup cost (RSC, %), Floor space used (FSU, sq. ft.) alternative 7 is highly preferred over alternative 4 and for Increase in quality (IQ) alternative 4 is same as alternative 7. From the comparison results of flexible manufacturing system 4 and 7, for two criteria 4 is slightly preferred and for one criteria 4 is highly preferred while flexible manufacturing system 7 is slightly preferred for one criteria and highly preferred for two criteria and for one criteria both are same. So, flexible manufacturing system 7 is the best choice which is also the best choice from the ranking results of VIKOR, improved PROMETHEE and ELECTRE III.

These results are like those suggested by Karsak and Kuzgunkaya (2002) using the fuzzy multiple objective programming approach. However, it may be mentioned that the ranking depends on the judgements of relative importance made by the user. The ranking may change if the user assigns different relative importance values to the criteria's. The same is true with all these multi criteria decision making approaches.

Table 8.81: Comparative ranking of alternatives for FMS.

Alternatives	*VIKOR*	*PROMETHEE*	*ELECTRE*
1	5	3	3
2	4	5	6
3	7	7	4
4	2	2	2
5	3	4	5
6	6	6	7
7	1	1	1
8	8	8	8

Table 8.82: Comparison of alternative 7 and 4 of FMS.

Criteria	*Comparison*		
	7	*4*	*Preferred*
RLC	23	25	4, slightly
RWP	18	20	4, slightly
RSC	20	13	7, highly
IMR	0.5	0.745	4, highly
IQ	0.745	0.745	Same
CMC	1100	1200	7, slightly
FSU	3000	4000	7, highly

8.5. SUMMARY AND CONCLUSIONS

VIKOR method focuses on ranking and selecting from a set of alternatives and determines compromise solutions for a problem with conflicting criteria, which can help the decision makers to reach a final decision. PROMETHEE proceeds to a pair-wise comparison of alternatives in each single criterion to determine partial binary relations denoting the strength of preference of an alternative over the other. Ranking by PROMETHEE gives the same results as ranking by VIKOR, with measure S representing "group utility". Results by ELECTRE III are relatively same to the results by VIKOR. The similar results PROMETHE-VIKOR (based on S-measure) and VIKOR-ELECTRE are consistent with the discussion.

9

Design of Forward and Reverse Supply Chain Networks: A Simulation Based Approach

9.1. INTRODUCTION

In this chapter, design of forward and reverse supply chain of a lead acid recycling process of a battery recycling company XYZ Ltd. is presented. Inverter batteries are collected at battery collection center, from collection center they are transported to the recycling plant where recycling is done and finally converted it to new battery production and other lead products. Recycling involves processing used materials into new products to prevent waste of potentially useful materials, reduce the consumption of fresh raw materials, reduce energy usage, reduce air pollution (from incineration) and water pollution (from land filling)

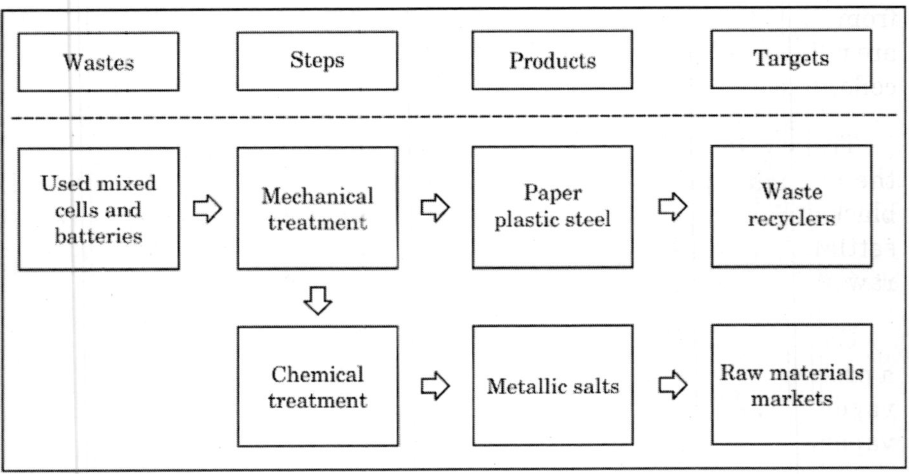

Fig. 9.1: Shows recycling process.

by reducing the need for "conventional" waste disposal, and lower greenhouse gas emissions as compared to virgin production. Recycling is a key component of modern waste reduction and is the third component of the "Reduce, Reuse, and Recycle" waste hierarchy as shown in Fig. 9.1.

The Lead-acid battery has led the way in recycling. The automotive industry should be given credit in organizing ways to dispose of spent car batteries. In the USA, 98% of all Lead acid batteries are recycled. In comparison, only one in six households in North America recycles batteries.

9.2. CURRENT FORWARD AND REVERSE SUPPLY CHAIN OF COMPANY

Most lithium batteries are non-rechargeable and are used in cameras, hearing aids and defense applications. For proper disposal, the batteries must first be fully discharged to consume the metallic lithium content. Battery Recycling Plant requires that the batteries be sorted according to chemistries. Some sorting must be done prior to the battery arriving at the recycling Plant. Nickel-cadmium, nickel-metal-hydride, lithium-ion and Lead acid are placed in designated boxes at the collection point. Battery recyclers claim that if a steady stream of batteries, sorted by chemistry, were available at no charge, recycling would be profitable. But preparation and transportation add to the cost.

The recycling process starts by removing the combustible material, such as plastics and insulation, with a gas fired thermal oxidizer. Gases from the thermal oxidizer are sent to the plant's scrubber where they are neutralized to remove pollutants. The process leaves the clean, naked cells, which contain valuable metal content.

The cells are then chopped into small pieces, which are heated until the metal liquefies. Non-metallic substances are burned off; leaving a black slag on top that is removed with a slag arm. The different alloys settle according to their weights and are skimmed off like cream from raw milk.

Cadmium is relatively light and vaporizes at high temperatures. In a process that appears like a pan boiling over, a fan blows the cadmium vapor into a large tube, which is cooled with water mist. This causes the vapors to condense and produces cadmium that is 99.95 percent pure.

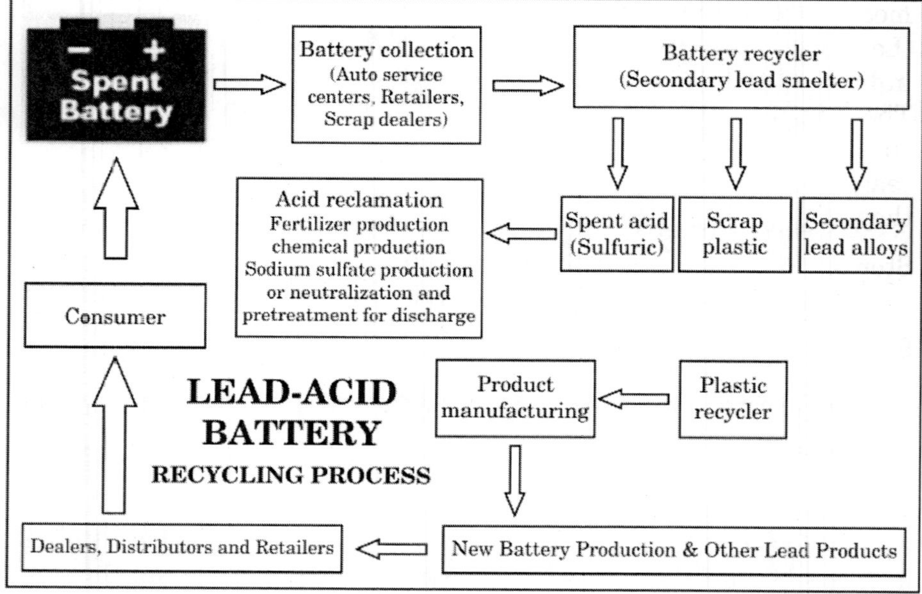

Fig. 9.2: Flowchart: Lead acid battery recycling process.

Current Battery recycling methods requires a high amount of energy. It takes six to ten times the amount of energy to reclaim metals from recycled batteries than it would through other means.

9.3. PROBLEMS WITH CURRENT RECYCLING MODEL

- Supply of used batteries is irregular.
- The company is not having any well-structured model of reverse logistics.
- Performance evaluation criteria's are not defined in the existing model of the company.
- Underutilization of existing workstations of the supply chain.
- To solve this kind of problems, we have developed an integrated model of the company, which is capable enough to predict the future performance of the company.

9.4. PYROMETALLURGICAL REFINING

Pyrometallurgical Process has been the most widely used for lead refining. Pyrometallurgical refining is performed in liquid phase, which means that the crude Lead must be melted to temperatures from 327°C (Lead fusion point) to 650°C. As a general trend, the process is performed in batches of 20 to 200 tons, according to the refining plant capacity. The chemical concept behind the refining process is the addition of specific reagents to the molten Lead at proper temperatures. These reagents will then remove the unwanted metals in a specific order as they are added selectively.

Steps of removal of unwanted metals in specific order.

Step I: Removal of Copper (Cu) from Molten lead.

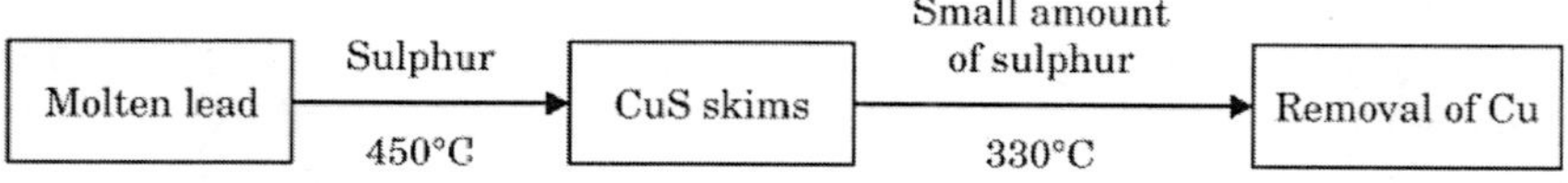

Step II: Removal of Tin (Sn) from Molten lead.

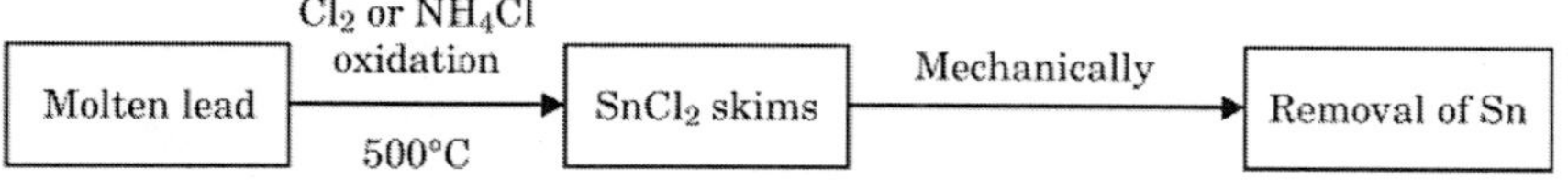

Step III: Removal of Arsenic (As) and Antimony (Sb).

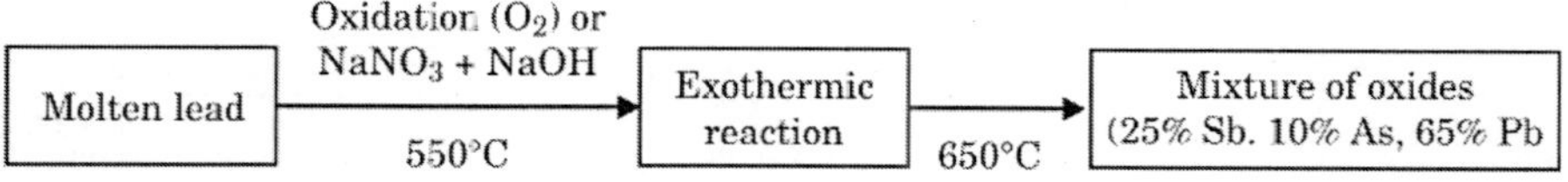

Step IV: Removal of Silver (Ag).

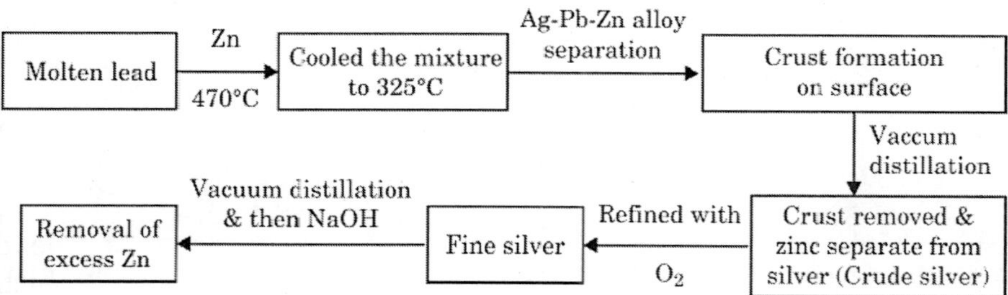

9.5. PERFORMANCE PARAMETERS OF THE COMPANY

Table 9.1: Different performance parameters of the company.

	Distance (Km)	*Speed (Km/hr)*	*Transfer time (Min)*	*Transfer cost (Rs)*
Council's	60	50	72	7
SBO's	60	50	72	7
Retailer's	42	55	45.8	5.4
Disassembly center	52	55	34.6	5.46

9.6. DESIGN OF THE CONCEPTUAL MODEL

At the current stage, the company is not using a well define structured approach to get EOL product in the company. Therefore, we have designed and developed an integrated model of forward and reverse logistics supply chain networks as shown in Fig. 9.3.

Prior to model Construction with the Arena simulation package, it is first fundamental to obtain an appropriate conceptual model, by which

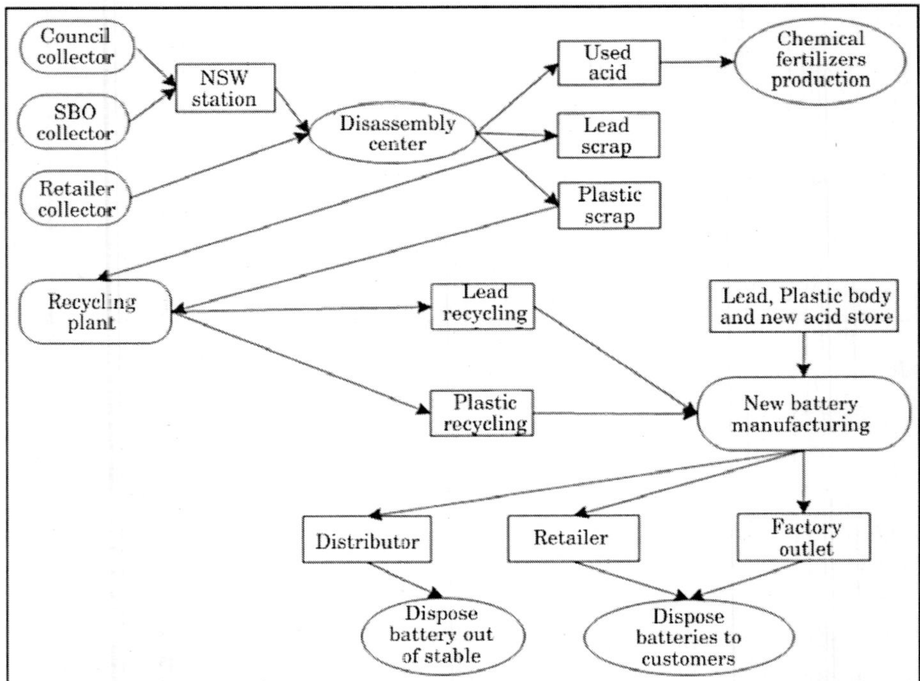

Fig. 9.3: Proposed integrated model of forward and reverse supply chain network for battery recycling.

the reverse logistics infrastructure can be described. When designing a network structure of reverse logistics, there are many factors to be considered. The factors include inverter batteries, the number and the type of participants in the system, battery collection center, battery recyclers, consumers, characteristics of the material flow and product characteristics. The retailers will deliver the inverter battery directly to the disassembly center. At the center, the parts that can be manufactured or recycled will be extracted and transferred either to the recycling facility or to remanufacturing plant.

9.7. SIMULATION

Simulation refers to a broad collection of methods and applications to mimic the behavior of real systems, usually on a computer with appropriate software. In fact, "simulation" can be an extremely general term since the idea applies across many fields, industries, and applications. These days, simulation is more popular and powerful than ever since computers and software are better than ever.

9.7.1. Modeling

Simulation, like most analysis methods, involves systems and models of them. Some examples of models and describe options for studying them to learn about the corresponding system.

- What is being modeled?
- How about just playing with the system?
- Sometimes you shouldn't play with the system
- Physical models
- Logical or Mathematical models
- What do you do with a logical model?

9.7.2. Simulation Languages

Special-purpose simulation languages like GPSS, Sim-script, SLAM, and SIMAN appeared on the scene some time later and provided a much better framework for the kinds of simulations many people do. Simulation languages have become very popular and are in wide use.

9.7.3. High-level Simulators

Several high-level "simulators" products emerged that are indeed very easy to use. They typically operate by intuitive graphical user interfaces, menus, and dialogs. We select from available simulation-modeling constructs, connect them and run the model along with a dynamic graphical animation of system components as they move around and change.

9.7.4. Where Arena Fits In

Arena combines the ease of use found in high-level simulators with the flexibility of simulation languages and even all the way down to general-purpose procedural language like the Microsoft Visual Basic programming system or C if we really want. It does this by providing alternative and interchangeable templates of graphical simulation modeling and analysis modules that we can combine to build a fairly wide variety of simulation models. For ease of display and organization, modules are typically grouped into panels to compose a template. By switching panels, we gain access a whole different set of simulation modeling constructs and capabilities. In most cases, modules from different panels can be mixed together in the same model.

In fact, the modules in Arena are composed of SIMAN components; we can create our own modules and collect them into our own templates for various classes of systems. For instance, Rockwell Software has built templates for general modeling, high-speed packaging, contact centers and other industries. Other people have built templates for their company in industries as diverse as mining, auto manufacturing, fast-food and forest-resource management. In this way, we don't have to compromise between modeling flexibility and ease of use.

Further Arena includes dynamic animation in the same work environment. It also provides integrated support, including graphics, for some of the statistical design and analysis issues that are part and parcel of a good simulation study.

9.7.5. Future Scope of Simulation

The rate of change in simulation has accelerated in the recent years, and there is every reason to believe that it will continue its rapid growth and cross the bridges to mainstream acceptance. Simulation software of new operating systems to provide greater ease of use, particularly for the first-time user. This trends must continue if simulation is to become a state-of-the-art tool resident on every systems-analysis computer. These new operating systems have also allowed for greater integration of simulation with other packages (like spreadsheets, databases, and word processors). It is now becoming possible to foresee the complete integration of simulation with other software packages that collect, store, and analyze system data at the front end along with software that helps control the system at the back end.

The Internet and intranets have had a tremendous impact on the way organizations conduct their business. Information-sharing in real time is not only possible, but is becoming mandatory. Simulation tools are being developed to support distributed model-building, distributed processing, and remote analysis of results. Soon, a set of enterprise-wide tools will be available to provide the ability for the employees throughout an organization to obtain answers to critical and even routine questions.

Today's simulation projects concentrate on the design or redesign of complex systems. They often must deal with complex system-control issues, which can lead to the development of new systems-control logic that is tested using the developed simulation.

With the rapid advances being made in computers and software, it is very difficult to predict much about the simulations of the distant future, but even now we are seeing the development and implementation of features such as automatic statistical analysis, software that recommends system changes, simulations totally integrated with system operating software, and yes, even virtual reality.

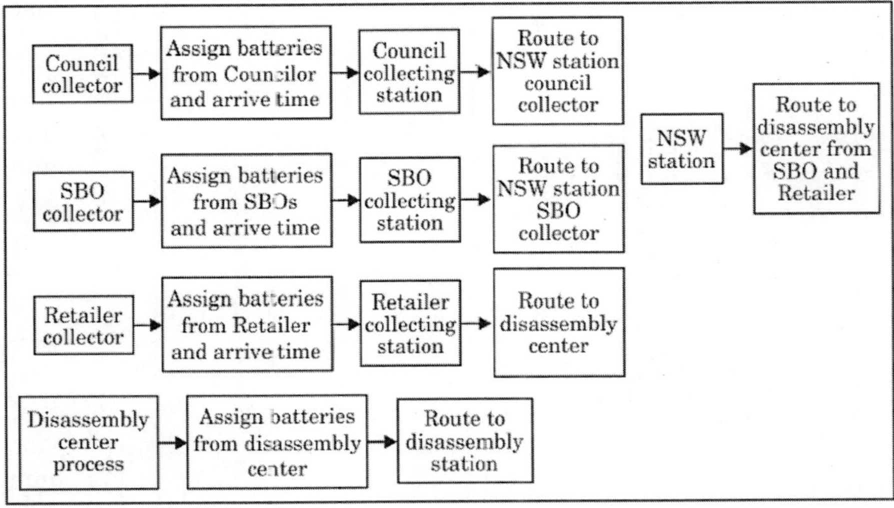

Fig. 9.4 (a): Design of integrated model along their routes.

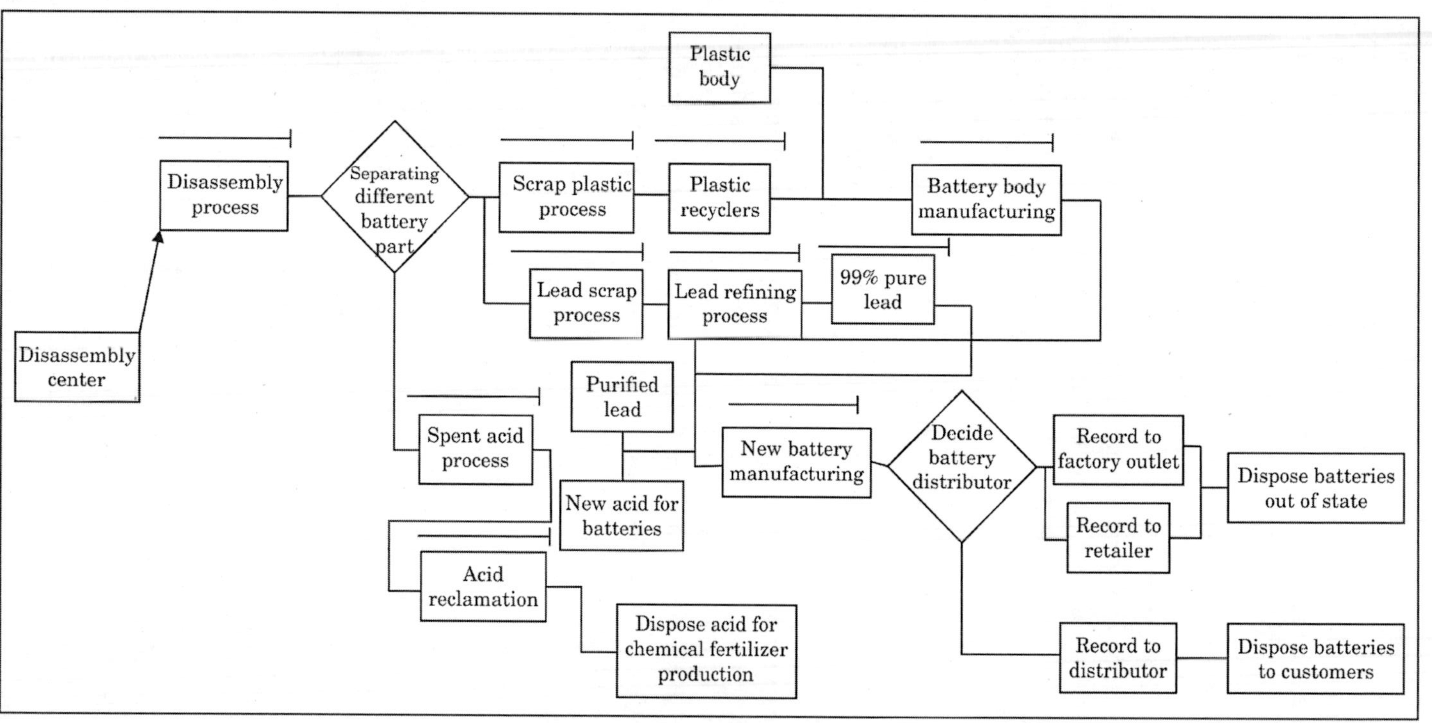

Fig. 9.4 (b): Design of integrated model along their process.

9.8. SIMULATION MODELING OF THE REVERSE LOGISTICS NETWORK

The Arena Simulation package by Rockwell Inc, which allows such modular design, was create the simulation model. Fig. 9.4a to 9.4b shows the structure of simulation models.

The starting point of simulation is in the "Creation and Assigning of Collector" sub model, where inverter batteries are created and then assign to the three collectors. The first process in the council sub-model is to determine the area of the council in which inverter batteries are discarded in accordance with the council's population. Then collection of EOL batteries is assigned to the council, retailer and SBOs area in the area by using the probability distribution. Inverter batteries are then delivered either to the NSW "Transfer station" sub-model if the collector is a council or a SBO or to the "Disassembly Plant" sub-model if the collector is a retailer. The inverter batteries coming from NSW station are brought in by the disassembly Plant where they are disassembled and separate different components of batteries and finally transfer lead scrap and plastic scrap to the recycling plant.

9.9. MODEL TRANSLATION

The system represents the final operation of manufacturing of batteries after recycling. The arriving parts are inverter batteries that are collected by council, SBOs, retailer.

The first units called, council collector with inter arrival times to our model being exponentially distributed with a mean 5 (all times are in minutes). Upon arrival, they are transferred to NSW - station with a route time 25 minutes.

The second units called, SBO collector with inter arrival times to our model being exponentially distributed with a mean 5. Upon arrival, they are transferred to NSW- station with a route time 25 minutes.

The third units called, retailer collector with inter arrival times to our model being exponentially distributed with a mean 5. Upon arrival, they are transferred to NSW- station with a route time 45 minutes. The inverter batteries collected at NSW-Station are transferred to Disassembly center. At Disassembly center, the batteries are dismantled and separate out into different parts by "N-way by chance" distribution where 70% lead scrap, 20% Plastic scrap and 10% spent acid is obtained. The spent acid is directly reclamation by acid reclamation process and the process time for the combined operation at acid reclamation follows a TRIA (5, 8, 10) distribution and used for chemical fertilizer production. 70% lead scrap is recycled by refining through pyrometallurgical process and the process time follows a TRIA (30, 35, 45) distribution and get 99% pure lead. Similarly, the plastic scrap is recycled by plastic recycling process and the process time follows a TRIA (15, 20, 30) distribution and manufacture outer bodies of batteries. There are stores from which we take the plastic body, purified lead, and new acid for manufacturing of new batteries by battery manufacturing process and the process time follows a TRIA (30, 35, 45) distribution. In the last, the decision how these batteries are distributed should be taken. It should be distributed through N-way by chance distribution i.e., 45% through distributor, 40% through retailer and remaining 15% through factory out let. Distributor dispose batteries to out of states while retailer and factory out let dispose batteries to customers.

We want to collect statistics in each area on entity, transfer time and transfer cost (councilor, SBOs, retailer and disassembler), process and resource utilization. We will initially run the simulation for four consecutive 8-hours shift or 1, 920 minutes.

9.10. VERIFICATION AND VALIDATION

Verification concerns with the operational model (whether it is performing properly). It is done to ensure that:

- The model is programmed correctly, and
- The model does not contain errors, oversights, or bugs.

In recycling process, model verification took place as a continuing process. In this case, the percentage error is at an acceptable level of less than 5%.

In this study, comparisons between the simulation and calculated results are made. The associated parameters are set constant for calculating the transportation costs by using deterministic data. This includes average truck load size, average speed of transporter, transporter cost per hour, and its travel distance. The transfer times and costs per part are calculated by using

$$\text{Transportation cost/part} = \frac{\text{Transportation time * Transportation cost/h}}{\text{No. of transported part per trip}}$$

In order to make sure that the model data accurately represent the actual process, hence enhance model validity, a confidence interval analysis was carried out. A 95% confidence interval was determined and the required number of simulation runs was decided by the following equation (Kelton *et al.*, 1991)

$$n = Z\alpha/2S^2/d^2$$

Where n is the number of desired replication, d is the accuracy expressed, Z is the critical value from the standard normal table at given confidence interval & S is the standard deviation desired.

9.11. LIMITATIONS OF WORK

XYZ Ltd. company manufactures different types of batteries but we are considering only one type of battery *i.e.*, invertors battery 150 AH (Amp. hour) for our work due to shortage of time and to reduce the complexity of the simulation model. It can extend the scope of simulation model to other products with some modification.

9.12. RESULTS AND DISCUSSION

Statistics from animation models are summarized in the Tables no. 9.1 to 9.17 and are self-explainable given below which focus on main performance indicators in our process that are: transfer time, transfer cost, process, total time for the process (VA), resource utilization and the number of entities processed in the process.

The influencing factors on the collection cost include the following:

***Collection Strategy*:** The collection strategies used in the model have a significant effect on the resulting transportation cost. In general, a collection strategy may involve varying degree of participation from all participants *e.g.*, manufacturer, consumer, collectors,

Recyclers, in reverse logistics. There are many factors, which influences the effectiveness of these collection strategies. Although some of these factors are country specific, researchers agree that the utilization of the existing structure simplifies the problem and makes it more economical. It is possible to further reduce the cost if the expertise and the infrastructure of NSW are fully utilized.

***Disassembly Plant*:** In this study, the low cost "center of gravity" with respect to incoming inverter batteries are used to find a good geographic location for the disassembly plant. The two factors affecting the result are the position of the nodes and their weight. In this study, the location of the disassembly plant was suitable because of its industrial environment. If the selected location falls within a residential area, the real location may have to be reviewed.

One or Bi-directional Delivery: The transportation cost per batteries displayed in the project is the traveling cost in one-direction, which is the case for most of the deliveries. If bi-directional deliveries can be used, for instance by using the trucks for delivery of new products, the "in truck time" and "in truck cost" could be halved.

***Inventory Cost at Stations*:** During the simulation, the inventory costs of the discarded spent batteries were calculated as a fixed cost according to the percentage of space they occupy at the stations. This is even though, in current practice, the holding cost of the spent batteries is assumed insignificant. If the calculations were carried out based on the current practice, the cost of collection would be significantly lower.

***Number of Reusable Components per White Good*:** It has assumed in the analysis that there is only one part per inverter batteries that can be recycled. If the average number of remanufactured parts per spent batteries is higher, it will have a significant effect on the transportation cost. For instance, if the average number of parts is two instead of one, the transportation cost from end consumer to the disassembly plant will be halved. This will not affect the transportation cost from the disassembly plant.

CATEGORY OVERVIEW

Recycling of Batteries
Table 9.1
Replication: 70 Time Units: Minutes

Key Performance Indicators

All Entities	*Average*
Non-Value Added Cost	0
Other Cost	0
Transfer Cost	Rs.4,470
Value Added Cost	Rs.7,454
Total Cost	Rs.11,925
System	Average
Total Cost	Rs.11,925
Number Out	128

ENTITY

Time

VA Time

Table 9.2: Value added time (va)

VA time	*Average (min.)*
Council battery collecting	116.33
SBO battery collecting	110.55
Retailer battery collecting	109.87
Disassembler collector	115.36

Transfer Time

Table 9.3: Transfer time

Transfer time	*Average (min.)*
Council battery collecting	70
SBO battery collecting	70
Retailer battery collecting	45
Disassembler collector	35

Total Time

Table 9.4: Total time

Total time	*Average (min.)*
Council battery collecting	186.33
SBO battery collecting	180.55
Retailer battery collecting	154.87
Disassembler collector	150.36

COST

VA Cost

Table 9.5: Value added cost (VA)

VA cost	*Average (Rs.)*
Council battery collecting	34.08
SBO battery collecting	41.85
Retailer battery collecting	83.24
Disassembler collector	33.84

Transfer Cost

Table 9.6: Transfer cost

Transfer cost	*Average (Rs.)*
Council battery collecting	1017.50
SBO battery collecting	1023.33
Retailer battery collecting	1530
Disassembler collector	808.75

Total Cost

Table 9.7: Total cost

Total cost	*Average (Rs.)*
Council battery collecting	1056.58
SBO battery collecting	1081.18
Retailer battery collecting	1639.25
Disassembler collector	869.59

Other

Table 9.8: Number of entities in

Number in	*Value*
Council battery collecting	60
SBO battery collecting	71
Retailer battery collecting	74
Disassembler collector	37

Table 9.9: Number of entities out

Number out	*Value*
Council battery collecting	30
SBO battery collecting	39
Retailer battery collecting	42
Disassembler collector	17

Table 9.10: Work in process

WIP	*Average*
Council battery collecting	23.66
SBO battery collecting	28.19
Retailer battery collecting	25.52
Disassembler collector	11.84

PROCESS

Total Time Per Entity

Table 9.11: Total time per entity

Total time per entity	*Average (min.)*
99% pure lead	15.09
Acid reclamation	7.63
Battery body manufacturing	36.37
Disassembly process	25.17
Lead refining process	37.85
Lead scrap process	24.73
New battery manufacturing	21.22
Plastic recyclers	25.37
Scrap plastic Process	20.21
Spent acid process	12.27

Cost Per Entity

Table 9.12: Total cost per entity

Total cost per entity	*Average (Rs.)*
99% pure lead	6.1345
Acid reclamation	3.3470
Battery body manufacturing	15.2166
Disassembly process	10.1872
Lead refining process	15.5970
Lead scrap process	9.8680
New battery manufacturing	8.5894
Plastic recyclers	10.6642
Scrap plastic process	8.5585
Spent acid process	5.4534

RESOURCE

Usage

Number Busy

Table 9.13: Number busy during resources

Number busy	*Average (%)*
Acid reclamation	0.4239
Battery manufacturing	0.8107
Body manufacturing	0.8636
Disassembly	0.7842
Lead scrap	0.7983
Plastic recycling	0.3575
Pure lead	0.2466
Pyrometallurgical process	0.9926
Scrap plastic	0.5500
Spent acid	0.6817

Queue

Table 9.14: Queue in the system

Queue length in the system	*Waiting time (min.)*
Disassembly center	38.00
New battery manufacturing	5.00
Lead recycling	45.00
Plastic recycling	44.00

9.13. COMPARISON OF PARAMETERS

After simulation of the proposed integrated model on Arena Rockwell software, the existing parameters of the company has compared with the results obtained after simulating. The comparisons are shown below:

Transfer Cost (Rs.)

Table 9.15: Comparison of transfer cost

	Existing cost (M)	*Cost after simulation (S)*	*Difference D = {(M-S)/M}*100*
Council's	7	6.78	3.14%
SBO's	7	6.8	2.85%
Retailer's	5.4	5.1	5.5%
Disassembly center	5.46	5.39	1.44%

Transfer Time (Min.)

Table 9.16: Comparison of transfer time

	Existing cost (M)	*Cost after simulation (S)*	*Difference D = {(M-S)/M}*100*
Council's	72	70	2.7%
SBO's	72	70	2.7%
Retailer's	45.8	45	1.74%
Disassembly center	34.6	35	1.15%

Resources Utilization

Table 9.17: Comparison of resource utilization

Number busy	*Existing (M) Average (%)*	*After simulation (S) Average (%)*
Acid reclamation	0.40	0.4239
Battery manufacturing	0.75	0.8107
Body manufacturing	0.78	0.8636
Disassembly	0.55	0.7842
Lead scrap	0.70	0.7983
Plastic recycling	0.25	0.3575
Pure lead	0.20	0.2466
Pyrometallurgical process	0.85	0.9926
Scrap plastic	0.42	0.5500
Spent acid	0.62	0.6817

9.14. SUMMARY AND CONCLUSIONS

Recycling process of lead acid battery in which waste batteries or spent batteries are collected at battery collection center through different suppliers from where they are transported to the recycling plant for obtaining secondary lead alloys, plastic scrap, spent acid. The spent acid is collected and transported to acid reclamation center and it is used in fertilizer production, chemical production, sodium sulphate production or neutralization and pretreatment for the discharge. The plastic scrap is further transported to the plastic recyclers for product manufacturing. The secondary lead alloys and the plastic product manufacturing are used in new battery production and other lead products. Finally, the dealers, distributors and retailers supply the new batteries and other lead products to the consumers according to their requirements. 98% of all Lead acid batteries are recycled in the USA. In comparison, only one in six households in North America recycles batteries.

The principal of reverse logistics has provided the basis, which investigates the take-back system for spent batteries collected from the consumers, and transported to the recycling plant. Designing of a collection network for the take-back of spent batteries for recycling require the integration of the network into an existing collection infrastructure. The simulation modeling allows the user to analyze the future performance of the network and to understand the complex relationship between the parties involved. At the current stage, the company is not using a well define structured approach to get EOL product in the company. Therefore, we have designed and developed an integrated model of forward and reverse supply chain network for the company. The presented model provided the process mapping of the company in a predictable manner and company can decide its future strategies regarding further use of EOL products for making new products in a cost effective and flexible manner.

10

Simulation Modeling of AGVs Based Production System in Job-Shop Manufacturing Environment

10.1. INTRODUCTION

Automated Guided Vehicles (AGVs) are often used in general manufacturing of products. AGVs can typically be found delivering raw materials, transporting work-in process, moving finished goods, removing scrap materials, and supplying packaging materials. These days, high competition is occurring in several production industries. The flexible manufacturing system (FMS) is designed to manufacture a variety of workplaces in a group of machines and other workstations connected by a material transport system, under computer control, such as an AGV system. A company is required to improve its production lines productivity by keeping the production line running smoothly. Strategies for scheduling become necessary. In addition, one type of job scheduling cannot optimize all objectives, such as machine utilization, works in progress, and throughput. In FMS, several problems arise in parts waiting to be processed by a machine. An AGV is used to transfer parts and products beside the production line, and is guided by markers on the floor. In job shop- manufacturing AGVs assembly-line vehicles used. The AGVs assembly-line vehicles are Variation of a light load transporter, for serial assembly processes, as the vehicle moves from one station to another succeeding assembly operations are performed, this kind provides flexibility for the manufacturing processes, lower expenses and ease of installation. The Automated Guided Vehicles (AGVs) is designed to perform their operations without direct human supervision. AGVs consist of one or more computer-controlled wheel based load carriers (normally battery- powered) that run on the plant floor or if out-of-doors on a paved area without require for an onboard operator or driver. Automated Guided Vehicles systems are fully automatic transport system using unmanned vehicles. Automated Guided Vehicles undamaged transport all kinds of products without human involvement within production, logistic, warehouse and distribution environments. This results in cost reduction and increase efficiency and productivity. Automated Guided Vehicle System to transport material from loading to unloading station. It is a highly flexible, intelligent and versatile material handling systems. A very flexible solution for the problem of integrating a new automated transportation line into an existing transportation environment by using transportation environment by using automated guided vehicle. Automated Guided Vehicles is a driverless vehicle, electric motor, battery powered. The programming capabilities are destination, path selection, positioning, and collision avoidance. One reason is that as manufacturers strive to become more competitive, they are adopting flexible manufacturing systems (FMS). These systems integrate automated material handling systems, robots, numerically controlled machine tools, and automated inspection stations. FMS offer high capital utilization and reduced direct labor costs in addition to lower work-in-process inventory and shorter lead times. Because the systems are flexible, they are more responsive to changes in production requirements. These systems offer high product quality and increased productivity.

Flexible manufacturing systems can benefit from the linkage with AGVs. While robots are often highlighted as saving billions in production costs, at some plants-including steel and other metals plants-automated material-handling systems have made the biggest inroads. Today, there are hundreds of instances of computer-controlled systems designed to handle and transport materials, many of which have replaced conventional human-driven platform trucks. Although only a single component of a flexible manufacturing system, automated material handling systems have advantages of their own. These include a reduction in damage to in-process materials, simplified inventory tracking and production scheduling, increased safety, and the need for fewer personnel than in conventional systems.

A variety of analytical methods have been planned by researchers for the design and control of AGVs. Simulation is used to compare the performance of tandem AGV system with that of conventional AGV path systems. Numbers of Simulation studies have been conducted to evaluate the effect of different parameters such as number of AGVs, the number of pallets, buffer sizes, bi-directional flows, dispatching rules etc.

The commendable safety record of Automated Guided Vehicles is due to a number of factors. They have numerous safety devices to prevent accidents such as flashing lights, physical emergency bumpers, audible devices, passive infrared and sonar devices. In addition, ANSI regulations specify layout requirements, stopping distances and other safety related items to insure safe implementation of Automated Guided Vehicles system.

Computer simulations are often used in planning complex transportation systems. Facilities may require pathways, wire-guidance systems, automatic cranes, and additional computer software and hardware to run the entire AGV system. Some AGV systems even use laser scanners as guidance systems.

10.2. PROBLEM EXPLANATION

The present work includes more number of machines and jobs, utilization of machines and AGVs in the manufacturing systems. The main objectives of this paper are study of AGVs in manufacturing environment, Sequencing of machines and jobs to improve utilization of AGVs system and Simulation modeling of designed hypothetical AGVs based system in job-shop manufacturing environment. Simulation of AGVs in job- shop manufacturing environment model in Arena Rockwell software simulation environment. Consider a Flexible Manufacturing System (FMS) producing four types of jobs, Job1, Job2, Job3 and Job4 and six machines, operation sequence by job types in Fig. 10.1. Consider of jobs arriving batches are 40% of type Job1, 30% of type Job2, 20% of type Job and 10% of type Job4. At first a job arrives at the arrival station and from there a job is departed passing through sequence of manufacturing operations. During this process a sequence consists of a subset of J1, J2, J3, J4, J5, J6 operations. Jobs of time between arrivals are 25 minutes Random (Expo). Operations plan are displayed in Table 10.2 which are showing the processing times and sequence of operation. Two AGV running at a constant speed of 100 feet/minute transport jobs among locations. Each AGV can carry only one job at a time. When a job is completed at a location, the job is placed into an output buffer, AGV is requested and the job waits for an arrival of AGV. When a job is transported to the next location, it is placed in a FIFO input buffer. Ultimately, when the J6 process on machine is completed, the finished job departed from the shop exit.

10.3. DESIGN OF LAYOUT OF JOB-SHOP MANUFACTURING ENVIRONMENT

Consider flexible manufacturing systems producing four types of jobs, like job1, job2, job3, and job4 and six machines, operation sequence by job types in Fig. 10.1. In this layout consider arrival station, M1 workstation, M2 workstation, M3 workstation, M4 workstation, M5 workstation, M6 workstation, and shop exit, J1 process on machine M1, J2 process on machine M2, J3 process on machine M3, J4 process on machine M4, J5 process on machine M5, and J6 process on machine M6. Distances among the workstations are given in Table 10.1 and operation plan for job by types in given Table 10.2.

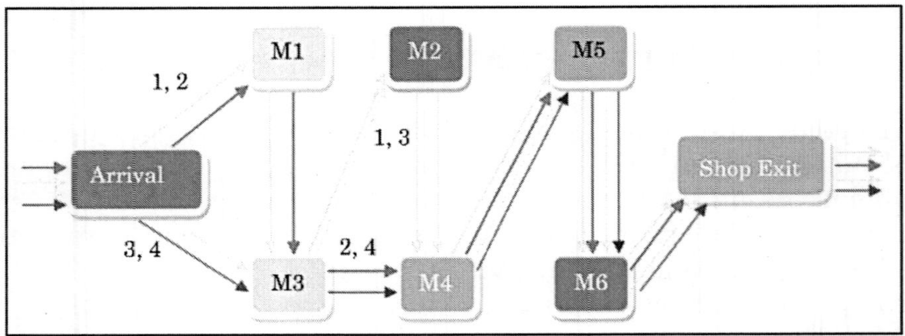

Fig. 10.1: Layout of operation sequence of FMS.

Table 10.1: The distances among locations

Sl. no.	*Beginning station*	*Ending station*	*Distance*
1.	Arrival	M1	140
2.	Arrival	M2	140
3.	M1	M2	200
4.	MI	M3	150
5.	M1	M4	250
6.	M1	M6	400
7.	M2	M3	250
8.	M2	M4	150
9.	M2	M5	400
10.	M3	M4	200
11.	M3	M5	150
12.	M3	M6	250
13.	M4	M5	250
14.	M4	M6	150
15.	M5	M6	200

Table 10.1 *(Contd...)*

Table 10.1 *(Contd...)*

Sl. no.	*Beginning station*	*Ending station*	*Distance*
16.	Arrival	M3	250
17.	Arrival	M4	250
18.	Arrival	M5	400
19.	Arrival	M6	400
20.	M2	M6	300
21.	M5	M1	300
22.	Shop exit	M5	400
23.	Shop exit	M6	400
24.	Shop exit	Arrival	660
25.	Shop exit	M1	500
26.	Shop exit	M2	500
27.	Shop exit	M3	300
28.	Shop exit	M4	300

Table 10.2: Operation plan for job by type

Job type	*Operation sequences*	*Processing time (minutes)*
Job1	M1	20
	M2	15
	M3	13
	M4	10
	M5	8
	M6	12
Job2	M1	10
	M3	9
	M4	8
	M5	4
	M6	7
Job3	M2	12
	M3	8
	M4	7
	M5	3
	M6	10
Job4	M3	11
	M4	14
	M5	6
	M6	4

Assumptions made during the modeling system are

- The FMS works for 24 hours a day in 3 shifts at 8 hours each.
- The free AGV stays at the destination until requested by another station.
- AGV speed is same both loaded and unloaded.

The objectives of the present work are to Simulation modeling of AGVs in manufacturing environment using ARENA Simulation software and find that

To study the Sequencing approaches, Utilization of AGV and Utilization of Machines.

10.4. SIMULATION MODELING OF AGVs IN MANUFACTURING ENVIRONMENT

The given system is modeled using ARENA Simulation software as shown in Figs. 10.2 and 10.3. The ARENA models for the job consisting of four main segments are Job Arrivals, Job Transportation, Job Processing, and Job Departure. Create Jobs and Assign Job Type and Sequence are called arrival section of jobs. In advanced Transfer template panel, each sequence consists of sequence name and a series of steps (Step column), listed in order of processing, each module data is shown in dialog box. Job entities are created in the create module, called Create Jobs, whose dialog box from arena software. The time between arrival sections specifies batch inter-arrival times to be uniformly distributed. An arriving job entity next enters the assign module is called assign type and sequence from arena software. Here, a job entity is assigned a type by sampling it from a discrete distribution, and saving the type code (1, 2, 3, or 4) in its type attribute. The operations sequences for job types are specified in the Sequence module from advanced transfer template panel, whose dialog spreadsheet is displayed, four sequences (row entities) are defined here, one for each job type. This part includes the arrival section of the jobs, Job arrival segment from arena software.

10.5. SIMULATION MODEL TRANSLATION

In advanced Transfer template panel, each sequence consists of sequence name (Name column) and a series of steps (Step column), listed in order of processing. To specify steps, the modeler clicks the button under the steps column and pops up the steps dialog spreadsheet. The five steps of type job1 job processing are displayed in the middle spreadsheet in dialog box. Each step is a row entry specifying the location name and associated values. Clicking the corresponding button pops up the associated Assignments dialog spreadsheet. Job Transportation segment includes jobs arriving to arrival, requesting an AGV and transporting jobs in manufacturing system. Job entities will be transported to the shop floor to start the first step in their operations

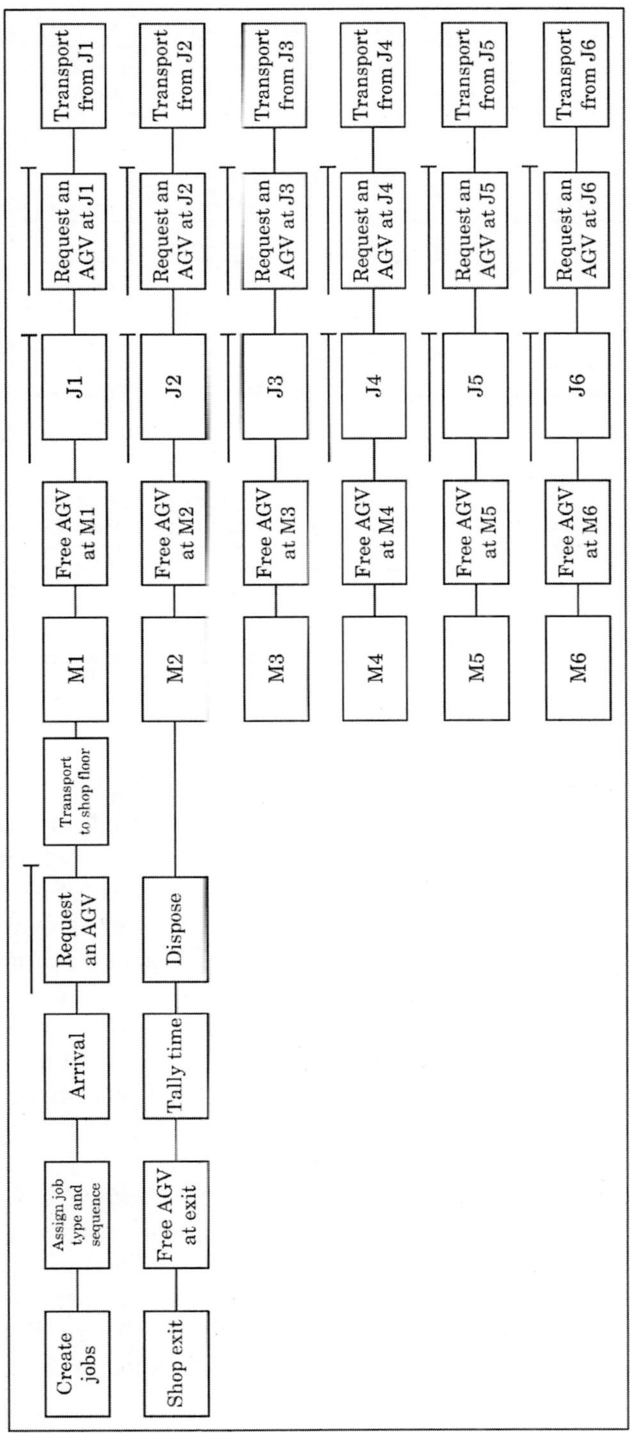

Fig. 10.2: Arena model for the job-shop manufacturing.

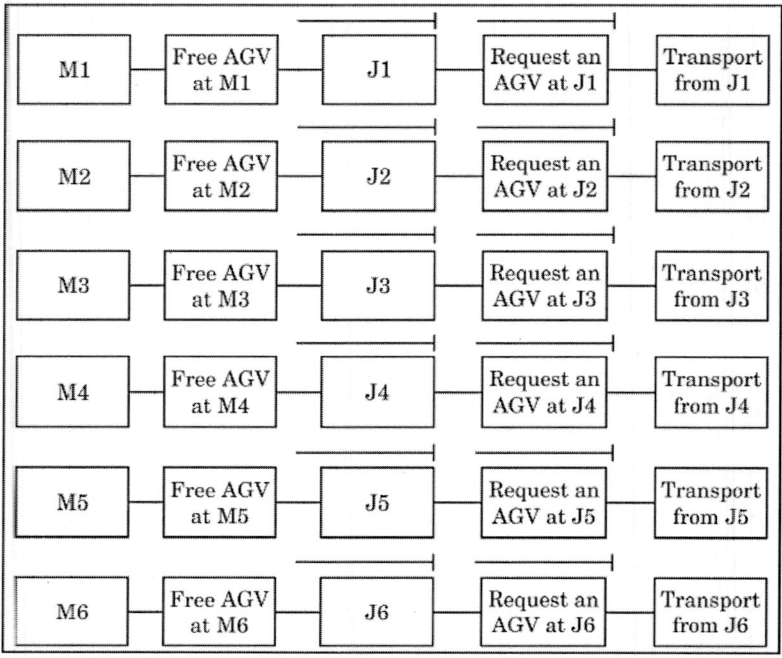

Fig. 10.3: Job processing segment

sequence. In this system locations are modeled as station modules. Every job entity proceeds to the station module, is called Arrival, to model its physical arrival at the job-shops arrival. If multiple transporters are available, the modeler can specify how to select one in the Selection Rule field. Such selection may be cyclical, random, preferred order as listed in the transporter module, smallest distance, largest distance, or a specific transporter. Here the selection rule requests the transporter nearest to the arrival. A job entity enters the Request module from the advanced transfer template, is called request an AGV in the dialog box. The transporter name field indicates a request for an AGV transporter. After that the save Attribute field specifies that the ID of the select transporter be saved in the agv_ID attribute of the requesting job entity. The saved ID will be used in due time to free that particular AGV. The entity location field indicates the location of the requesting entity, and the Velocity field specifies the transporter's velocity, which is 100 feet/ minute in our case. Finally, job entities requesting transportation at the same Request module are instructed in the Queue Name field to wait in the queue, is called Agv. Queue, until a transporter becomes available.

As soon as a job entity grabs an AGV, it proceeds to the Transporter module is called transport to shop floor, which dialog box is displayed. The Transporter name and Unit number field specify the type and ID of the selected transporter, which here is an AGV whose ID is kept in an agv_ID attribute of the requesting job entity. The transporter per job destination is specified in the entity destination type field as the by sequence option, including that the destination is determined by the job entity's sequence number. This field may also specify a Station module name, using the station option. It can also specify an attribute or expression. The job entity and the transporter move as a grouped entity at a velocity of 100 feet/minute as specified in the velocity field. Job processing segment includes the actual processing of jobs. The job processing segment encompasses sets of Station modules, each modeling an operation in the sequence, from J1 to J6. Since all sets have the same structure (except for name), only the M1 operation logic is explained here. When a job entity is transported to the J1 process on M1, it enter the station module is called M1 station. It then proceeds to the free module is called Free AGV at M1, whose dialog box is displayed. Here the Transporter name and Unit number field specify an AGV to be freed for use by other job entities, using the agv_ID attribute of the free job entity. Job entities move from one operation to another according to their specified sequences. It should be noted out that Arena Simulation handles all sequencing details at runtime. The internal Arena attribute be keeps track of each job entity's step number in its sequence at any time a sequential transport is requested, Arena increments to be attribute and indexes into the appropriate steps module spreadsheet to determine the end location and travel time. This segment includes the transport of finished jobs from manufacturing system to outer surface. Arena computes travel times of transporter among station modules based on their distances and transporter speeds.

The ARENA model is simulated for a period of one month. Parameters like Replication length and hour per day etc. While running in simulation, we can movement of entities (jobs) through different facilities, waiting for processing, transfer between machines etc. The process time flow as assign DISC (0.4, 1, 0.7,2, 0.9, 3, 1, 4). The following parameter ware specified:

(1) Length of each simulation run = 43200 minutes (one month).

(2) Number of independent simulation runs = 10 and 20 replications.

After completing the Simulation, report is generated automatically. From the report, we can find job flow times, Job delays at operations location, Machine utilization and AGVs utilization of the system.

10.6. RESULTS AND DISCUSSION

Statistics from animation model is summarized in the Tables 10.3 to 10.7 and in Figs 10.4 to 10.7 as given below are self-explainable, which focus on main performance indicators in our process that are transfer time, value added time, wait time, total time for the process, resource utilization and the number of entities processed in the process. Utilization of AGVs is display after the simulation run.

Table 10.3: VA time per entities

VA time per entity	*Average (minute)*	
	Replication 10	***Replication 20***
J1	15.69	15.69
J2	14.00	14.00
J3	10.59	10.59
J4	9.17	9.18
J5	5.59	5.59
J6	9.31	9.31

Table 10.4: Total time per entities

Process	*Average (minute)*	
	Replication 10	***Replication 20***
J	22.48	22.71
J2	15.80	15.74
J3	12.77	12.78
J4	9.67	9.68
J5	5.69	5.68
J6	10.78	10.75

Table 10.5: Number of jobs in process.

Process	*Average*	
	Replication 10	***Replication 20***
J1	1229	1217
J1	1044	1043
J1	1739	1736
J1	1739	1736
J1	1738	1735
J1	1738	1735

Table 10.6: Number of jobs out process

Process	*Average*	
	Replication 10	*Replication 20*
J1	1228	1217
J2	1044	1042
J3	1739	1736
J4	1739	1736
J5	1738	1735
J6	1738	1735

Table 10.7: Scheduled utilization of process

Process	*Average*	
	Replication 10	*Replication 20*
J1 process	44.61	44.23
J2 process	33.85	33.77
J3 process	42.66	42.57
J4 process	36.94	36.92
J5 process	22.52	22.46
J6 process	37.47	37.36

Trends of accumulated times are generated after simulation run of Replication 10 given below:

Accum VA Time	*Average*	*Half width*	*Minimum average*	*Maximum average*
J1	19264.00	554.19	18310.00	21050.00
J2	14620.80	366.92	13992.00	15555.00
J3	18429.10	480.90	17614.00	19858.00
J4	15957.30	425.30	15137.00	17174.00
J5	9729.30	250.14	9307.00	10484.00
J6	16187.00	406.82	15532.00	17352.00

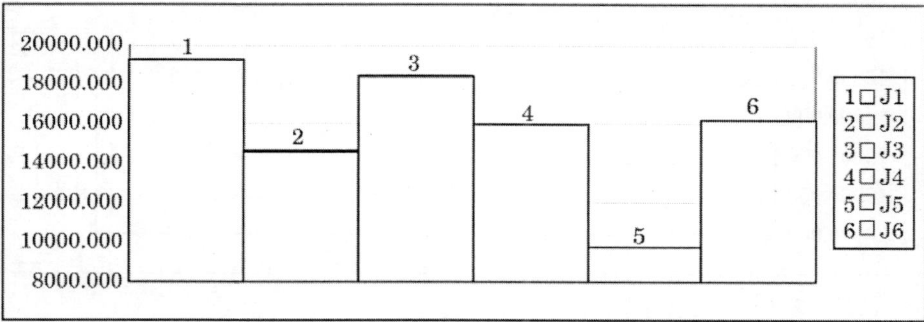

Fig. 10.4: Accumulated VA time (replication 10).

Accum VA Time	*Average*	*Halfwidth*	*Minimum average*	*Maximum average*
J1	8370.86	1007.28	6316.28	10938.59
J2	1874.20	210.14	1446.80	2357.40
J3	3806.77	455.26	3148.48	5082.10
J4	867.72	112.87	685.34	1222.69
J5	164.36	27.93	114.16	261.22
J6	2552.05	194.04	2196.22	3042.17

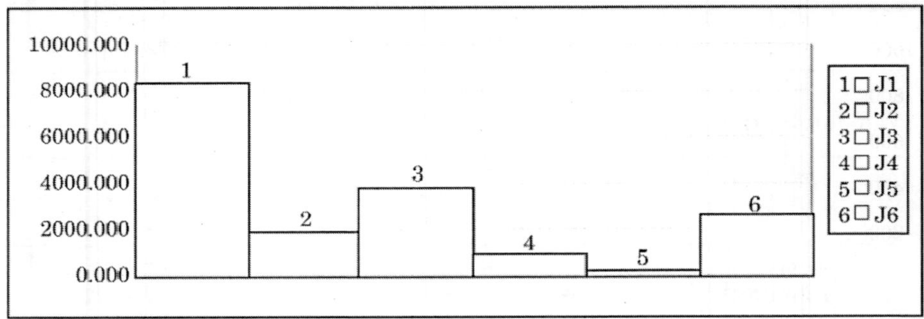

Fig. 10.5: Accumulated wait time (replication 10).

Trends of accumulated times are generated after simulation run of Replication 20 given below:

Accum VA Time	*Average*	*Halfwidth*	*Minimum average*	*Maximum average*
J1	19101.50	287.46	18310.00	21050.00
J2	14583.30	208.28	13992.00	15555.00
J3	18387.65	262.23	17614.00	19858.00
J4	15946.35	237.64	15137.00	17174.00
J5	9702.40	136.96	9307.00	10484.00
J6	16136.80	222.13	15483.00	17352.00

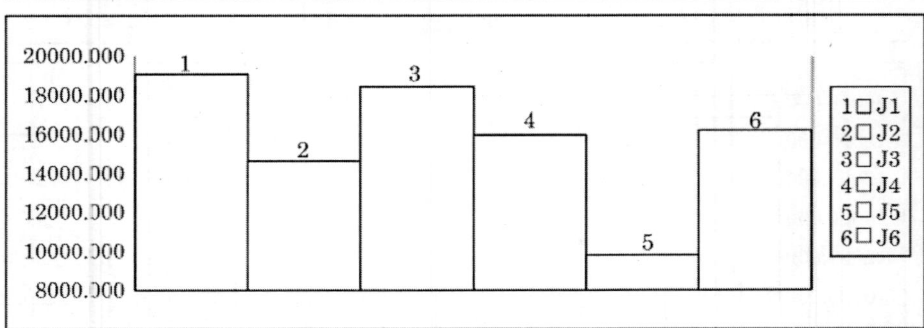

Fig. 10.6: Accumulated VA time (replication 20).

Accum VA Time	Average	Halfwidth	Minimum average	Maximum average
J1	8553.98	562.02	6316.28	10938.59
J2	1827.91	117.32	1446.80	2357.40
J3	3815.01	230.91	3148.48	5082.10
J4	870.72	54.69	685.34	1222.69
J5	161.94	13.49	114.16	261.22
J6	2523.08	101.06	2196.22	3042.17

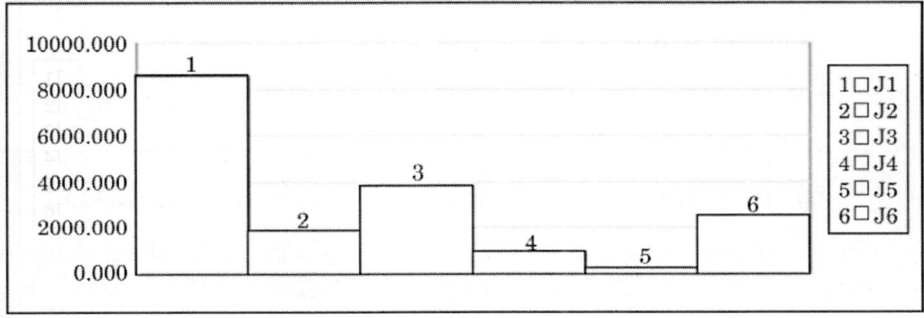

Fig. 10.7: Accumulated wait time (replication 20).

Jobs Delay at Operation

After simulation of proposed integrated model on Arena Rockwell software, comparison the machines delay operation time of replication 10 and 20 are given below in Table 10.8 and in Fig. 10.8.

Table 10.8: Job delay at operation on machine with waiting times.

No. of replication	Machine	Waiting time (minute)
10	M1	6.7902
	M2	1.7907
	M3	2.1812
	M4	0.4971
	M5	0.0940
	M6	1.4656
20	M1	7.0140
	M2	1.7519
	M3	2.1932
	M4	0.5007
	M5	0.0930
	M6	1.4527

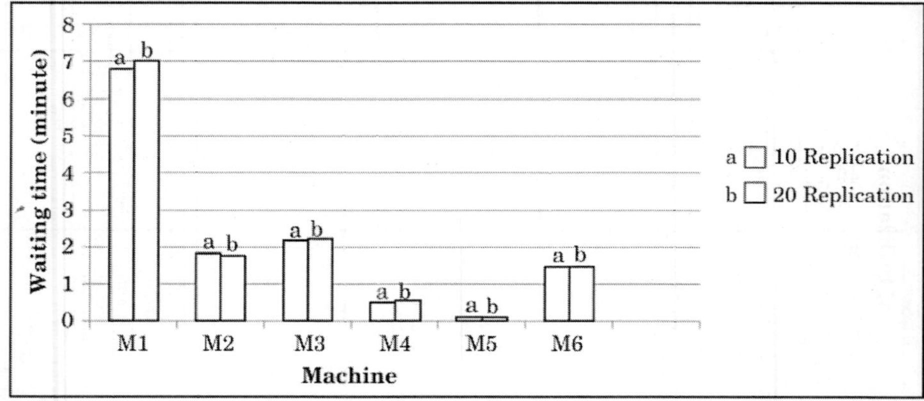

Fig. 10.8: Comparing machine with waiting times of replication 10 and 20.

In the given table there is not much variation of waiting time of each machine and hence system is stabilized. When AGVs has introduced in the system waiting time of each machine has greatly improved. The waiting of machine is reduced, the waiting time of jobs also reduced; and finally manufacturing process time will be reduced.

Resources Utilization

After simulation of proposed integrated model on Arena Rockwell software, comparison of the resource utilization of machines and AGV replication 10 and 20 are given below in Table 10.9.

Table 10.9: Resource utilization of machines and AGV.

No. of replication	*Machine and AGV*	*% of Utilization*
10	M1	44.61
	M2	33.85
	M3	42.66
	M4	36.94
	M5	22.52
	M6	37.47
	AGV	55.48
20	M1	44.23
	M2	33.77
	M3	42.57
	M4	36.92
	M5	22.46
	M6	37.36
	AGV	55.35

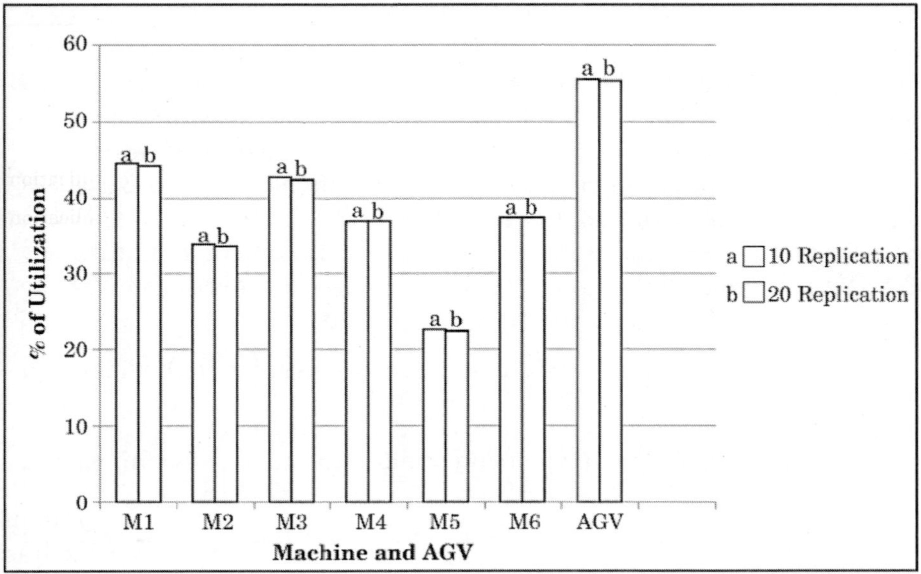

Fig. 10.9: Comparing resource utilization of machines and AGVs.

The trend between percentage utilization and machine, AGV is shown in Fig. 9. In this system numbers of replications are changes but there is little variation in the percentage of utilization in machines and AGVs in both replications and hence system is stabilized. Sequencing is the main problem in the system so job handling with defined path and hence utilization of Automated Guided Vehicles (AGVs) and machine are improved. Therefore, production in the manufacturing process increased. The resources utilization of machines and AGVs are given in Table 10.9.

10.7. CONCLUSIONS AND FUTURE SCOPE

The objective of the study has been basically to review the current literature and research works related to Simulation of Automated Guided Vehicles (AGVs) in job-shop manufacturing environment. The researcher has been facing the problems of scheduling, dispatching, sequencing of Automated Guided Vehicles (AGVs) in the job-shop manufacturing environment. In this research, resources utilization of machines and AGVs performance has been increased by reducing waiting times of machines and jobs in FMS. The given system is modeling using ARENA Simulation Software and results are generated. The analyzed the results, number of replications are changes but waiting times of each machines is almost same as shown in Table 10.8. The waiting times of each machine are little variation and hence our system is stabilized. Percentage utilization of resources of Machines and AGVs are increased as shown in Fig. 10.9. Hence the performance of job production is also improved. Therefore, efficiency of whole system is increased significantly. In this system number of machines and jobs increase, and utilization of machines are improved shown in Table 10.9. Utilization of AGVs is increases as 55.48% shown in the Table 10.9. Transportation efficiency of AGVs is increased and job production is also improved, thereby the efficiency of whole system improved significantly.

The present study can be extended in the broad scope in the following context

- The model could be developed for more number of machines and jobs.
- Efficiency of the AGVs may be further improved in manufacturing system.
- Applying the dispatching/ scheduling of jobs, machines and AGV system.
- Applicability on AGV systems of approaches for vehicle routing and scheduling in physical distribution.

Bibliography

Ablhamid, R.K., Budi Santosh M. and Muslim, M. (2013). "Decision Making and Evaluation System for Employee Recruitment Using Fuzzy Analytic Hierarchy Process". *International Refereed Journal of Engineering and Science* (IRJES), 2(7): 24–31.

Agarwal G. and Lokesh Vijayvargy (2012). "Green Supplier Assessment in Environmentally Responsive Supply Chains through Analytical Network Process" *Proceedings of International multi–conference for engineers & computer scientists*, Volume II, Hongkong.

Agrawal A. and Shankar R. (2002). "Analysing alternatives for improvement in supply chain performance". *Work–Study*, 51(1): 32–37.

Alinezad A., Seif A. and Esfandiari N. (2013). "Supplier evaluation and selection with QFD and FAHP in a pharmaceutical company". *Int. J. Adv. Manuf. Technology,* 68: 355–364.

AlKhidir T. and Zailani, S. (2009). "Going Green in supply chain towards Environmental Sustainability". *Global Journal of Environmental Research*, 3(3): 246–251.

Amid A., Ghodsypour, S.H. and O'brien, C. (2011). "A weighted max–min model for fuzzy multi–objective supplier selection in a supply chain". *Int. J. Production Economics,* 131: 139–145.

Anand G. and Kodali, R. (2008). "Selection of lean manufacturing systems using the PROMETHEE". *Journal of Modelling in Management*, 3(1): 40–70.

Asghar A., Meysam and Ramezani, A. (2013). "Green Supply Chain Management Evaluation in Publishing Industry Based on Fuzzy AHP *Approach". Journal of Logistics Management*, 2(1): 9–14.

Attri Rajesh (2013). "Interpretive Structural Modelling (ISM) approach: An Overview" *Research Journal of Management Sciences*, 2(2): 3–8.

Awasthi A., Chauhan S.S. and Goyal S.K. (2010). "A fuzzy multi-criteria approach for evaluating environmental performance of suppliers". *Int. J. Production Economics*, 126: 370–378.

Ba Dere B., Chan K., Morrow R., Prasitnarit A., Seliger G. and Skerlos S. (2003). "Economic and environmental characteristics of global cellular telephone remanufacturing". *In: Proceedings of the IEEE international symposium on electronics & environment*, Boston.

Bai and Sarkis J. (2010). Integrating sustainability into supplier selection with grey system and rough set methodologies. *International Journal of Production Economics,* 24(1): 252–264.

Benjamin T. Hazen (2011). "Strategic reverse logistics disposition decisions: From theory to practice". *Int. J. of Logistics Systems and Management,* 10(3): 275–292.

Berman S., Schechtman E. and Edan Y. (2009). "Evaluation of automatic guided vehicle systems". *Robotics and Computer–Integrated Manufacturing,* 25: 522–8.

Bhaduria Shubham and Jayant Arvind (2017). "Development of ANN Models for Demand Forecasting". *American Journal of Engineering Research (AJER)*, 6(11): 142–147.

Bharti R.V., Giri and Jayant A. (2015). "Green Supply Chain Management Strategy Selection by Analytical Network Process (ANP) Approach: A Case Study". *Journal of Material Science and Mechanical Engineering* (JMSME), 2(12): 8–13. ISSN: 2393–9109.

Bowen F., Cousins, P., Lamming, R. and Faruk, A. (2001). "The role of supply management capabilities in green supply". *Production and Operations Management,* 10(2): 174–189

Brans J.P., Mareschal, B. and Vincke P. (1984). "PROMETHEE: A new family of outranking methods in multi-criteria analysis". *Proceedings of Operational Research,* Amsterdam, North Holland, 84: 477–490.

Cooper W.W., Seiford L.M. and Zhu J. (2008). *"Handbook of Data Envelopment Analysis"* Kluwer Academic.

Dargi A., Ali A., Masoud. R., Galankashia and Akarte A.M.M. (2001). "Web based casting supplier evaluation using analytic hierarchy process". *Journal of the Operational Research Society*, 52(5): 511–522.

Dickson G.W. (1966). "An analysis of vendor selection systems and decisions". *Journal of Purchasing,* 2(1): 5–17.

Dowlastshahi S. (2005). "A strategic framework for the design and implementation of remanufacturing operations in reverse logistics". *International Journal of Production Research*, 43(16): 3455–3480.

Emiliani M.L. (2010). "Historical lessons in purchasing and supply relationship management". *Journal of Management History,* 16(1): 116–136.

Ernest Benedito and Albert Corominas (2011). "Optimal production and storage capacities in a system with reverse logistics and periodic demand". *J. of Logistics Systems and Management,* 10(3): 340–360[7].

Fawcett S.E. and Fawcett S.A. (1995). "The firm as a value–added system: Integrating logistics, operations and purchasing". *Int. Journal of Physical Distribution & Logistics Management*, 25(5): 24–4.

Fleischmann M. (2001). "Quantitative models for reverse logistics" *Berlin: Springer Verlag*, pp. 11–5.

Franke C., Basdere, B., Ciupek M. and Seliger S. (2004). "Remanufacturing of Mobile Phones — Capacity, Program and Facility Adaptation Planning". *Omega,* 27: 77–101.

Gant R.M. (1996) "Prospering in dynamically – competitive environments: Organizational capability as knowledge integration". *Organizational Science*, 7(4): 375–387. doi:10.1287/orsc.7.4.375

Gibbon P. (1997). "Saviours and Punks: The Political Economy of the Nile Perch Marketing Chain in Tanzania" CDR Working Paper 97.3, *Danish Institute for International Studies. http://dlc.dlib.indiana.edu/archive/00002866/01/gibbon.pdf.*

Giri Vikrant and Jayant Arvind (2016). "Development of multi–criteria decision making (MCDM) support models using hybrid grey relational analysis (HGRA)" unpublished *M.Tech. Dissertation* in SLIET Deemed University, Longowal, Punjab.

Grisi R.M., Guerra, L. and Naviglio, G. (2010). "Supplier performance evaluation for green supply chain management". *Business Performance Measurement and Management*, Part 4: 149–163.

Guneri A., Yucel A. and Ayyildiz G. (2009). "An integrated fuzzy – lp approach for a supplier selection problem in supply chain management". *Expert Systems with Applications,* 36: 9223–9228.

Handfield R., Walton S.V., Sroufe R. and Melnyk S.A. (2002). "Applying environmental criteria to supplier assessment: A study in the application of the analytical hierarchy process". *European Journal of Operational Research*, 141(1): 70–87.

Hentschel C., Seliger G. and Zussman E. (1995). "Grouping of used products for cellular recycling systems". *Annals of the CIRP*, 41(1): 11–4.

Hsu C.W. and Hu A.H. (2008). "Green Supply Chain Management in the Electronic Industry". *International Journal of Science and Technology*, 5(2): 205–216. ISSN: 1735–1472.

Hsu C.W. and Hu A.H. (2009). "Applying hazardous substance management to supplier selection using analytic network process". *Journal of Cleaner Production,* 17: 255–264.

Humphreys P., McIvor R. and Chan F. (2003). "Using case–based reasoning to evaluate supplier environmental management performance". *Expert Systems with Applications*, 25: 141–153.

Jain V., Tiwari M. K. and Chan F.T.S. (2004). Evaluation of the supplier performance using an evolutionary fuzzy – based approach. *Journal of Manufacturing Technology Management: Special Issue on Logistics and Supply Chain Management with Artificial Intelligence Techniques– Part 1*, 15(8): 735–744.

Jayant A. and Paul Virender (2013). "Analytical Network Process (ANP) in Selection of Green Supplier: A Case Study of Automotive Industry". *International Scientific Journal on Science Engineering & Technology*, 17(05): 453–465.

Jayant Arvind (2014). "Strategic Decision Modeling of Reverse Logistics Systems: Selection of Recovery Operations for EOL products". *Research Journal of Economics & Business Studies* (Singapore), 3(12): 42–62.

Jayant Arvind (2016). "Selection of Reverse Logistics Service Provider (RLSP) Using Analytical Network Process (ANP): A Case Study of an Automotive Company". *International Journal of Analytic Hierarchy Process*, 8(1): 131–160.

Jayant Arvind and Azhar Md. (2014). "Analysis of Barriers to Implement Green Supply Chain Management (GSCM) Practices: An Interpretive Structural Modelling (ISM) Approach". *Procedia Engineering*, 97: 2157–2166.

Jayant Arvind and Kumar Sanjiv (2011). "Simulation Modelling of Integrated Supply Chain Logistics Networks" *Proceedings of the International Conf. on Soc–ProS. AISC* 131: 87–97. springerlink.com © Springer India 2012.

Jayant Arvind and Singh Priya (2015). "Application of AHP–VIKOR Hybrid MCDM Approach for 3PL Selection: A Case Study". *International Journal of Computer Applications (IJCA)*, 125(5): 4–11. ISSN: 0975 – 8887.

Jayant Arvind, Gupta P. and Garg S.K. (2012). "Simulation modeling of outbound logistics of supply chain: A Case Study of telephone company". *International Journal of Industrial Engineering*, 19(2): 90–100.

Jayant Arvind, Gupta P. and Garg S.K. (2013). "Reverse Logistics Network Design for Spent Batteries: A Simulation Study". *International Journal of Logistics System and Management*", 18(3): 343–365.

Jayant Arvind, Gupta P. and Garg S.K. (2014). "Simulation modeling and analysis of network design for closed–loop supply chain: A case study of battery industry". *Procedia Engineering*, 97: 2213–2221.

Jayant Arvind, Khan M. and Kumar V. (2014). "Multi–Criteria Supplier Selection Using Fuzzy–AHP Approach: A Case Study of Manufacturing Company". *International Journal of Research in Mechanical Engineering & Technology,* 4(3): 73–79. ISSN 2249–5770

Jeong B.H. and Randhawa S.U. (2001). "A multi–attribute dispatching rule for automated guided vehicle systems". *International Journal of Production Research*, 39: 2817–2832.

Jharkharia S. and Shankar R. (2005). "IT enablement of supply chains: Understanding the barriers". *Journal of Enterprise Information Management*, 18(1): 11–27.

Kannan G., Devika K. and NoorulHaq A. (2010). "Analysing Supplier Development criteria for an Automobile Industry". *Industrial Management & Data Systems*, 110(1): 43–62.

Kaplan R.S. and Norton D.P. (1992). "The balanced scorecard — Measures that drive performance". *Harvard Business Review*, 70(1): 71–79.

Kara S., Rugrungrang F. and Kaebernick H. (2007). "Simulation modeling of reverse logistics networks". *International Journal of Production Economics*, 106: 61–69.

Karsak E.E. and Kuzgunkaya O. (2002) "A fuzzy multiple objective programming approach for the selection of a flexible manufacturing system". *International Journal of Production Economics,* 79: 101–111.

Khidir Al and Zailani S. (2009). "Going Green in supply chain towards Environmental Sustainability". *Global Journal of Environmental Research*, 3(3): 246–251

Kim C.W., Tanchoco J.M.A. and Koo P.H. (1999). "AGV dispatching based on workload balancing". *International Journal of Production Research*, 37: 4053–4066.

Koo P.H. and Jang J. (2002). "Vehicle travel time models for AGV systems under various dispatching rules". *International Journal of Flexible Manufacturing Systems*, 14: 249–261.

Kull T.J. and Talluri S. (2008). "A supply risk reduction model using integrated multi-criteria decision making". *IEEE Transactions on Engineering Management*, 55(3): 409–419.

Kumar S. and Malegeant P. (2006). "Strategic alliance in a closed–loop supply chain, a case of manufacturer and eco–non–profit organization". *Technovation*, 26(10): 1127–1135.

Kumar Uttam and Jayant Arvind (2015). "Simulation Modeling of AGVs in Job Shop Manufacturing Environment". *Journal of Aeronautical and Mechanical Engineering (JAAE)*, 2(8): 67–69. ISSN: 2393–8587.

Lee A.H.I., Kang H.Y., Hsu C.F. and Hung H.C. (2009). "A green supplier selection model for high–tech industry". *Expert Systems with Applications*, 36: 7917–7927.

Lettice F., Wyatt C. and Evan, S. (2010). "Buyer–supplier partnerships during product design and development in the global automotive sector: Who invests in what and when? *International Journal of Production Economics*, 127(2): 309–319. doi: 10.1016/j.ijpe.2009.08.007

Mallidis, I. and Vlachos, D. (2010). "A Framework for green supply chain management". *In: 1st Olympus international conference on supply chain, proceedings*, pp. 1–2.

Mandal A. and Deshmukh S.G. (1994). "Vendor selection using interpretive structural modelling (ISM)". *International Journal of Operations and Production Management*, 14(6): 52–59. doi:10.1108/01443579410062086.

Meade L. and Sarkis J. (1998). "Strategic analysis of logistics and supply chain management systems using the analytical network process". *Transp. Res. – E*, 34(3): 201–215.

Meade L.M. and Sarkis J. (1998). "Strategic analysis of logistics and supply chain management systems using the analytic network process". *Logistics and Transportation Review*, 34(2): 201–15.

Meade L.M., Liles, D. and Sarkis J. (1997). "Justifying strategic alliances and partnering: A prerequisite for virtual enterprising". *Omega: The International Journal of Management Science*, 25: 29–42.

Mendoza A., Jose A. and Ventura (2012). "Analytical models for supplier selection and order quantity allocation". *Applied Mathematical Modelling*, 36: 3826–3835.

Michaels R., Kumar A. and Samu S. (1995). "Activity–specific role stress in purchasing". *International Journal of Purchasing and Materials Management*, 31(1): 11–19.

Min H. and Zhou G. (2002). "Supply chain modelling: Past, present and future". *Computers & Industrial Engineering*, 43: 231–249.

Monczka R.M., Trent R.J. and Callahan T.J. (1983). "Supply base strategies to maximize supplier performance". *International Journal of Physical Distribution and Logistics Management*, 23(4): 42–54.

Monda Sandeep and Mukherjee Kampan (2012). "Simulation of tyre retreading process – An Indian case study". *Int. J. of Logistics Systems and Management*, 13(4): 525–539.

Morgan J. (1996). "Supplier integration: A new level of supply chain management Purchasing". 120(1): 110–113.

Morgan J. (1996b). "Good may not be good enough". *Purchasing*, 120(7): 38–41.

Mudgal R.K., Shankar R., Talib P. and Raj T. (2010). "Modelling the barriers of green supply chain practices: An Indian perspective". *Int. Journal of Logistics Systems and Management*, 7(1): 81–107. Electronic copy available at: *http://inderscience.metapress.com/link.asp?id=8343t48231372825*

Narasimhan R. (1983). "Analytical approach to supplier selection". *Journal of Purchasing and Materials Management*, 19(4): 27–32.

Nimawat D. and Namdev V. (2012). "An Overview of Green Supply Chain Management in India". *Research Journal of Recent Sciences*, 16: 77–82.

Noci G. (1997). "Designing green vendor rating systems for the assessment of a supplier's 18 environmental performance". *European Journal of Purchasing & Supply Management*, 3(2): 103–114.

Nydick R.L. and Hill R.P. (1992). "Using the analytic hierarchy process to structure the supplier selection procedure". *International Journal of Purchasing and Materials Management*, 28(2): 31–36.

Olugu E.U., Wong K.Y. and Shaharoun A.M. (2010). "A Comprehensive Approach in Assessing the Performance of an Automobile closed loop Supply Chain". *Sustainability*, 2: 871–879. doi:10.3390/su2040871

Patton W.E. (1997). "Individual and joint decision making in industrial vendor selection". *Journal of Business Research*, 38: 115–122.

Perron G.M. (2005). "Barriers to Environmental Performance Improvements in Canadian SMEs" *Dalhousie University Press Release*, Canada.

Porter M.E. and Van de Linde C. (1995). "Green and Competitive". *Harvard Business Review,* 24: 120–134.

Raj T., Shankar R. and Suhaib M. (2010). "An ISM approach for modelling the enablers of flexible manufacturing system: The case for India". *International Journal of Production Research,* 46(24): 6883–6912.

Rajhans K. and Jayant Arvind (2017). "Development of Decision Support System for Vendor Selection Using AHP–VIKOR Based Hybrid Approach". *American Journal of Engineering Research (AJER),* 06(12): 195–208. ISSN (e): 2320–0847,

Ravi V. and Shankar R. (2005). "Analysis of interactions among the barriers of reverse logistics". *International Journal of Technological Forecasting & Social Change*, 72(8): 1011–1029.

Revell A. and Rutherfoord R. (2003). "UK environmental policy and the small firm: Broadening the focus". *Business Strategy and the Environment,* 12: 26–35.

Saaty T.L. (1990). "The analytic hierarchy processes" New York, NY: McGraw–Hill.

Saaty T.L. (1996). "Decision making with dependence and feedback: The analytic network process" Pittsburgh, PA: RWS Publications.

Saaty T.L. (2005). "Theory and applications of the analytic network process: Decision making with benefits, opportunities, costs and risks" USA: RWS Publications.

Saaty Thomas L. (1999). "Fundamentals of the Analytic Network Process" Proceedings of *ISAHP Japan*, pp. 12–14.

Saaty Thomas L. and Niemira P. Michael (2004). "An analytical network process model for financial – crisis forecasting". *Int. J. Forecasting,* 20: 573–587.

Sabuncuoglu (1998). "Study of scheduling rules of flexible manufacturing systems: A simulation approach". *International Journal of Production Research*, 36: 527–546.

Sage A. (1977). "Interpretive Structural Modelling: Methodology for Large Scale Systems" New York: McGraw–Hill, pp. 91–164.

Saini Yaman and Jayant Arvind (2017). "Performance Evaluation of Suppliers and Manufacturer using Data Envelopment Analysis (DEA) Approach". *International Journal of Advanced in Management, Technology and Engineering Sciences,* 7(11): 263–270.

Sajan T., John R. and Sridharan (2013). "Modelling and analysis of network design for a closed–loop supply chain". *Int. J. of Logistics Systems and Management*, 14(3): 329–352.

Sarkis J. (1995). "Manufacturing strategy and environmental consciousness". *Technovation,* 15(2): 79–97.

Sarkis J. (1998). "Evaluating environmentally conscious business practices". *European Journal of Operational Research,* 107: 159–74.

Sarkis J. (2003). "A strategic decision framework for green supply chain management". *Journal of Cleaner Production*, 11: 397–409

Sarkis J. and Talluri S. (2002). "A model for strategic supplier selection". *Journal of Supply Chain Management*, 38: 18–28.

Schultmann Engel and Rentz (2003). "Closed loop supply chain for spent batteries". *Interfaces,* 33(6).

Seliger G., Ba dere B., Ciupek M. and Franke C. (2003). "Remanufacturing of cellular phones", *In*: *CIRP Seminar on Life Cycle Engineering*, Copenhagen.

Seliger G., Ba dere B., Keil T. and Rebafka U. (2002). "Innovative processes and tools for disassembly" *Annals of the CIRP*, 51(1): 37–40.

Setaputra Robert and Mukhopadhyay Samar K. (2010). "A framework for research in reverse logistics". *Int. J. of Logistics Systems and Management*, 7(1): 19 – 55

Shah A. and Nejad S.S. (2013). "Developing a new model using Fuzzy AHP and TOPSIS methods in supplier selection problem in Supply Chain Management – A case study of SADRA Company in IRAN". *Supply Chain Management Journal*, 4(1): 26–44.

Shang J. and Sueyoshi T. (1995). "A unified framework for the selection of a flexible manufacturing system". *European Journal of Operational Research*, 85: 297–315.

Sharma B.P., Singh M.D. and Neha (2012). "Modeling the Knowledge Sharing Barriers using an ISM Approach". *International Conference on Information and Knowledge Management*, 45: 233–238.

Sharma S. (2000). "Managerial interpretations and organizational context as predictors of corporate choice of environmental strategy". *Academy of Management Journal*, 43(4): 681–697.

Shiang T.L. (2008). "A fuzzy DAE/AR approach to the selection of manufacturing system". *Computer and Industrial Engineering*, 54: 66–76.

Shyur H.J. and Shih H.S. (2006). "A hybrid MCDM model for strategic vendor selection". *Mathematical and Computer Modelling*, 44: 749–761.

Singh Priya and Jayant Arvind (2015). "Selection of FMS: A Synthesis of 3MCDM Approaches". *Journal of Material Science and Mechanical Engineering (JMSME),* 2(8): 15–20. ISSN: 2393–9109.

Sinha Sanjeev, Shankar Ravi and Pichai Taneerananon (2010). "Modeling and case study of reverse logistics for construction aggregates". *Int. J. of Logistics Systems and Management*, 6(1): 39 –59.

Sorayaei A., Bolboli S. and Zahra Atf (2014). "Assessment and prioritization of business processes with the ability to outsource municipal fuzzy AHP approach: Case Study of Amir kola Municipality". *American Journal of Engineering Research (AJER)*, 3(4): 260–264.

Talluri S. and Baker R.C. (2010). "A multi–phase mathematical programming approach for effective supply chain". *OMEGA*, 67(5): 78–90.

Talluri S., Whiteside M.M. and Seipel S.J. (2000). "A non–parametric stochastic procedure for FMS evaluation". *European Journal of Operational Research*, 124: 529–538.

Tian–tian D., Yi–sheng, F.Y. and Yu M. (2014). "Macro–site selection of wind/solar hybrid power station based on ELECTRE–II". *Renewable and Sustainable Energy Reviews*, 3: 194–204.

Torres B., Nones S., Morques S. and Evgenio R. (2004). "A Theoretical Approach for Green Supply Chain Management" Federal University DO RIO GRANDE, Industrial Engineering Program, NATAL–BRAZIL.

Treleven M. (1987) "Single sourcing: a management tool for the quality supplier" *Journal of Purchasing and Materials Management*, 21,19–24, spring.

Tsai W. and Ghoshal S. (1998). "Social capital and value creation: The role of intra firm networks". *Academy of Management Journal*, 41(4): 464–476.

V. Ravi, Shankar Ravi and Tiwari M.K. (2005). "Analysing alternatives in reverse logistics for end–of–life computers: ANP and balanced scorecard approach". *Computers & Industrial Engineering,* 48: 327–356.

Warfield J.W. (1974) "Developing interconnected matrices in Structural modelling" *IEEE Transcript on Systems, Men and Cybernetics*, 4(1), 51–81.

Weber C.A., Current J.R. and Benton W.C. (1991). "Vendor selection criteria and methods". *European Journal of Operational Research,* 50(1): 2–18.

Westkamper E., Alting L. and Arndt G. (2000). "Life cycle management and assessment, approaches and visions towards sustainable manufacturing". *Annals of the CIRP,* 49(2): 501–26.

Wu W.Y., Sukoco B.M., Li C.Y. and Chen S.H. (2009). "An integrated multi–objective decision–making process for supplier selection with bundling problem". *Expert Systems with Applications*, 36: 2327–2337.

Xanthopoulos E. Iakovou (2010). "A strategic methodological optimization framework for the design.

Yu Lin, C. (2007). "Adoption of green supply in Taiwan logistic industry". *Journal of Management Study,* 12: 90–98.

Zhu Q , Sarkis J. and Lai K.H. (2007). "Green supply chain management: Pressures, practices and performance within the Chinese automobile industry". *Journal of Cleaner Production*, 15: 1041–1052.

Subject Index

L

M

N

P

R

S

T

U

V

W